SIGNIFICANT CHANGES TO THE

INTERNATIONAL BUILDING CODE®

2015 EDITION

CENGAGE Learning®

Australia • Brazil • Japan • Korea • Mexico • Singapore • Spain • United Kingdom • United States

Significant Changes to the International Building Code® 2015 Edition

International Code Council

Douglas W. Thornburg, AIA; John R. Henry, P.E.; Jay Woodward

Cengage Learning Staff:

Executive Director of Professional Technology and Trades Training Solutions: **Taryn Zlatin McKenzie**

Product Manager: **Vanessa Myers**

Associate Content Developer: **Jenn Wheaton**

Director of Marketing: **Beth A. Lutz**

Senior Marketing Manager: **Marissa Lavigna**

Marketing Communications Manager: **Nicole McKasty-Stagg**

Senior Production Director: **Wendy Troeger**

Production Director: **Patty Stephan**

Senior Content Project Manager: **Stacey Lamodi**

Senior Art Director: **Benjamin Gleeksman**

ICC Staff:

Executive Vice President and Director of Business Development: **Mark A. Johnson**

Senior Vice President, Business and Product Development: **Hamid Naderi**

Vice President, Education and Certification: **Doug Thornburg**

Director, Products and Special Sales: **Suzane Nunes**

Senior Marketing Specialist: **Dianna Hallmark**

Cover images courtesy of:
© Marko Bradic/Shutterstock;
© Leungchopan/Shutterstock;
© VLADJ55/Shutterstock

Library of Congress Control Number: 2014937025

ISBN: 978-1-305-25471-8

ICC World Headquarters
500 New Jersey Avenue, NW
6th Floor
Washington, D.C. 20001-2070
Telephone: 1-888-ICC-SAFE (422-7233)
Website: **http://www.iccsafe.org**

Cengage Learning
20 Channel Center Street
Boston, MA 02210
USA

Cengage Learning is a leading provider of customized learning solutions with office locations around the globe, including Singapore, the United Kingdom, Australia, Mexico, Brazil, and Japan. Locate your local office at: **international.cengage.com/region**

Cengage Learning products are represented in Canada by Nelson Education, Ltd.

Visit us at **www.ConstructionEdge.cengage.com**
For more learning solutions, please visit our corporate website at **www.cengage.com**

Printed in China
3 4 5 6 7 18 17 16 15

Contents

PART 4
Means of Egress
Chapter 10 141

PART 5
Accessibility
Chapter 11 177

PART 6
Building Envelope, Structural Systems, and Construction Materials
Chapters 12 through 26 199

Preface

The purpose of *Significant Changes to the International Building Code® 2015 Edition* is to familiarize building officials, fire officials, plans examiners, inspectors, design professionals, contractors, and others in the construction industry with many of the important changes in the 2015 *International Building Code®* (IBC®). This publication is designed to assist those code users in identifying the specific code changes that have occurred and, more important, understanding the reasons behind the changes. It is also a valuable resource for jurisdictions in their code-adoption process.

Only a portion of the total number of code changes to the IBC are discussed in this book. The changes selected were identified for a number of reasons, including their frequency of application, special significance, or change in application. However, the importance of those changes not included is not to be diminished. Further information on all code changes can be found in the *Code Changes Resource Collection,* available from the International Code Council® (ICC®). The resource collection provides the published documentation for each successful code change contained in the 2015 IBC since the 2012 edition.

This book is organized into seven general categories, each representing a distinct grouping of code topics. It is arranged to follow the general layout of the IBC, including code sections and section number format. The table of contents, in addition to providing guidance in use of this publication, allows for quick identification of those significant code changes that occur in the 2015 IBC.

Throughout the book, each change is accompanied by a photograph, an application example, or an illustration to assist and enhance the reader's understanding of the specific change. A summary and a discussion of the significance of the changes are also provided. Each code change is identified by type, be it an addition, modification, clarification, or deletion.

The code change itself is presented in a format similar to the style utilized for code-change proposals. Deleted code language is shown with a strike-through, whereas new code text is indicated by underlining. As a result, the actual 2015 code language is provided, as well as a comparison with the 2012 language, so the user can easily determine changes to the specific code text.

As with any code-change text, *Significant Changes to the International Building Code 2015 Edition* is best used as a study companion to the 2015 IBC. Because only a limited discussion of each change is provided, the code itself should always be referenced in order to gain a more comprehensive understanding of the code change and its application.

The commentary and opinions set forth in this text are those of the authors and do not necessarily represent the official position of the ICC. In addition, they may not represent the views of any enforcing agency, as such agencies have the sole authority to render interpretations of the IBC. In many cases, the explanatory material is derived from the reasoning expressed by the code-change proponent.

Comments concerning this publication are encouraged and may be directed to the ICC at significantchanges@iccsafe.org.

About the *International Building Code®*

Building officials, design professionals, and others involved in the building construction industry recognize the need for a modern, up-to-date building code addressing the design and installation of building systems through requirements emphasizing performance. The *International Building Code* (IBC), in the 2015 edition, is intended to meet these needs through model code regulations that safeguard the public health and safety in all communities, large and small. The IBC is kept up to date through the open code-development process of the International Code Council (ICC). The provisions of the 2012 edition, along with those code changes approved through 2013, make up the 2015 edition.

The ICC, publisher of the IBC, was established in 1994 as a nonprofit organization dedicated to developing, maintaining, and supporting a single set of comprehensive and coordinated national model building construction codes. Its mission is to provide the highest-quality codes, standards, products, and services for all concerned with the safety and performance of the built environment.

The IBC is 1 of 15 International Codes® published by the ICC. This comprehensive building code establishes minimum regulations for buildings systems by means of prescriptive and performance-related provisions. It is founded on broad-based principles that make possible the use of new materials and new building designs. The IBC is available for adoption and use by jurisdictions internationally. Its use within a governmental jurisdiction is intended to be accomplished through adoption by reference, in accordance with proceedings establishing the jurisdiction's laws.

About the Authors

Douglas W. Thornburg, AIA, CBO
International Code Council
Vice-President, Education and Certification

Douglas W. Thornburg is currently Vice-President of Education and Certification for the International Code Council (ICC), where he provides administrative and technical leadership for the ICC education and certification programs. Prior to employment with ICC in 2004, he spent nine years as a code consultant and educator on building codes.

Formerly Vice-President/Education for the International Conference of Building Officials (ICBO), Doug also continues to develop and present building code seminars nationally and has developed numerous educational texts and resource materials. He was presented with ICC's inaugural Educator of the Year Award in 2008, recognizing his outstanding contributions in education and training.

A graduate of Kansas State University and a registered architect, Doug has over 33 years of experience in building code training and administration. He has authored a variety of code-related support publications, including the *IBC Handbook* and *Significant Changes to the International Building Code.*

John R. Henry, P. E.

John R. Henry is the former Principal Staff Engineer with the International Code Council (ICC) Business and Product Development Department, where he was responsible for the research and development of technical resources pertaining to the structural engineering provisions of the *International Building Code* (IBC). John also developed and presented technical seminars on the structural provisions of the IBC. He has a broad range of experience that includes structural design in private practice, plan-check engineering with consulting firms and building department jurisdictions, and 17 years as an International Conference of Building Officials (ICBO)/ICC Staff Engineer. John graduated with honors from California State University in Sacramento with a Bachelor of Science Degree in Civil Engineering and is a Registered Civil Engineer in the State of California. He is a member of the Structural Engineers Association of California (SEAOC) and is an ICC Certified Plans Examiner. John has written several articles on the structural provisions of the IBC that have appeared in *Structure* magazine and *Structural Engineer* magazine's Code Series. He is also the coauthor of the *2012 IBC Handbook* and coauthor with S. K. Ghosh, PhD, of the *IBC Handbook—Structural Provisions.*

Jay Woodward
International Code Council
Senior Staff Architect

Jay Woodward is a senior staff architect with the ICC's Business and Product Development department. With more than 31 years of experience in building design, construction, code enforcement, and instruction, Jay's experience provides him with the ability to address issues of code application and design for code enforcement personnel as well as architects, designers, and contractors. Jay has previously served as the Secretariat for the ICC A117.1 standard committee, ICC's *International Energy Conservation Code,* and the *International Building Code's* Fire Safety Code Development committee.

A graduate of the University of Kansas and a registered architect, Jay has also worked as an architect for the Leo A. Daly Company in Omaha, Nebraska; as a building plans examiner for the City of Wichita, Kansas; and as a senior staff architect for the International Conference of Building Officials (ICBO) prior to working for the ICC. He is also author of *Significant Changes to the A117.1 Accessibility Standard 2009 Edition.*

About the ICC

The International Code Council is a member-focused association dedicated to helping the building safety community and construction industry provide safe, sustainable and affordable construction through the development of codes and standards used in the design, build and compliance process. Most U.S. communities and many global markets choose the International Codes. ICC Evaluation Service (ICC-ES), a subsidiary of the International Code Council, has been the industry leader in performing technical evaluations for code compliance, fostering safe and sustainable design and construction.

Headquarters:
500 New Jersey Avenue, NW, 6th Floor
Washington, DC 20001-2070

Regional Offices:
Birmingham, AL; Chicago. IL; Los Angeles, CA

1-888-422-7233
www.iccsafe.org

PART 1

Administration

Chapters 1 and 2

- **Chapter 1** Administration
- **Chapter 2** Definitions

The provisions of Chapter 1 address the application, enforcement, and administration of subsequent requirements of the code. In addition to establishing the scope of the *International Building Code* (IBC), the chapter identifies which buildings and structures come under its purview. A building code, as with any other code, is intended to be adopted as a legally enforceable document to safeguard health, safety, property and public welfare. A building code cannot be effective without adequate provisions for its administration and enforcement. Chapter 2 provides definitions for terms used throughout the IBC. Codes, by their very nature, are technical documents, and as such, literally every word, term, and punctuation mark can add to or change the meaning of the intended result. ■

101.2
Exempt Residential Accessory Structures

111.1
Change of Use or Occupancy

202
Definition of Horizontal Exit

202
Definition of Platform

202
Definition of Private Garage

202
Definition of Treated Wood

101.2

Exempt Residential Accessory Structures

CHANGE TYPE: Modification

CHANGE SUMMARY: Modifications to the *International Residential Code* (IRC) provisions have been reflected in the exception to IBC Section 101.2 such that the limiting height of an IRC structure accessory to a dwelling unit or townhouse has increased from two stories to three stories above grade plane.

2015 CODE: **101.2 Scope.** The provisions of this code shall apply to the construction, alteration, relocation, enlargement, replacement, repair, equipment, use and occupancy, location, maintenance, removal and demolition of every building or structure or any appurtenances connected or attached to such buildings or structures.

> **Exception:** Detached one- and two-family dwellings and multiple single-family dwellings (townhouses) not more than three stories above grade plane in height with a separate means of egress, and their accessory structures not more than three stories above grade plane in height, shall comply with the *International Residential Code.*

CHANGE SIGNIFICANCE: As a general rule, the IBC applies to any structure undergoing construction, alteration, relocation, enlargement, replacement, use and occupancy, repair, maintenance, removal, or demolition. However, the exception to Section 101.2 indicates that the IRC is to be applied to specified residential buildings, along with their accessory structures. The IRC has previously limited the area and height of such accessory structures through the definition of "accessory structure" in Section R202. That definition has been deleted, in effect limiting the

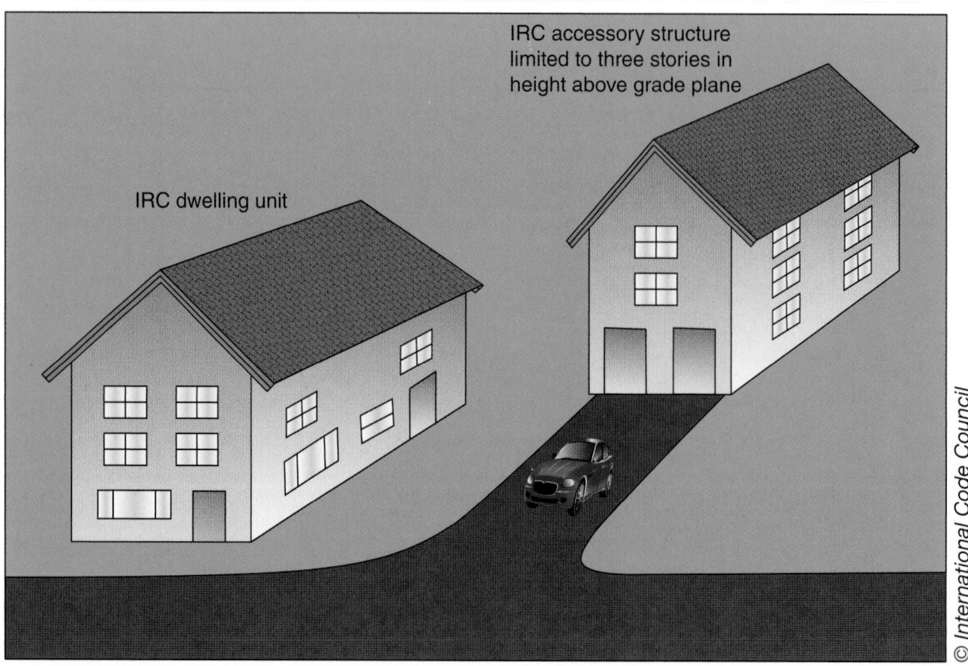

Residential accessory structure

height of an IRC accessory structure to the dwelling unit/townhouse limit of three stories above grade plane. The modifications in the IRC provisions have been reflected in the exception to IBC Section 101.2.

The IRC has previously limited accessory buildings to 3000 square feet and two stories in height. The 2015 IRC no longer contains these limitations, but rather only limits the height of an accessory structure to the scoping height of those dwellings and townhouses regulated by the IRC. As a side note, the IRC no longer places a floor area limitation on accessory structures. It was determined that the more appropriate approach to limiting the size of a residential accessory structure is through local zoning ordinances.

111.1

Change of Use or Occupancy

CHANGE TYPE: Modification

CHANGE SUMMARY: A change in a building's use, or a portion of a building's use, with no change in its occupancy classification now requires that a new certification of occupancy be issued by the building official.

2015 CODE: **111.1 Use and Occupancy.** A building or structure shall not be used or occupied, and a change in the existing use or occupancy classification of a building or structure or portion thereof shall not be made, until the building official has issued a certificate of occupancy therefor as provided herein. Issuance of a certificate of occupancy shall not be construed as an approval of a violation of the provisions of this code or of other ordinances of the jurisdiction.

> **Exception:** Certificates of occupancy are not required for work exempt from permits under Section 105.2.

CHANGE SIGNIFICANCE: The tool that the building official utilizes to control the uses and occupancies of the various buildings and structures within the jurisdiction is the certificate of occupancy. The IBC makes it unlawful to use or occupy a building or structure unless a certificate of occupancy has been issued for that use. The provisions of Section 111.1 have previously only required the issuance of a new certificate of occupancy where the building undergoes a change in occupancy classification. Additional code text now specifies that a change in use, without a change in occupancy, also requires that a new certification of occupancy be obtained.

There is a key distinction in the application of the IBC concerning the terms "use" and "occupancy." Most buildings have multiple uses, but often only contain a single occupancy classification. For example, an office building may have business areas, small storage areas, and small assembly uses, as well as support areas such as restrooms and mechanical equipment rooms. Although multiple uses occur in the building, it only

Outpatient surgery center

contains a single occupancy, Group B. It is anticipated that all of the hazards that are anticipated as part of the building's function can be effectively addressed through classification as a single-occupancy group.

The modification in application of Section 111.1 is not intended to address minor changes in use that are typical when new ownership or tenancy occurs. It will be applicable where change in the use results in a significant change to the hazards involved. As an example, where a Group B business office building undergoes a change of use to a Group B ambulatory care facility, additional fire- and life-safety safeguards are necessary in order to address the increased hazard due to the presence of healthcare recipients incapable of self-preservation. Although both types of facilities are classified as Group B occupancies, there is a distinct difference in their uses and, as such, a new certificate of occupancy must be issued to address the change in use.

202

Definition of Horizontal Exit

CHANGE TYPE: Modification

CHANGE SUMMARY: The definition of "horizontal exit" now focuses on the compartmentalization aspect of using a horizontal exit, rather than on the path of egress travel.

2015 CODE: **202.** ~~**EXIT, HORIZONTAL.** A path of egress travel from one building to an area in another building on approximately the same level, or a path of egress travel through or around a wall or partition to an area on approximately the same level in the same building, which affords safety from fire and smoke from the area of incidence and areas communicating therewith.~~

HORIZONTAL EXIT. An exit component consisting of fire-resistance rated construction and opening protectives intended to compartmentalize portions of a building thereby creating refuge areas that afford safety from the fire and smoke from the area of fire origin.

CHANGE SIGNIFICANCE: As established by its definition, a means of egress consists of three separate and distinct parts: the exit access, the exit, and the exit discharge. The "exit" portion of the means-of-egress system can include a number of different components, one of which is a horizontal exit. The definition of "horizontal exit" has been modified to more accurately describe its role as an exit component.

The previous definition indicated that a horizontal exit was a path of egress travel, which was not accurate. In fact, the horizontal exit concept is based upon the creation of a refuge area where exit travel is no longer independently regulated for the individuals passing through the horizontal exit's fire-resistance-rated separation. The definition now focuses on the compartmentalization aspect of using a horizontal exit, with the recognition that a fire-resistance-rated separation and appropriate opening protectives are keys to its use as an exit component. In addition, the portion of the definition that described travel to and from areas on approximately the same level has been deleted because the specifics of the various acceptable building configurations are best addressed in Section 1026.

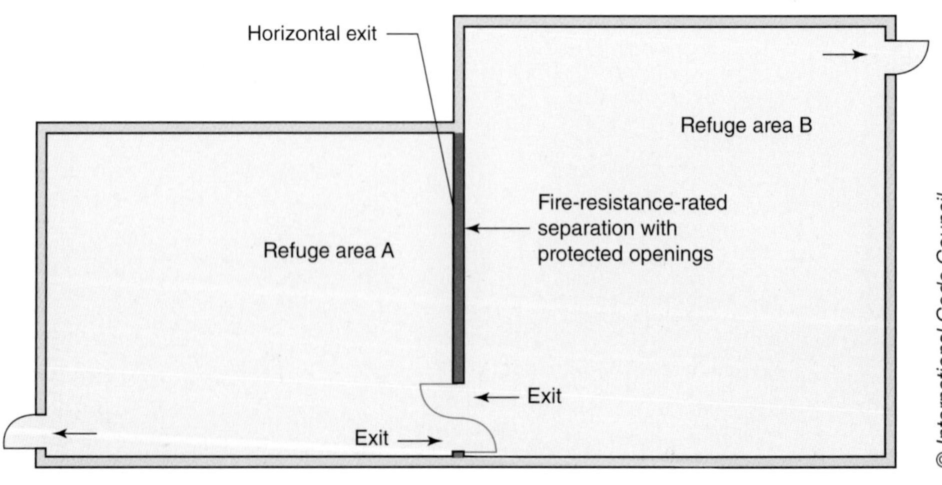

Horizontal exits

202

Definition of Platform

CHANGE TYPE: Clarification

CHANGE SUMMARY: By definition, horizontal sliding curtains are now specifically permitted on a raised performance area regulated as a platform.

2015 CODE: 202. PLATFORM. A raised area within a building used for worship, the presentation of music, plays or other entertainment; the head table for special guests; the raised area for lecturers and speakers; boxing and wrestling rings; theater-in-the-round stages; and similar purposes wherein, other than horizontal sliding curtains, there are no overhead hanging curtains, drops, scenery or stage effects other than lighting and sound. A temporary platform is one installed for not more than 30 days.

CHANGE SIGNIFICANCE: The distinctions between the definitions of a stage and a platform are very important because of the requirements applicable to each element. The primary difference between a stage and a platform is the presence of overhead hanging curtains, drops, scenery, and other effects that a stage contains. The amount of combustible materials associated with a platform is typically much less than for a stage. Thus, the fire-severity potential is much lower. The allowance for horizontal sliding curtains at platforms has been clarified, as they are now specifically permitted by definition.

In order to be regulated as a platform, a raised performance area cannot have overhead hanging curtains. It has previously been unclear as to whether or not horizontal sliding curtains are included in this prohibition. Although it has been widely interpreted that horizontal sliding curtains are not considered as overhead hanging curtains, it was deemed important for consistency in application that the issue be addressed directly. Therefore, the definition of a platform now clearly states that horizontal sliding curtains are permitted.

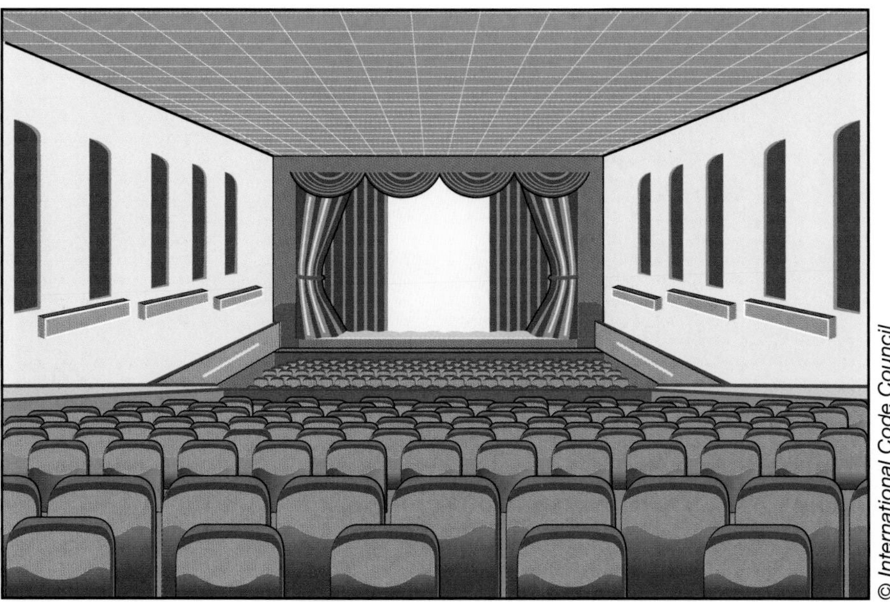

Platform with horizontal sliding curtains

202

Definition of Private Garage

CHANGE TYPE: Addition

CHANGE SUMMARY: Motor vehicles stored in a "private garage" are now limited through a new definition to only those vehicles used by tenants of the building or buildings on the same premises as the garage.

2015 CODE: **202.** **PRIVATE GARAGE.** A building or portion of a building in which motor vehicles used by the tenants of the building or buildings on the premises are stored or kept, without provisions for repairing or servicing such vehicles for profit.

CHANGE SIGNIFICANCE: There are fundamentally two types of parking garages regulated by the IBC, private garages and public garages. Although there has previously been no specific definition for either type of garage, the basis for both classifications has been Section 406.3 addressing private garages and carports. Those parking structures that fell outside of the scope of Section 406.3 were then considered as public parking garages. The primary difference between private and public garages has been the size of the facility, rather than the use. With the new definition of "private garage," the use of the garage is now also a determining factor in how a garage is to be classified.

A private garage, in addition to its limit on floor area, is now limited as to its use. Motor vehicles stored in a private garage are limited to only those vehicles used by tenants of the building or buildings on the same premises as the garage. In addition, the repair and/or servicing of vehicles for business purposes cannot occur within a private garage.

While the new definition places additional restrictions on those structures classified as Group U private garages, the controlling factor in their classification will typically continue to be their size. Additional changes to the provisions regulating private garages and carports are addressed in the discussion of significant changes to Section 406.3.

Private garages serving an assisted living facility

202

Definition of Treated Wood

CHANGE TYPE: Clarification

CHANGE SUMMARY: The definition of "treated wood" has been revised to clarify that approved treatment methods by other than pressure are acceptable.

2015 CODE: 202. TREATED WOOD. Wood ~~and wood-based materials~~ products that ~~use vacuum-pressure impregnation processes~~ are conditioned to enhance fire retardant or preservative properties.

Fire-Retardant-Treated Wood. ~~Pressure-treated lumber and plywood~~ Wood products that, when impregnated with chemicals by a pressure process or other means during manufacture, exhibit reduced surface-burning characteristics and resist propagation of fire.

Preservative-Treated Wood. ~~Pressure-treated~~ Wood products that, conditioned with chemicals by a pressure process or other means, exhibit reduced susceptibility to damage by fungi, insects or marine borers.

CHANGE SIGNIFICANCE: Treated wood is frequently mandated where wood materials are used in various applications. Both fire-retardant-treated wood and preservative-treated wood are specified throughout the IBC. By definition, treated wood has previously been described as wood and wood-based materials that are enhanced through the process of vacuum pressure impregnation. However, pressure treatment is not the only method permitted by the code for treated wood. Sections 2303.2 and 2303.2.2 both indicate that means other than pressure treatment are acceptable for fire-retardant-treated wood. Preservative-treated wood can be pressure treated or treated by a number of other methods indicated in the AWPA standards referenced in Section 2303.1.9. The new definitions now provide consistency with conditions established in Chapter 23.

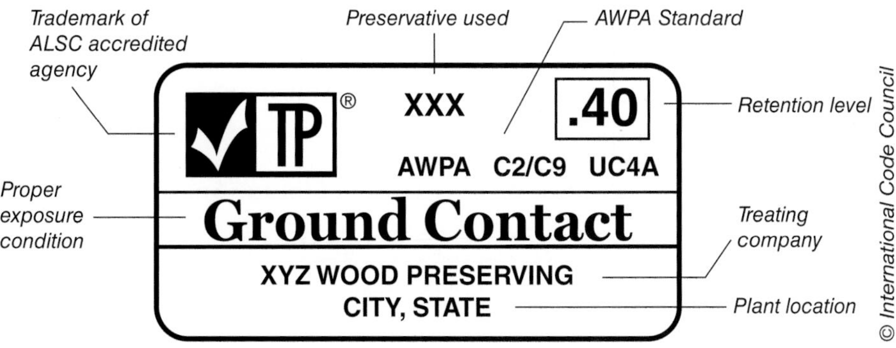

Example of treated wood quality mark

PART 2

Building Planning

Chapters 3 through 6

The application of the *International Building Code* to a structure is typically initiated through the provisions of Chapters 3, 5, and 6. Chapter 3 establishes one or more occupancy classifications based upon the anticipated uses of a building. The appropriate classifications are necessary to properly apply many of the code's non-structural provisions. The requirements of Chapter 6 deal with classification as to construction type, based on a building's materials of construction and the level of fire resistance provided by such materials. Limitations on a building's height and area, set forth in Chapter 5, are directly related to the occupancies it houses and its type of construction. Chapter 5 also provides the various methods available to address conditions in which multiple uses or occupancies occur within the same building. Chapter 4 contains special detailed requirements based on unique conditions or uses that are found in some buildings. ■

406.3.1

Private Garage Floor-Area Limitation

406.3.2

Private Parking Garage Ceiling Height

407.2.5

Group I-2 Shared Living Spaces

407.2.6

Group I-2 Cooking Facilities

407.5

Maximum Size of Group I-2 Smoke Compartments

410.3.5

Horizontal Sliding Doors at Stage Proscenium Opening

412.7

Travel Distance in Aircraft Manufacturing Facilities

423.3

Storm Shelters Serving Critical Emergency Operations Facilities

423.4

Storm Shelters Serving Group E Occupancies

503

General Building Height and Area Limitations

TABLES 504.3, 504.4

Building Height and Number of Stories

505.2.3, EXCEPTION 2

Mezzanine Openness

TABLE 506.2

Building Area

507.1

Basements in Unlimited Area Buildings

507.9

Group H-5 in Unlimited Area Buildings

TABLE 509

Fire Protection from Incidental Uses

510.2

Horizontal Building Separation

TABLE 601, FOOTNOTE D

One-Hour Substitution

602.4

Type IV Member Size Equivalencies

602.4.2

Cross-Laminated Timber in Exterior Walls

603.1, ITEM 26

Wall Construction of Freezers and Coolers

304.1

Food Processing Facilities and Commercial Kitchens

CHANGE TYPE: Modification

CHANGE SUMMARY: Small food processing establishments and commercial kitchens not associated with dining facilities are now considered as Group B occupancies.

2015 CODE: **304.1 Business Group B.** Business Group B occupancy includes, among others, the use of a building or structure, or a portion thereof, for office, professional or service-type transactions, including storage of records and accounts. Business occupancies shall include, but not be limited to, the following:

Airport traffic control towers
Ambulatory care facilities
Animal hospitals, kennels and pounds
Banks
Barber and beauty shops
Car wash
Civic administration
Clinic, outpatient
Dry cleaning and laundries: pickup and delivery stations and self-service
Educational occupancies for students above the 12th grade
Electronic data processing
Food processing establishments and commercial kitchens not associated with restaurants, cafeterias and similar dining facilities not more than 2500 square feet (232 m^2) in area.

Carry-out business with commercial kitchen

Laboratories: testing and research

Motor vehicle showrooms

Post offices

Print shops

Professional services (architects, attorneys, dentists, physicians, engineers, etc.)

Radio and television stations

Telephone exchanges

Training and skill development not in a school or academic program (this shall include, but not be limited to, tutoring centers, martial arts studios, gymnastics and similar uses regardless of the ages served, and where not classified as a Group A occupancy).

CHANGE SIGNIFICANCE: Facilities used for food processing and/or preparation have traditionally been considered as Group F-1 occupancies unless directly related to a dining activity. The Group F-1 classification has been applied to large-scale operations, such as food processing plants; moderate-scale uses, such as catering operations; and small-scale establishments, such as bakeries, carry-out pizza tenants and other uses that are open to the public. It is not uncommon for such small-scale food processing facilities to occur in mixed-occupancy buildings with retail sales, offices and restaurant tenants. For this reason, these establishments have sometimes been classified as Group M retail sales or Group B business occupancies.

Classifying such establishments as Group F-1 occupancies is now considered inappropriate where the floor area of the building or tenant space is relatively small. In addition, a Group M classification is considered not fully representative of the hazards involved with the food processing/public occupancy activity. Therefore, a Group B classification is to be applied where the facility does not exceed 2500 square feet in floor area. This classification also assumes the facility is not used for assembly purposes, such as a café or bar. This allowance is also extended to commercial kitchens such as those used for catering operations. Where the floor area exceeds the 2500-square-foot threshold, then a classification of Group F-1 continues to be appropriate.

304.1

Training and Skill Development Facilities

CHANGE TYPE: Clarification

CHANGE SUMMARY: The Group B classification for training and skill development uses has been clarified to address the ages of the occupants using the facility, the occupant load limitation where the facility is used for assembly purposes, and the types of permitted uses.

2015 CODE: **304.1 Business Group B.** Business Group B occupancy includes, among others, the use of a building or structure, or a portion thereof, for office, professional or service-type transactions, including storage of records and accounts. Business occupancies shall include, but not be limited to, the following:

Airport traffic control towers

Ambulatory care facilities

Animal hospitals, kennels and pounds

Banks

Barber and beauty shops

Car wash

Civic administration

Clinic, outpatient

Dry cleaning and laundries: pickup and delivery stations and self-service

Educational occupancies for students above the 12th grade

Electronic data processing

Food processing establishments and commercial kitchens not associated with restaurants, cafeterias and similar dining facilities not more than 2500 square feet in area.

Music center

Laboratories: testing and research

Motor vehicle showrooms

Post offices

Print shops

Professional services (architects, attorneys, dentists, physicians, engineers, etc.)

Radio and television stations

Telephone exchanges

Training and skill development not in a school or academic program (this shall include, but not be limited to, tutoring centers, martial arts studios, gymnastics and similar uses regardless of the ages served, and where not classified as a Group A occupancy).

CHANGE SIGNIFICANCE: Various types of facilities are designed to teach and train persons outside of an academic school program. These types of uses, classified as Group B occupancies, include various skills, trades and technical programs for persons with a business, as well as those outside of a formal business setting that are not a part of a school or a degree program. Although the nature of the uses may cause consideration as Group E or Group M occupancies, the degree of anticipated hazards is most accurately addressed under the Group B classification.

The scope of such facilities assigned the Group B classification has previously been inconsistently applied regarding three key issues: (1) the ages of the occupants using the facility, (2) the occupant load limitation where the facility is used for assembly purposes, and (3) the types of uses considered to be classified as Group B. Additional code language now clarifies the original intent of the classification criteria.

Many skill development facilities are devoted to the training of children of various ages. The question of whether or not such training should fall under the Group E classification has now been specifically addressed by expressing that the Group B classification is to be applied regardless of age. Where the number of occupants in an assembly setting, of any age, is such that the occupant load triggers a Group A classification, typically 50 or more occupants, then the Group B classification is inappropriate. Training rooms and similar assembly rooms typically classified as Group A occupancies will maintain that classification. The types of uses anticipated in this classification category include tutoring centers, martial arts studios and gymnastics studios, as well as music and art development facilities.

306.2

Food Processing Facilities and Commercial Kitchens

CHANGE TYPE: Modification

CHANGE SUMMARY: A classification of Group F-1 is now applied only to larger-sized food processing facilities and commercial kitchens not associated with dining facilities.

2015 CODE: 306.2 Moderate-Hazard Factory Industrial, Group F-1. Factory industrial uses which are not classified as Factory Industrial F-2 Low Hazard shall be classified as F-1 Moderate Hazard and shall include, but not be limited to, the following:

Aircraft (manufacturing, not to include repair)
Appliances
Athletic equipment
Automobiles and other motor vehicles
Bakeries
Beverages: over 16-percent alcohol content
Bicycles
Boats
Brooms or brushes
Business machines
Cameras and photo equipment
Canvas or similar fabric
Carpets and rugs (includes cleaning)
Clothing
Construction and agricultural machinery
Disinfectants
Dry cleaning and dyeing

Commercial catering facility

Electric generation plants
Electronics
Engines (including rebuilding)
Food processing establishments and commercial kitchens not associated with restaurants, cafeterias and similar dining facilities more than 2500 square feet (232 m^2) in area
Furniture
Hemp products
Jute products
Laundries
Leather products
Machinery
Metals
Millwork (sash and door)
Motion pictures and television filming (without spectators)
Musical instruments
Optical goods
Paper mills or products
Photographic film
Plastic products
Printing or publishing
Recreational vehicles
Refuse incineration
Shoes
Soaps and detergents
Textiles
Tobacco
Trailers
Upholstering
Wood; distillation
Woodworking (cabinet)

CHANGE SIGNIFICANCE: Food processing facilities and commercial kitchens not directly associated with dining facilities have traditionally been considered as Group F-1 occupancies due to the moderate-level hazards that are often encountered. Establishments where food is prepared in a commercial kitchen for carry-out purposes have also been sometimes considered as Group F-1 occupancies. Consistent with a modification to the Group B classification category, the floor area of the facility is now the determining factor in the proper occupancy classification of the use.

Where the food processing facility or establishment, or where the commercial kitchen not directly associated with dining activities, has a floor area exceeding 2500 square feet, a Group F-1 classification is warranted. Where the floor area of such a use does not exceed the 2500-square-foot limitation, a Group B classification is to be applied.

308.3

Group I-1 Occupancy Classification

CHANGE TYPE: Modification

CHANGE SUMMARY: The uses permitted in a Group I-1 custodial care facility have been expanded to include care recipients who may need a limited degree of verbal or physical assistance if responding to a fire or other emergency situation.

2015 CODE: 308.3 Institutional Group I-1. ~~This~~ Institutional Group I-1 occupancy shall include buildings, structures or portions thereof for more than 16 persons, excluding staff, who reside on a 24-hour basis in a supervised environment and receive custodial care. ~~The persons receiving care are capable of self preservation.~~ Buildings of Group I-1 shall be classified as one of the occupancy conditions specified in Sections 308.3.1 or 308.3.2. This group shall include, but not be limited to, the following:

Alcohol and drug centers
Assisted living facilities
Congregate care facilities
~~Convalescent facilities~~
Group homes
Halfway houses
Residential board and ~~custodial~~ care facilities
Social rehabilitation facilities

308.3.1 Condition 1. This occupancy condition shall include buildings in which all persons receiving custodial care who, without any assistance, are capable of responding to an emergency situation to complete building evacuation.

Assisted living facility

308.3.2 Condition 2. This occupancy condition shall include buildings in which there are any persons receiving custodial care who require limited verbal or physical assistance while responding to an emergency situation to complete building evacuation.

CHANGE SIGNIFICANCE: Institutional Group I-1 occupancies include those uses where individuals receiving custodial care on a 24-hour basis live in a residential environment. Historically, the Group I-1 classification identified those facilities where the care recipients do not require any staff assistance should a fire or other emergency exist that would require the occupants to evacuate the building or relocate within the building. Types of uses included in this category include halfway houses, assisted living facilities and group homes. The uses permitted in a Group I-1 occupancy have been expanded to include care recipients who may need a limited degree of verbal or physical assistance if responding to a fire or other emergency situation.

Most state custodial care licensing agencies allow for the housing of occupants in assisted living facilities, residential care facilities and group homes even though such occupants may require some limited assistance with evacuation. In order to gain consistency between the IBC and what the various states allow, the custodial care provisions of the IBC have been modified. The revised provisions allow both care recipients who require limited assistance with evacuation, as well as those who do not, to reside in a Group I-1 custodial care facility. Through the use of "Condition" classifications 1 and 2, the differences in the evacuation capabilities of the residents can be appropriately addressed.

Condition 1 custodial care facilities reflect the 2012 IBC Group I-1 classification where the residents are capable of responding to an emergency without the assistance of others. Condition 2 has been added to address facilities where the residents are receiving custodial care but may require some assistance with evacuation. Four more stringent requirements addressing story limitations, smoke barriers, sprinkler protection and smoke detection have been instituted for Condition 2 occupancies due to the concerns regarding evacuation potential. It is anticipated that most Group I-1 custodial care facilities will be classified as Condition 2 unless the permit application or submittal documents identify the applicable licensing regulations that limit the resident type to Condition 1. Similar changes occurred addressing custodial care facilities with a limited number of care recipients as regulated under the Group R-4 occupancy classification.

An additional revision to the list of Group I-1 occupancies is the deletion of the term "convalescent facilities." The term was deemed to be outdated and was similarly deleted from the listing of Group R-4 uses.

308.4

Group I-2 Occupancy Classification

© International Code Council

Group I-2, Condition 1 nursing home

CHANGE TYPE: Modification

CHANGE SUMMARY: Two basic conditions of Group I-2 medical care uses that have previously been regulated together as a single category have been created, dividing the classification into short-term care facilities, such as hospitals, and long-term care facilities, such as nursing homes.

2015 CODE: 308.4 Institutional Group I-2. ~~This~~ Institutional Group I-2 occupancy shall include buildings and structures used for medical care on a 24-hour basis for more than five persons who are incapable of self-preservation. This group shall include, but not be limited to, the following:

Foster care facilities
Detoxification facilities
Hospitals
Nursing homes
Psychiatric hospitals

308.4.1 Occupancy Conditions. Buildings of Group I-2 shall be classified as one of the occupancy conditions indicated in Sections 308.4.1.1 or 308.4.1.2.

308.4.1.1 Condition 1. This occupancy condition shall include facilities that provide nursing and medical care but do not provide emergency care, surgery, obstetrics or inpatient stabilization units for psychiatric or detoxification, including but not limited to nursing homes and foster care facilities.

308.4.1.2 Condition 2. This occupancy condition shall include facilities that provide nursing and medical care and could provide emergency care, surgery, obstetrics or in-patient stabilization units for psychiatric or detoxification, including but not limited to hospitals.

CHANGE SIGNIFICANCE: Group I-2 occupancies have historically been considered as facilities used for medical care on a 24-hour basis, where the care recipients are typically incapable of self-preservation. A "protect-in-place" philosophy is applied to these occupancies due to the inability of the occupants to evacuate efficiently and under their own control. Hospitals and nursing homes are the two primary types of facilities that are regulated under the Group I-2 classification. The revision to Section 308.4 now provides two categories of medical care uses that have previously been regulated together as a single category.

Due to the diversification of how medical care is provided in the five characteristic uses as established in Section 308.4, the Group I-2 classification has now been subdivided into two basic categories: Condition 1, long-term care (nursing homes), and Condition 2, short-term care (hospitals). Changes in how care is delivered include a general increase in the ratio of floor area per patient in hospitals due to the increase in diagnostic equipment and the movement toward single-occupant patient rooms, as well as a trend to provide more residential-type arrangements in nursing homes, such as group/suite living and cooking facilities. Although most applicable code requirements will continue to apply to both medical care conditions classified as Group I-2, the division of uses does allow for varying provisions based upon the type of care being provided. Detoxification facilities and those facilities where patients receive psychiatric treatment will also be classified as either Condition 1 or 2, depending upon the extent of care provided.

310.5

Group R-3 Lodging Houses

CHANGE TYPE: Modification

CHANGE SUMMARY: Lodging houses are now specifically defined in Chapter 2 and are typically permitted to be constructed in accordance with the *International Residential Code* (IRC) if they contain no more than five guest rooms.

2015 CODE:

SECTION 202
DEFINITIONS

GUEST ROOM. A room used or intended to be used by one or more guests for living or sleeping purposes.

LODGING HOUSE. A one-family dwelling where one or more occupants are primarily permanent in nature and rent is paid for guest rooms.

310.5 Residential Group R-3. Residential Group R-3 occupancies where the occupants are primarily permanent in nature and not classified as Group R-1, R-2, R-4 or I, including:

Buildings that do not contain more than two dwelling units
Boarding houses (nontransient) with 16 or fewer occupants
Boarding houses (transient) with 10 or fewer occupants
Care facilities that provide accommodations for five or fewer persons receiving care

Bed-and-breakfast lodging

Congregate living facilities (nontransient) with 16 or fewer occupants

Congregate living facilities (transient) with 10 or fewer occupants

Lodging houses with five of fewer guest rooms

310.5.2 Lodging Houses. Owner-occupied lodging houses with five or fewer guest rooms shall be permitted to be constructed in accordance with the *International Residential Code.*

CHANGE SIGNIFICANCE: Residential uses, both transient and nontransient, are considered as Group R occupancies. Common characteristics of Group R buildings include use by persons for living and sleeping purposes, relatively low potential fire severity and the worst fire record of all structure fires. Lodging houses are now specifically defined in Chapter 2 and are typically permitted to be constructed in accordance with the IRC if they contain no more than five guest rooms and are owner-occupied.

Two new definitions have been added to Chapter 2 related to the classification of lodging houses. A lodging house is defined as a single-family dwelling where at least one of the occupants is a permanent resident and guest rooms are available for payment. A guest room is simply a room used for transient sleeping and/or living purposes. The most common example of such a use is a small bed-and-breakfast facility.

Lodging houses with more than five guest rooms will typically continue to be classified as Group R-1 occupancies as transient living facilities with more than 10 occupants. Where there are 10 or fewer occupants, or where there are five or fewer guest rooms, a classification of Group R-3 is appropriate. The primary change in application occurs where the lodging house is owner-occupied. Although by definition a lodging house will be occupied by a permanent resident, it does not specify that the permanent occupant be the owner of the building. New Section 310.5.2 now permits lodging houses with five or fewer guest rooms to be constructed in accordance with the IRC where the single-family residence is occupied by the owner.

The allowance provided by new provisions recognizes that single-family dwellings with limited guests create little if any additional hazard beyond those structures regulated under the IRC. Although a similar allowance was introduced to the scoping provisions of the 2012 IRC, it did not extend to the 2012 IBC, which continued to regulate such buildings as Group R-3 occupancies. Section 310.5.2 now provides consistency between the IBC and IRC.

310.6

Group R-4 Occupancy Classification

CHANGE TYPE: Modification

CHANGE SUMMARY: The uses permitted in a Group R-4 custodial care facility have been expanded to include care recipients who may need a limited degree of verbal or physical assistance while responding to a fire or other emergency situation.

2015 CODE: 310.6 Residential Group R-4. ~~This~~ Residential Group R-4 occupancy shall include buildings, structures or portions thereof for more than five but not more than 16 persons, excluding staff, who reside on a 24-hour basis in a supervised residential environment and receive custodial care. Buildings of Group R-4 shall be classified as one of the occupancy conditions specified in Sections 310.6.1 or 310.6.2. The persons receiving care are capable of self-preservation. This group shall include, but not be limited to, the following:

Alcohol and drug centers
Assisted living facilities
Congregate care facilities
~~Convalescent facilities~~
Group homes
Halfway houses
Residential board and ~~custodial~~ care facilities
Social rehabilitation facilities

Group R-4 occupancies shall meet the requirements for construction as defined for Group R-3, except as otherwise provided for in this code.

310.6.1 Condition 1. This occupancy condition shall include buildings in which all persons receiving custodial care, without any assistance, are capable of responding to an emergency situation to complete building evacuation.

Group R-4 halfway house

310.6.2 Condition 2. This occupancy condition shall include buildings in which there are any persons receiving custodial care who require limited verbal or physical assistance while responding to an emergency situation to complete building evacuation.

CHANGE SIGNIFICANCE: Residential Group R-4 occupancies include those uses where individuals receiving custodial care on a 24-hour basis live in a residential environment. They differ from those facilities classified as Group I-1 occupancies in only one aspect: there are fewer individuals receiving custodial care in the facility. Group R-4 occupancies have no more than 16 individuals receiving care in a supervised residential environment. Historically, the Group R-4 classification was limited to those facilities where the care recipients require no staff assistance should a fire or other emergency exist that would require evacuation or relocation of the occupants. Types of uses included in this category are halfway houses, assisted living facilities, residential board and care facilities, alcohol and drug centers, social rehabilitation facilities and group homes. The uses permitted in a Group R-4 occupancy have been expanded to include care recipients who may need a limited degree of verbal or physical assistance while responding to a fire or other emergency situation.

Through the use of the "Condition 1" and "Condition 2" classifications, the differences in the evacuation capabilities of the residents can be appropriately addressed. Condition 1 custodial care facilities reflect the 2012 IBC Group R-4 classification where the residents are capable of responding to an emergency without the assistance of others. Condition 2 has been added to address facilities where the residents are receiving custodial care but may require some assistance with evacuation. Further information can be found in the discussion of significant changes to Section 308.3.

311.1.1

Classification of Accessory Storage Spaces

CHANGE TYPE: Modification

CHANGE SUMMARY: Storage rooms less than 100 square feet in floor area are not to be classified as Group S, but rather as the same occupancy as the portion of the building to which they are accessory.

2015 CODE: **311.1.1 Accessory Storage Spaces.** A room or space used for storage purposes that is less than 100 square feet (9.3 m^2) in area and accessory to another occupancy shall be classified as part of that occupancy. The aggregate area of such rooms or spaces shall not exceed the allowable area limits of Section 508.2.

CHANGE SIGNIFICANCE: The classification approach to storage rooms is similar to that of other support areas within a building. Where the hazard level within the storage area is such that the provisions for the general building use do not adequately address the risks posed by the storage use, the storage room is to be classified as a Group S occupancy, typically Group S-1. There has been a historic consensus that small storage areas pose a limited concern above that anticipated due to the major use of the building, thus allowing such small storage spaces to be included as part of the major classification. However, there has never been a specific threshold established in the code addressing this issue. The code now recognizes that storage rooms less than 100 square feet in floor area are not to be classified as Group S, but rather as the same occupancy as the portion of the building to which they are accessory.

It is important to note that Section 311.1 does not mandate that storage rooms of 100 square feet or more be classified as Group S occupancies. The classification of a storage room, much like other uses within a building, continues to be based upon the need to classify uses based upon the hazard level they pose. It continues to be acceptable for the building official to classify storage spaces of 100 square feet or more as part of the occupancy in which they are located based upon the hazard level created. However, where the storage space is less than 100 square feet, it is now clear that a Group S classification is inappropriate.

The reference to the allowable area limits of Section 508.2 is somewhat vague, but it can be assumed that it is intended that the aggregate of all storage rooms that are less than 100 square feet in floor area cannot exceed 10 percent of the floor area of the story nor more than the tabular allowable area for nonsprinklered buildings as established in Section 508.2.

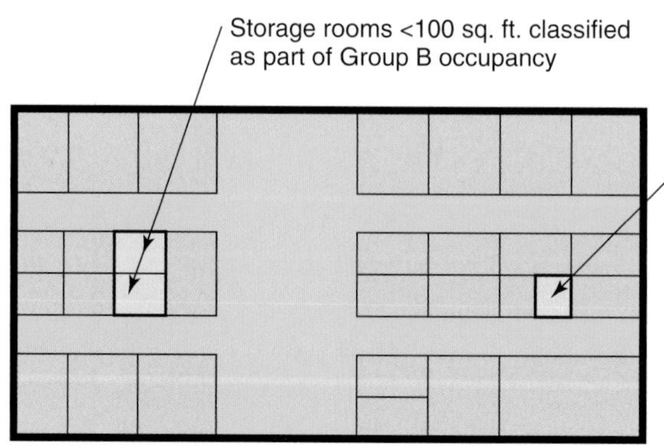
Storage rooms <100 sq. ft. classified as part of Group B occupancy
Storage room ≥100 sq. ft. classified as Group B or Group S as determined by building official
Example: Office suite with several small storage rooms

Accessory storage spaces

403.1, Exception Items 3 and 5

Applicability of High-Rise Provisions

© International Code Council

High-rise buildings

CHANGE TYPE: Clarification

CHANGE SUMMARY: Clarified code text now indicates that Group H-1 occupancies, as well as several specified types of Group H-2 and H-3 occupancies, are not required to comply with the high-rise provisions.

2015 CODE: 403.1 Applicability. High-rise buildings shall comply with Sections 403.2 through 403.6.

> **Exception:** The provisions of Sections 403.2 through 403.6 shall not apply to the following buildings and structures:
>
> 1. Airport traffic control towers in accordance with Section 412.3.
> 2. Open parking garages in accordance with Section 406.5.
> 3. ~~Buildings with~~ The portion of a building containing a Group A-5 occupancy in accordance with Section 303.6.
> 4. Special industrial occupancies in accordance with Section 503.1.1.
> 5. Buildings with:
>   - 5.1. A Group H-1 occupancy,
>   - 5.2. A Group H-2 occupancy in accordance with Sections 415.8, 415.9.2, 415.9.3 or 426.1; or,
>   - 5.3. A Group H-3 occupancy in accordance with Section 415.8.

CHANGE SIGNIFICANCE: Special fire- and life-safety requirements are applicable to those buildings of moderate or greater height that meet the definition of "high-rise buildings" in Chapter 2 of the IBC. The hazards created by the unique characteristics of high-rise buildings are addressed through a set of general and specific provisions that are applicable to buildings of considerable height. Certain buildings have traditionally been exempted from compliance with the high-rise provisions, among them buildings that contain a Group H-1, H-2 or H-3 occupancy. The code does not prohibit those specific occupancies from being in a high-rise building, but rather simply exempts them from compliance with Sections 403.2 through 403.6. The modification does not alter this application of the code, but it does directly indicate that only Group H-1 occupancies, as well as those specified types of Group H-2 and H-3 occupancies, are not required to comply with the high-rise provisions.

Group H-1 occupancies must be in a building used for no other purposes; therefore, a mixed-occupancy condition is prohibited. They are also limited to one story in height. For these reasons, it has been recognized that the special provisions of Section 403 have little, if any, relationship to buildings classified as Group H-1, and therefore need not apply. The reference to Section 415.8 for Group H-2 and H-3 occupancies addresses the need for such occupancies to be in detached buildings where specified hazardous materials are expected to exceed specified amounts. Consistent with the limits placed on Group H-1 occupancies, detached buildings classified as Group H-2 or H-3 are limited to one story in height, and as such, the high-rise building provisions have no real application.

403.1, Exception Items 3 and 5 continues

403.1, Exception Items 3 and 5 continued

Buildings containing one of three special types of Group H-2 occupancies are also exempted from compliance with the special high-rise requirements of Sections 403.2 through 403.6. Included are buildings where grain processing and storage conditions produce combustible dusts, liquefied petroleum gas facilities, and dry cleaning plants. The reasoning for the exemptions is consistent with that for the other listed Group H occupancies.

404.5, Exception

Atrium Smoke Control in Group I Occupancies

CHANGE TYPE: Modification

CHANGE SUMMARY: Smoke control is now required in atriums in Group I-2 occupancies, as well as those in Group I-1 occupancies classified as Condition 2, that connect two stories.

2015 CODE: **404.5 Smoke Control.** A smoke control system shall be installed in accordance with Section 909.

> **Exception:** In other than Group I-2, and Group I-1, Condition 2, smoke control is not required for atriums that connect only two stories.

© International Code Council

Group I-2, Condition 2 hospital

CHANGE SIGNIFICANCE: A major component of the life-safety system for an atrium is the required smoke control system. The control of any smoke that may be present in the atrium is important in maintaining a tenable environment for those occupants who may enter the atrium during means-of-egress travel. An exception had previously eliminated the requirement for smoke control in those atriums that connect only two stories. The exception has now been modified in a manner where it no longer applies to atriums in Group I-2 occupancies, as well as those Group I-1 occupancies classified as Condition 2.

Although the concept of omitting smoke control protection for two-story atriums is consistent with other provisions throughout the code that provide allowances for two-story conditions, it has now been deemed as unacceptable in specific healthcare applications. The exposure of care recipients who are incapable of self-preservation to a large unprotected vertical opening condition is seen as an unacceptable risk. The restriction also provides consistency with federal requirements that would not allow such an opening without either smoke control or passive separation. The exception is applicable to Group I-1, Condition 2 occupancies, as well as those uses classified as Group I-2. Persons in buildings classified in one of these groups are expected to need some degree of assistance from others in order to respond effectively to an emergency situation. This need for assistance often results in delayed relocation or evacuation for the care recipients, justifying the requirement for smoke control where two-story atriums occur in these two types of institutional occupancies.

404.9, 404.10

Egress Travel through an Atrium

© International Code Council

Egress travel through an atrium

CHANGE TYPE: Clarification

CHANGE SUMMARY: The three distinct travel distance conditions that could potentially occur for areas open to an atrium are now each addressed individually in order to clarify their application.

2015 CODE: **404.9 Exit Access Travel Distance.** ~~In other than the lowest level of the atrium, where the required means of egress is through the atrium space, the portion of exit access travel distance within the atrium space shall be not greater than 200 feet (60 960 mm). The travel distance requirements for areas of buildings open to the atrium and where access to the exits is not through the atrium, shall comply with the requirements of Section 1016.~~ Exit access travel distance for areas open to an atrium shall comply with the requirements of this section.

404.9.1 Egress Not Through the Atrium. Where required access to the exits is not through the atrium, exit access travel distance shall comply with Section 1017.

404.9.2 Exit Access Travel Distance at the Level of Exit Discharge. Where the path of egress travel is through an atrium space, exit access travel distance at the level of exit discharge shall be determined in accordance with Section 1017.

404.9.3 Exit Access Travel Distance at Other Than the Level of Exit Discharge. Where the path of egress travel is not at the level of exit discharge from the atrium, that portion of the total permitted exit access travel distance that occurs within the atrium shall be not greater than 200 feet (60 960 mm).

404.10 Interior Exit Stairways. A maximum of 50 percent of interior exit stairways are permitted to egress through an atrium on the level of exit discharge in accordance with Section 1028.

CHANGE SIGNIFICANCE: Of the special requirements in Section 404 applicable to atriums, there is limited focus on means-of-egress issues. The general provisions of Chapter 10 have been applicable under all conditions except for travel distance limitations on egress travel that occurs within the atrium space at other than the atrium level. Although there are no technical changes to the provisions, Section 404.9 has been reformatted to clarify the intent. The three distinct travel distance conditions that could potentially occur are now each addressed individually. Where travel does not occur through the atrium, or where travel within the atrium occurs only at the level of exit discharge, the general provisions of Section 1017 continue to apply. Where means-of-egress travel occurs at other than the level of exit discharge, the limitation of 200 feet of travel within the atrium also continues to apply.

A new provision addresses the extension of interior exit stairways through an atrium at the level of exit discharge. It has been clarified that if the atrium complies with the interior exit discharge provisions established in Exception 1 of Section 1028, a maximum of 50 percent of the interior exit stairways may egress through the atrium space at the discharge level.

406.3.1

Private Garage Floor-Area Limitation

CHANGE TYPE: Modification

CHANGE SUMMARY: A Group U private garage is now limited to a maximum floor area of 1000 square feet; however, multiple Group U private garages are permitted in the same building where they are compartmentalized by minimum 1-hour fire separations.

2015 CODE: 406.3.1 Classification. ~~Buildings or parts of buildings~~ Private garages and carports shall be classified as Group U occupancies. ~~because of the use or character of the occupancy~~ Each private garage shall be not greater than 1000 square feet (93 m^2) in area. ~~or one story in height except as provided in Section 406.3.2. Any building or portion thereof that exceeds the limitations specified in this section shall be classified in the occupancy group other than Group U that it most nearly resembles.~~ Multiple private garages are permitted in a building where each private garage is separated from the other private garages by 1-hour fire barriers in accordance with Section 707, or 1-hour horizontal assemblies in accordance with Section 711, or both.

~~**406.3.2 Area Increase.** Group U occupancies used for the storage of private or pleasure-type motor vehicles where no repair work is completed or fuel is dispensed are permitted to be 3,000 square feet (279 m^2) where the following provisions are met~~

1. ~~For a mixed occupancy building, the exterior wall and opening protection for the Group U portion of the building shall be as required for the major occupancy of the building. For such a mixed occupancy building, the allowable floor area of the building shall be as permitted for the major occupancy contained therein.~~
2. ~~For a building containing only a Group U occupancy, the exterior wall shall not be required to have a fire-resistance rating and the area of openings shall not be limited where the fire separation distance is 5 feet (1524 mm) or more.~~

406.3.1 continues

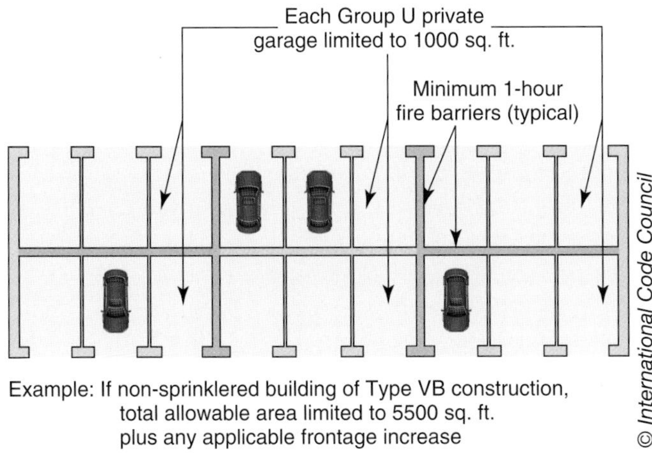
Each Group U private garage limited to 1000 sq. ft.
Minimum 1-hour fire barriers (typical)
Example: If non-sprinklered building of Type VB construction, total allowable area limited to 5500 sq. ft. plus any applicable frontage increase
© International Code Council

Private garage floor-area limits

406.3.1 continued

~~More than one 3,000-square-foot (279 m2) Group U occupancy shall be permitted to be in the same structure, provided each 3,000-square-foot (279 m^2) area is separated by fire walls complying with Section 706.~~

CHANGE SIGNIFICANCE: Although uncommon, fire hazards related to motor vehicle parking are a concern. Where the amount of floor area devoted to vehicle parking is relatively small, the code recognizes the limited hazard level by establishing a limited set of safeguards. Several significant issues have been addressed in the modifications to these provisions as they address private garages and carports.

An important factor in the revised application of the private garage provisions is the new definition of "private garage" in Section 202. In order to apply the provisions of Section 406.3, the garage must house motor vehicles used by the tenants of a building on the same premises. A more comprehensive review of the new definition can be found in the discussion of significant changes to Section 202. In addition to the new scoping limit provided by the definition, there are a number of other technical modifications applicable in the regulation of private garages.

A Group U private garage is now limited to a maximum floor area of 1000 square feet. Although the 1000-square-foot limit has always been the base requirement, an often-utilized area increase to a maximum of 3000 square feet was previously permitted. The allowance for an area increase has been deleted, effectively capping the permissible floor area at 1000 square feet per garage. A new provision establishes a method for multiple Group U private garages to be located in the same building. Where each private garage is separated from the other private garages in the building by minimum 1-hour fire barriers, 1-hour horizontal assemblies or both, multiple garages are permitted provided the total floor area of the building does not exceed the allowable building area for a Group U occupancy as established in Sections 503 and 506. As a part of the code revisions, carports are no longer limited in floor area by Section 406.3. As a Group U occupancy, they will, however, continue to be limited based on the general allowable area provisions of Chapter 5.

406.3.2

Private Parking Garage Ceiling Height

CHANGE TYPE: Clarification

CHANGE SUMMARY: The allowance for a 7-foot ceiling height previously permitted for public garages has now been extended to private garages and carports.

2015 CODE: **406.3.2 Clear Height.** In private garages and carports, the clear height in vehicle and pedestrian traffic areas shall be not less than 7 feet (2134 mm). Vehicle and pedestrian areas accommodating van-accessible parking shall comply with Section 1106.5.

CHANGE SIGNIFICANCE: Occupiable spaces require a minimum ceiling height of 7 feet 6 inches based upon the interior space dimension provisions of Section 1208.2. In addition, Section 1003.2 also mandates a minimum 7-foot, 6-inch ceiling height for any portion of the means of egress. Although Exception 7 to Section 1003.2 permits a reduction in minimum height in vehicular and pedestrian traffic areas in parking garages, the exception has previously only been applicable to public garages. Section 406.4.1 requires a minimum clear height of 7 feet in open parking garages and enclosed parking garages. The allowance for a 7-foot ceiling height has now been extended to private garages and carports.

Special provisions applicable to private garages and carports are provided in Section 406.3. In the past, a minimum required ceiling height was not established for these limited-size parking facilities. Therefore, the general height of 7 feet 6 inches was the default requirement for minimum clear ceiling height. The new provisions for private garages and carports are now fully consistent with those for public garages, allowing for a minimum ceiling height of 7 feet. Reference is also made to Section 1106.5 for minimum ceiling heights at vehicle and pedestrian traffic areas accommodating van accessible parking.

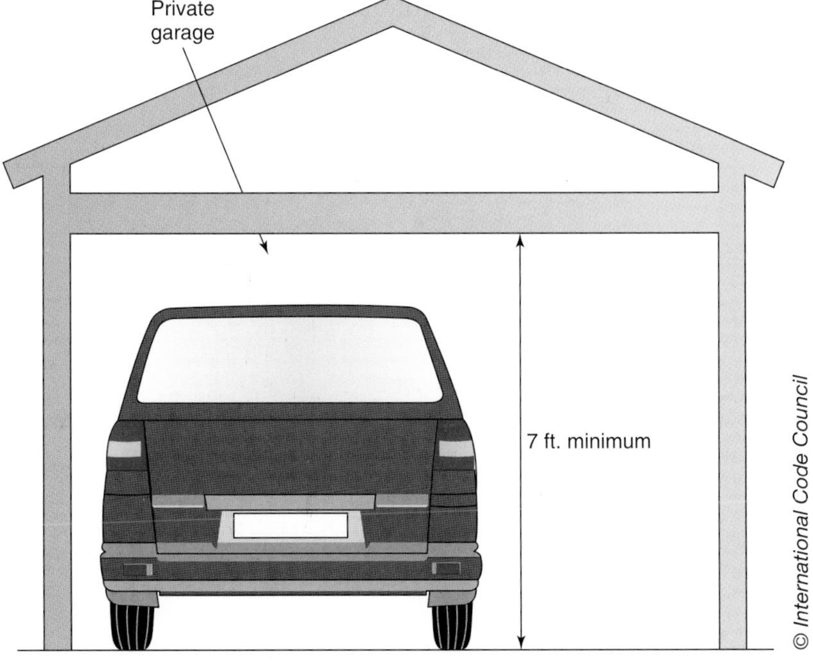

Private garage ceiling height

407.2.5

Group I-2 Shared Living Spaces

CHANGE TYPE: Addition

CHANGE SUMMARY: Shared living spaces, group meeting areas, and multipurpose therapeutic spaces are now permitted to be open to corridors in Group I-2, Condition 1 nursing homes provided five specific conditions are met.

2015 CODE: **407.2.5 Nursing Home Housing Units.** In Group I-2, Condition 1, occupancies, in areas where nursing home residents are housed, shared living spaces, group meeting or multipurpose therapeutic spaces shall be permitted to be open to the corridor, where all of the following criteria are met:

1. The walls and ceilings of the space are constructed as required for corridors.
2. The spaces are not occupied as resident sleeping rooms, treatment rooms, incidental uses in accordance with Section 509, or hazardous uses.
3. The open space is protected by an automatic fire detection system installed in accordance with Section 907.
4. The corridors onto which the spaces open, in the same smoke compartment, are protected by an automatic fire detection system installed in accordance with Section 907, or the smoke compartment in which the spaces are located is equipped throughout with quick-response sprinklers in accordance with Section 903.3.2.
5. The space is arranged so as not to obstruct access to the required exits.

CHANGE SIGNIFICANCE: Corridors in Group I-2 occupancies are intended to provide a direct egress path adequately separated from hazards in adjoining spaces by smoke partitions. Such protection is intended to

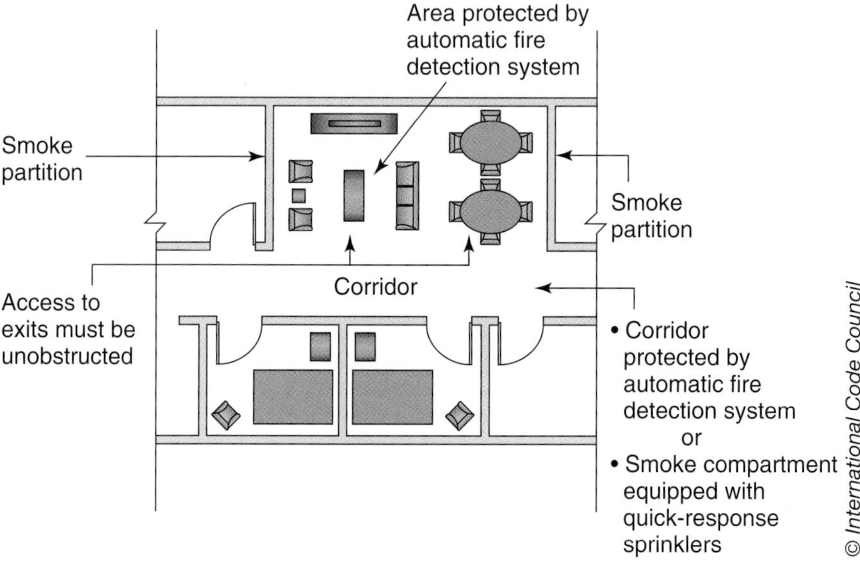

Group I-2 shared living space

provide a relatively smoke-free environment for the relocation of patients during a fire emergency. However, in hospitals, nursing homes and other Group I-2 occupancies, necessary modifications have been provided to facilitate the primary functioning of these types of healthcare facilities. These modifications recognize the special needs of these occupancies to provide the most efficient and effective healthcare services. An additional allowance applicable only to Group I-2, Condition 1 nursing homes has been included in the 2015 IBC, which now permits shared living spaces, group meeting areas or multipurpose therapeutic spaces to be open to the corridor provided five specific conditions are met.

In nursing home environments, residents are encouraged to spend time outside of their rooms. By providing a variety of shared living spaces open to the circulation/means-of-egress system, socialization and interaction are encouraged. Further, being able to preview activities that are occurring helps to encourage joining and allows reluctant participants to join at their own pace. Finally, a more open plan allows staff to more easily monitor residents throughout the day. For these reasons, the required physical separation of shared resident spaces from corridors has been eliminated.

In order to address the concerns of having common resident spaces open to the corridor system, several conditions have been established. The walls and ceilings of the shared spaces must be constructed in the same manner as required for corridor construction. This will result in surrounding construction equivalent to smoke partitions as set forth in Section 710. The shared spaces are limited in use, as they cannot be occupied as sleeping rooms, treatment rooms, incidental uses or hazardous uses. As expected, access to the required exits cannot be obstructed due to the arrangement of the spaces.

From a fire protection standpoint, the open space must be protected by an automatic fire detection system. In addition, the corridors into which the space open, within the same smoke compartment, must also be protected by an automatic fire detection system, or, as an alternative, the entire smoke compartment in which the spaces are located is to be protected throughout with quick-response sprinklers. Through the application of these conditions, the openness desired for Group I-2, Condition 1 nursing homes can be safely achieved.

407.2.6

Group I-2 Cooking Facilities

CHANGE TYPE: Addition

CHANGE SUMMARY: A room or space containing a cooking facility with domestic cooking appliances is now permitted to be open to the corridor in a Group I-2, Condition 1 nursing home provided 13 specific conditions are met.

2015 CODE: **407.2.6 Cooking Facilities.** In Group I-2 Condition 1, occupancies, rooms or spaces that contain a cooking facility with domestic cooking appliances shall be permitted to be open to the corridor where all of the following criteria are met:

1. The number of care recipients housed in the smoke compartment is not greater than 30.
2. The number of care recipients served by the cooking facility is not greater than 30.
3. Only one cooking facility area is permitted in a smoke compartment.
4. The types of domestic cooking appliances permitted are limited to ovens, cooktops, ranges, warmers and microwaves.
5. The corridor is a clearly identified space delineated by construction or floor pattern, material or color.
6. The space containing the domestic cooking facility shall be arranged so as not to obstruct access to the required exit.
7. A domestic cooking hood installed and constructed in accordance with Section 505 of the *International Mechanical Code* is provided over the cooktop or range.

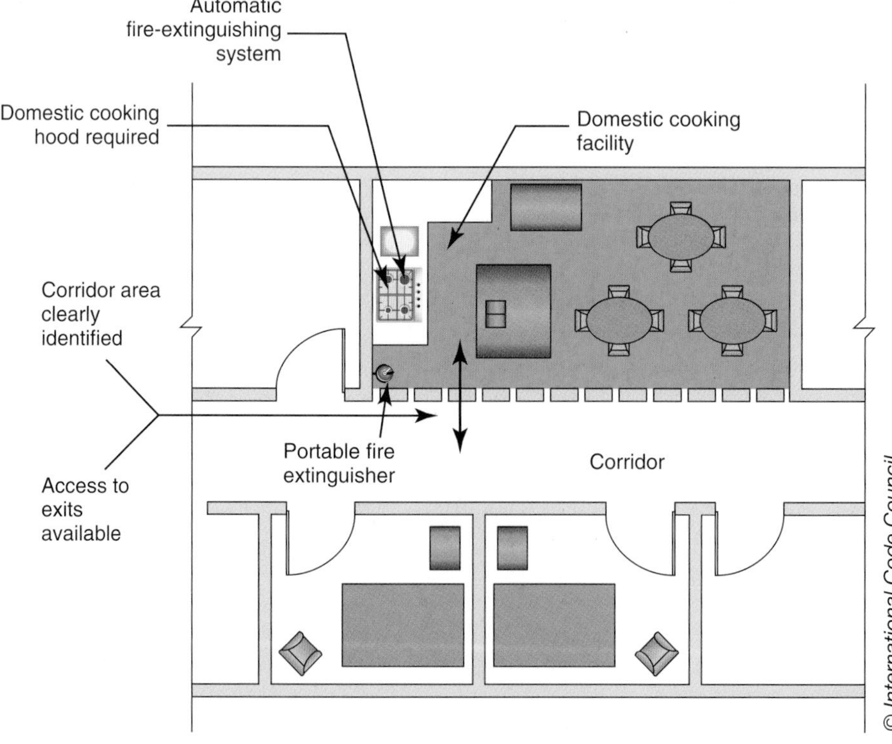

Group I-2 domestic cooking facilities

8. The domestic cooking hood provided over the cooktop or range shall be equipped with an automatic fire-extinguishing system of a type recognized for protection of domestic cooking equipment. Pre-engineered automatic extinguishing systems shall be tested in accordance with UL 300A and listed and labeled for the intended application. The system shall be installed in accordance with this code, its listing and the manufacturer's installation instructions.
9. A manual actuation device for the hood suppression system shall be installed in accordance with Sections 904.12.1 and 904.12.2.
10. An interlock device shall be provided such that upon activation of the hood suppression system, the power or fuel supply to the cooktop or range will be turned off.
11. A shut-off for the fuel and electrical power supply to the cooking equipment shall be provided in a location that is accessible only to staff.
12. A timer shall be provided that automatically deactivates the cooking appliances within a period of not more than 120 minutes.
13. A portable fire extinguisher shall be installed in accordance with Section 906 of the *International Fire Code.*

CHANGE SIGNIFICANCE: Nursing home means-of-egress corridors are intended to provide a direct egress path adequately separated from hazards in adjoining spaces through the use of smoke partitions, providing for a relatively smoke-free environment during the relocation of patients during a fire emergency. However, necessary modifications have been provided to facilitate the primary functioning of nursing homes and other types of healthcare facilities. These modifications recognize the special needs of these occupancies to provide the most efficient and effective healthcare services possible. An additional allowance applicable to Group I-2, Condition 1 nursing homes has been included in the 2015 IBC, which now permits a room or space containing a cooking facility with domestic cooking appliances to be open to the corridor provided 13 specific conditions are met.

As nursing homes transition from traditional models, the need for open and shared resident spaces is very important. A part of that group environment is a functioning kitchen that can also serve as the hearth of the home. Instead of a large, centralized, institutional kitchen where all meals are prepared and delivered to a central dining room or the resident's room, the new "household model" nursing home uses de-centralized kitchens and small dining areas to create the focus and feeling of home. Allowing kitchens that serve a small, defined group of residents to be open to common spaces and corridors is viewed as critically important to enhancing the feeling and memories of home for older adults. The code now allows spaces that contain domestic cooking facilities to be open to the corridor system provided 13 specific conditions are met.

Because unattended cooking equipment is a leading cause of fires in residential facilities, it is important that necessary safeguards be put in place to address the hazards involved. In addition, limitations on occupancy are necessary to allow for efficient egress should an emergency condition occur. It is expected that each smoke compartment functions as a distinct residential cooking and dining unit; therefore, no more than 30 care recipients can be housed in the smoke compartment or served by

407.2.6 continues

407.2.6 continued

the cooking facility. Only one cooking facility is permitted in each such smoke compartment and it is limited to domestic ovens, cooktops, ranges, warmers and microwave ovens.

It is important that the means of egress within and through the cooking area be unobstructed, obvious and easily identifiable. Some means shall be provided at the floor surface to clearly identify the path of travel through the cooking area, such as a contrasting floor pattern, material or color. In addition, access to required exits shall be maintained by arranging the cooking space in a manner that will not obstruct egress travel.

Additional conditions deal primarily with the fire protection and mechanical systems related to the cooking activity. An IMC-compliant domestic hood must be installed over the cooktop or range, protected by an appropriate fire-extinguishing system. A manual actuation device, along with an interlock device, shall be provided for the hood suppression system. Other safeguards required include the installation of an emergency shutoff for the fuel and electrical power supply, a timer that automatically deactivates the cooking appliances and the installation of a portable fire extinguisher.

407.5

Maximum Size of Group I-2 Smoke Compartments

CHANGE TYPE: Modification

CHANGE SUMMARY: The maximum allowable smoke compartment size for Group I-2, Condition 2 hospitals and similar occupancies has been increased to 40,000 square feet.

2015 CODE: 407.5 Smoke barriers. Smoke barriers shall be provided to subdivide every story used by persons receiving care, treatment or sleeping and to divide other stories with an occupant load of 50 or more persons, into no fewer than two smoke compartments. Such stories shall be divided into smoke compartments with an area of not more than 22,500 square feet (2092 m^2) in Group I-2, Condition 1, and not more than 40,000 square feet (3716 m^2) in Group I-2, Condition 2, and the ~~travel~~ distance of travel from any point in a smoke compartment to a smoke barrier door shall be not greater than 200 feet (60 960 mm). The smoke barrier shall be in accordance with Section 709.

CHANGE SIGNIFICANCE: Evacuation of a Group I-2 occupancy is often virtually impossible in the event of a fire, particularly in multistory structures. Horizontal relocation, on the other hand, is possible with a properly trained staff. As a result, the code makes provisions for compartmentalization on each story, so that if necessary, care recipients can be moved from one compartment to another. The size of each compartment has historically been limited to 22,500 square feet in floor area, with a maximum 200-foot distance of travel from any point in the compartment to a smoke barrier door. Although the provisions are unchanged for nursing homes and similar Group I-2, Condition 1 occupancies, the maximum allowable smoke compartment size for Group I-2, Condition 2 hospitals and similar occupancies has been increased to 40,000 square feet.

407.5 continues

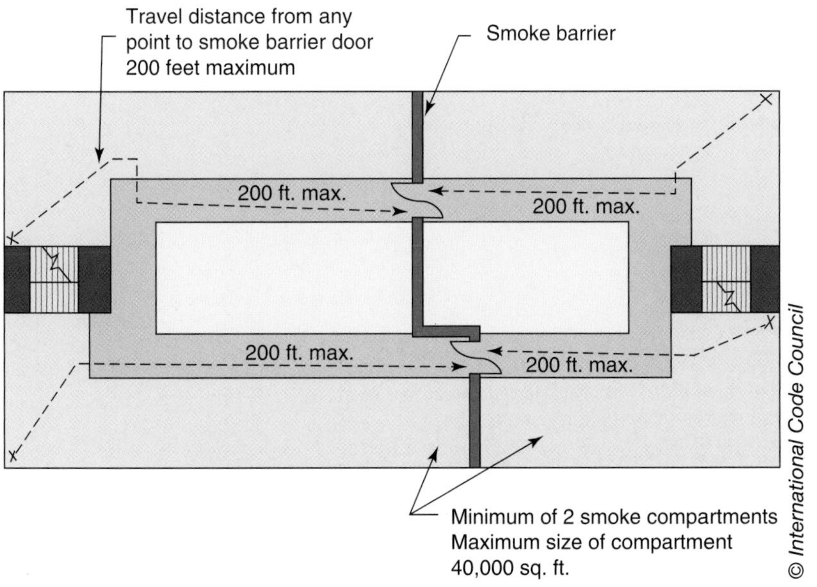

Travel distance from any point to smoke barrier door 200 feet maximum

Smoke barrier

200 ft. max.

200 ft. max.

200 ft. max.

200 ft. max.

Minimum of 2 smoke compartments Maximum size of compartment 40,000 sq. ft.

Group I-2, Condition 2 smoke compartments

407.5 continued

The increase in compartment size is supported by two basic issues: (1) recent increases in allowable travel distances without a corresponding increase in permitted smoke compartment size, and (2) floor area increases of functional patient areas with no increase in the occupant load within the compartment. At the advent of smoke compartmentalization in health-care facilities, compartment size was only limited by travel distance. The earlier travel distance limitations of 150 feet were extrapolated to a compartment area limit of 22,500 square feet (150 feet by 150 feet) when compartment size was first mandated. The current travel distance limit of 200 feet for Group I-2 occupancies is the basis for the increase in maximum compartment size to 40,000 square feet (200 feet by 200 feet).

The increase in the allowable smoke compartment size is also supported due to an increase in the size of patient treatment areas in hospitals. The primary reason for the increase is the equipment and utilities necessary for the treatment of a patient, such as patient monitoring, gases and diagnostic equipment, while maintaining staff access to the patients. In addition, patient treatment areas are increasing in size for a variety of reasons. It was determined that the current operational considerations required an increase to the smoke compartment size to match contemporary requirements, delivery of care and technologies. Because similar considerations are not necessarily present in nursing homes, the increase in compartment size was not considered applicable for Group I-2, Condition 1 occupancies.

Horizontal Sliding Doors at Stage Proscenium Opening

CHANGE TYPE: Addition

CHANGE SUMMARY: An additional method of stage proscenium opening protection has now been provided that permits the use of horizontal sliding doors having a minimum fire protection rating of 1 hour.

2015 CODE: 410.3.5 Proscenium Curtain. Where a proscenium wall is required to have a fire-resistance rating, the stage opening shall be provided with a fire curtain complying with NFPA 80<u>, horizontal sliding doors complying with Section 716.5.2 having a fire protection rating of at least 1 hour,</u> or an approved water curtain complying with Section 903.3.1.1 or, in facilities not utilizing the provisions of smoke-protected assembly seating in accordance with Section 1029.6.2, a smoke control system complying with Section 909 or natural ventilation designed to maintain the smoke level not less than 6 feet (1829 mm) above the floor of the means of egress.

CHANGE SIGNIFICANCE: Where a stage height exceeds 50 feet, a minimum 2-hour proscenium wall is required to provide a fire separation between the stage and the assembly seating area. The stage opening must also be appropriately protected in order to maintain the integrity of the separation. Because the opening in a fire-resistance-rated stage proscenium wall is typically too large to be protected with any usual type of fire protective, the code has historically required that it be protected with a fire curtain. Recent alternative methods permitted by the code include the use of a water curtain or assembly seating protection through a smoke control system or natural ventilation. An additional method of proscenium opening protection has now been included that permits the use of horizontal sliding doors having a minimum fire protection rating of 1 hour.

410.3.5 continues

Stage proscenium opening

410.3.5 continued Horizontal sliding doors have long been identified and approved by the code as a component of the means of egress when in compliance with eight criteria specified in Section 1010.1.4.3 addressing special-purpose horizontal sliding, accordion and folding doors. Such doors are often used to create fire separations at expansive openings where wall construction is not desirable. The provisions of Section 716.5.2, addressing the testing of fire door assemblies other than side-hinged and pivoted swinging doors, indicate that horizontal sliding fire assemblies are to be tested in accordance with NFPA 252 or UL 10B.

412.7

Travel Distance in Aircraft Manufacturing Facilities

Aircraft manufacturing facility

CHANGE TYPE: Modification

CHANGE SUMMARY: The travel distance allowances for aircraft manufacturing facilities have been significantly increased based upon a combination of the manufacturing area's height and floor area.

2015 CODE: **412.7 Aircraft Manufacturing Facilities.** In buildings used for the manufacturing of aircraft, exit access travel distances indicated in Section 1017.1 shall be increased in accordance with the following:

1. The building shall be of Type I or II construction.
2. Exit access travel distance shall not exceed the distances given in Table 412.7.

TABLE 412.7 Aircraft Manufacturing Exit Access Travel Distance

Height (feet)[b]	Manufacturing Area (sq. ft.)[a]					
	≥150,000	≥200,000	≥250,000	≥500,000	≥750,000	≥1,000,000
≥25	400	450	500	500	500	500
≥50	400	500	600	700	700	700
≥75	400	500	700	850	1,000	1,000
≥100	400	500	750	1,000	1,250	1,500

For SI: 1 foot = 304.8 mm

a. Contiguous floor area of the aircraft manufacturing facility having the indicated height.

b. Minimum height from finished floor to bottom of ceiling or roof slab or deck.

412.7.1 Ancillary Areas. Rooms, areas and spaces ancillary to the primary manufacturing area shall be permitted to egress through such area having a minimum height as indicated in Table 412.7. Exit access travel distance within the ancillary room, area or space shall not exceed that indicated in Table 1017.2 based on the occupancy classification of that ancillary area. Total exit access travel distance shall not exceed that indicated in Table 412.7.

CHANGE SIGNIFICANCE: Aircraft manufacturing facilities are classified by Section 306.2 as Group F-1 occupancies. As such, there is often difficulty for larger facilities to be in compliance with the 250-foot maximum travel distance for sprinklered Group F-1 occupancies without incorporating exit passageways or horizontal exits into the building's means-of-egress system. The use of either exit component has been considered somewhat problematic. Due to the compartmentalized nature of horizontal exits, they do not lend themselves to aircraft production processes or movement of the completely assembled aircraft. For similar reasons, exit passageways are generally installed below the floor of the manufacturing level. The use of underground passageways during a fire event or other emergency in such a large, high-volume space is widely

412.7 continues

412.7 continued

viewed as generally contrary to human nature. Once aware of an event, employees typically evacuate the building instinctively at the level with which they are most familiar. It is also relatively common for occupants to want to move away from the point of origin of a fire due to a person's sensory awareness within the entire open space. Given the fact that occupants sense safety as they move away from a fire incident, it is counterintuitive to enter an underground area unless as a final resort.

In spite of these observations, it is important that it can be demonstrated that such large-volume spaces are able to provide a tenable environment for the evacuation or relocation of building occupants. In support of this requirement, smoke and temperature fire modeling was conducted using the National Institute of Standards and Technology Fire Dynamics Simulator (NIST FDS) computer program. Results of the fire modeling activity, based on conservative assumptions, were used to establish the maximum travel distance increases as set forth in Table 412.7.

The travel distance allowances for aircraft manufacturing facilities are therefore based on a combination of building features; the minimum height from the finished floor to the bottom of the ceiling, roof slab or roof deck above; and the contiguous floor area of the aircraft manufacturing facility having the indicated height. The extended maximum travel distances can range well above the 250-foot limit established in Table 412.7, with a maximum of 1500 feet in structures over 1 million square feet with a clear manufacturing area height of at least 100 feet.

Ancillary spaces within or adjacent to the manufacturing area are permitted to egress through the manufacturing area having a minimum height as established by Table 412.7. The portion of travel within the ancillary spaces is limited to the travel distances set forth in Table 1017.2 based upon the occupancy classification of the ancillary space. The overall travel distance cannot exceed the distance indicated in Table 412.7.

423.3

Storm Shelters Serving Critical Emergency Operations Facilities

CHANGE TYPE: Addition

CHANGE SUMMARY: The construction of complying storm shelters is now required in critical emergency operations facilities where such facilities are located in geographical areas where the shelter design wind speed for tornadoes is at its highest.

2015 CODE: **423.3 Critical Emergency Operations.** In areas where the shelter design wind speed for tornadoes in accordance with Figure 304.2(1) of ICC 500 is 250 MPH, 911 call stations, emergency operation centers and fire, rescue, ambulance and police stations shall have a storm shelter constructed in accordance with ICC 500.

> **Exception:** Buildings meeting the requirements for shelter design in ICC 500.

CHANGE SIGNIFICANCE: In order to establish minimum requirements for structures and spaces designated as hurricane, tornado or combination shelters, the IBC first referenced ICC 500, *ICC/NSSA Standard for the Design and Construction of Storm Shelters*, in the 2009 edition. However, in the 2009 and 2012 editions of the IBC, there are no scoping provisions to enact the technical requirements. Although the IBC did not mandate that storm shelters be provided, it did regulate their design if they were constructed. The 2015 IBC now provides scoping provisions that mandate the construction of complying storm shelters in critical emergency operations facilities where such facilities are located in geographical areas where the shelter design wind speed for tornadoes is at its highest, 250 miles per hour.

423.3 continues

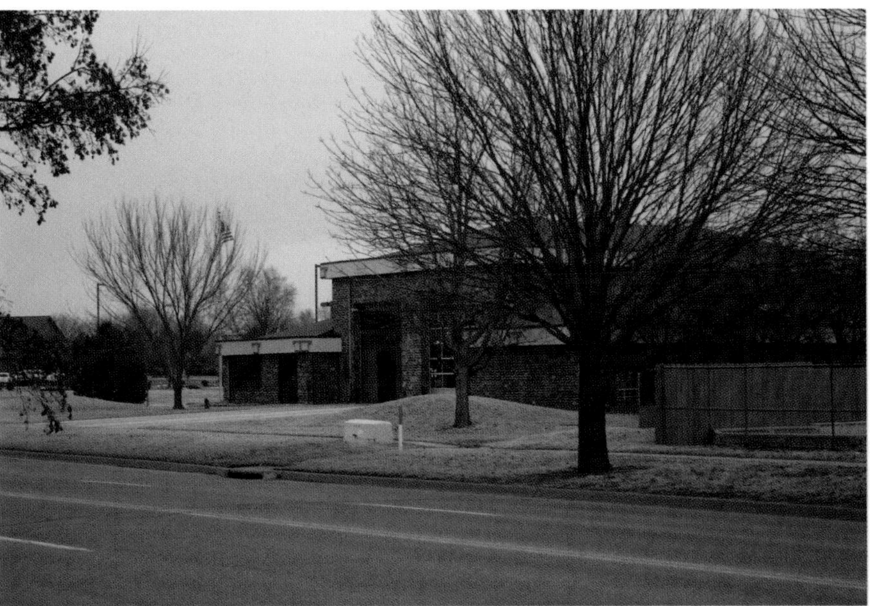

Fire station

423.3 continued

Critical facilities, such as emergency operations centers, fire and police stations, and buildings with similar functions, are essential for the delivery of vital services or the protection of a community. It is important to protect the occupants of such critical facilities struck by tornadoes, as well as to maintain continuity of operations for those facilities. Emergency operation centers, as well as police and fire rescue facilities, are critical to disaster response because an interruption in their operation as a result of building or equipment failure may prevent rescue operations, evacuation, assistance delivery or general maintenance of law and order, which can have serious consequences for the community after a storm event. Section 423.3 now requires storm shelters in those critical emergency operations facilities located in those areas of the United States where the shelter design wind speed for tornadoes is established at 250 miles per hour. These areas include a substantial portion of the central United States, as shown in Figure 304.2(1) of ICC 500.

In addition to the necessary administrative and application provisions established in Chapter 1 and definitions provided in Chapter 2, the ICC 500 standard includes criteria for structural design, siting, occupancy, means of egress, access, accessibility and fire safety. In addition, essential features and accessories are regulated independently for tornado shelters and hurricane shelters. The exception indicates that a separate storm shelter space is not required where the entire building housing the critical emergency operations is designed to meet the shelter requirements of ICC 500.

423.4

Storm Shelters Serving Group E Occupancies

CHANGE TYPE: Addition

CHANGE SUMMARY: Storm shelters are now required in Group E occupancies located in those areas of the United States where the shelter design wind speed for tornadoes is at its highest.

2015 CODE: **423.4 Group E Occupancies.** In areas where the shelter design wind speed for tornadoes is 250 MPH in accordance with Figure 304.2(1) of ICC 500, all Group E occupancies with an aggregate occupant load of 50 or more shall have a storm shelter constructed in accordance with ICC 500. The shelter shall be capable of housing the total occupant load of the Group E occupancy.

Exceptions:

1. Group E day care facilities.
2. Group E occupancies accessory to places of religious worship.
3. Buildings meeting the requirements for shelter design in ICC 500.

CHANGE SIGNIFICANCE: In order to establish minimum requirements for structures and spaces designated as hurricane, tornado or combination shelters, the IBC first referenced ICC 500, *ICC/NSSA Standard for the Design and Construction of Storm Shelters*, in the 2009 edition. However, in the 2009 and 2012 editions of the IBC, there are no scoping provisions to enact the technical requirements. Although the IBC did not mandate that storm shelters be provided, it did regulate their design if they were constructed. The 2015 IBC now provides scoping provisions that mandate the construction of complying storm shelters in specified Group E educational occupancies.

423.4 continues

School gymnasium constructed as tornado shelter

423.4 continued

Due to unpredictable and often very short tornado warning time, there are many high-wind events where it is unfeasible to evacuate school buildings. Therefore, it is very important that refuge areas be provided that are capable of providing a near absolute level of protection for these vulnerable individuals. Section 423.4 now requires storm shelters in Group E occupancies located in those areas of the United States where the shelter design wind speed for tornadoes is at its highest. These areas, where the shelter design wind speed for tornadoes is 250 miles per hour, include a substantial portion of the central United States, as shown in Figure 304.2(1) of ICC 500.

In addition to the necessary administrative and application provisions established in Chapter 1 and the definitions provided in Chapter 2, the ICC 500 standard includes criteria for structural design, siting, occupancy, means of egress, access, accessibility, and fire safety. In addition, essential features and accessories are regulated independently for tornado shelters and hurricane shelters.

The IBC mandates that the shelter be designed to house the total occupant load assigned to the Group E occupancy. As a general requirement, the provisions apply to all Group E occupancies having an occupant load of 50 or more. However, there are two types of educational occupancies where the construction of a storm shelter is not required. Group E day-care facilities are exempt, as are Group E occupancies accessory to places of religious worship. In addition, a separate storm shelter space is not required where the entire Group E building is designed to meet the shelter requirements of ICC 500.

503

General Building Height and Area Limitations

CHANGE TYPE: Clarification

CHANGE SUMMARY: The provisions regulating building height and area limitations have been extensively revised with no change in technical application in order to provide an increased degree of user-friendliness and technical consistency.

2015 CODE:

SECTION 503
GENERAL BUILDING HEIGHT AND AREA LIMITATIONS

503.1 General. Unless otherwise specifically modified in Chapter 4 and this chapter, ~~the~~ building height, number of stories and building area shall not exceed the limits specified in ~~Table 503~~ Sections 504 and 506 based on the type of construction as determined by Section 602 and the occupancies as determined by Section 302 except as modified hereafter. Building height, number of stories and building area provisions shall be applied independently. Each portion of a building separated by one or more fire walls complying with Section 706 shall be considered to be a separate building.

CHANGE SIGNIFICANCE: The IBC regulates the size of buildings in order to limit to a reasonable level the magnitude of a fire that potentially may develop. The size of a building is controlled by its floor area and height, both of which are limited by the IBC. Although floor-area limitations are concerned primarily with property damage, life safety is enhanced as well by the fact than in the larger buildings there are typically more people at

503 continues

Large assembly occupancy

503 continued

risk during a fire. Height restrictions are imposed to primarily address egress concerns and fire department access limitations. The provisions regulating building height and area limitations have been extensively revised in order to provide an increased degree of user-friendliness and technical consistency.

Table 503, representing the unmodified base allowable area and height data, has historically been the starting point in the determination of all three limiting conditions; a building's allowable height in feet, its allowable height in stories above grade plane and its allowable floor area per story. The limiting values in Table 503 of the 2012 IBC have now been separated into three separate tables and placed in context at the appropriate technical sections for the design or review process. In addition, the allowable values within the new tables represent whether or not the building is sprinklered, and if it is sprinklered, the type of sprinkler system provided. Occupancy classification and construction type continue to be the other required variables necessary to determine the maximum allowable height and area of a building.

Integrating all three required variables into the three revised tables effectively eliminates or seamlessly incorporates many of the previous provisions and exceptions, reducing misinterpretation of the code's intent. In addition, the revised format uses new terminology, along with changes in organization, to improve on a somewhat confusing multistep process for value determination. Through an improved sequential format and technical consolidation, this process has been simplified, resulting in consistent allowable height and area determinations.

Because Code Change G101-12, as amended, extensively revised the allowable height and area provisions of Sections 503, 504 and 506, the entire code change text is too extensive to be included here. Refer to the *2015 IBC Code Changes Resource Collection* for the complete text and rationale of the code change.

Tables 504.3, 504.4

Building Height and Number of Stories

CHANGE TYPE: Clarification

CHANGE SUMMARY: In order to increase the degree of user-friendliness of the process by which the allowable building height provisions are determined, Table 503 has now been reformatted as Tables 504.3 (allowable height in feet) and 504.4 (allowable number of stories above grade plane), and any applicable sprinkler increase has been incorporated directly into the new tables.

2015 CODE: ~~**504.1 General.** The building height permitted by Table 503 shall be increased in accordance with Sections 504.2 and 504.3.~~

504.3 Height in Feet. The maximum height, in feet, of a building shall not exceed the limits specified in Table 504.3.

504.4 Number of Stories. The maximum number of stories of a building shall not exceed the limits specified in Table 504.4.

Tables 504.3, 504.4 continues

TABLE ~~503~~ 504.3[a] Allowable Building Heights ~~and Areas~~ in Feet Above Grade Plane

		Type of Construction								
		Type I		Type II		Type III		Type IV	Type V	
Occupancy Classification	See Footnotes	A	B	A	B	A	B	HT	A	B
A,B,E,F,M,S,U	NS[b]	UL	160	65	55	65	55	65	50	40
	S	UL	180	85	75	85	75	85	70	60
H-1, H-2, H-3, H-5	NS[c,d] S	UL	160	65	55	65	55	65	50	40

Note: UL = Unlimited; NS = Buildings not equipped throughout with an automatic sprinkler system; S = Buildings equipped throughout with an automatic sprinkler system installed in accordance with Section 903.3.1.1

(Only a portion of Table 504.3 is shown above.)

TABLE ~~503~~ 504.4[a,b] Allowable ~~Building Heights and Areas~~ Number of Stories Above Grade Plane

		Type of Construction								
		Type I		Type II		Type III		Type IV	Type V	
Occupancy Classification	See Foot-Notes	A	B	A	B	A	B	HT	A	B
A-1	NS	UL	5	3	2	3	2	3	2	1
	S	UL	6	4	3	4	3	4	3	2
A-2	NS	UL	11	3	2	3	2	3	2	1
	S	UL	12	4	3	4	3	4	3	2
A-3	NS	UL	11	3	2	3	2	3	2	1
	S	UL	12	4	3	4	3	4	3	2

Note: UL = Unlimited; NS = Buildings not equipped throughout with an automatic sprinkler system; S = Buildings equipped throughout with an automatic sprinkler system installed in accordance with Section 903.3.1.1

(Only a portion of Table 504.4 is shown above.)

Tables 504.3, 504.4 continued

EXAMPLE:

GIVEN: A fully sprinklered Group A-2 restaurant in a building of Type VB construction. Building has two stories above grade plane and is 32 feet in hight.

DETERMINE: If in compliance with maximum allowable building height in feet and stories above grade plane.

2015 IBC Procedure for Determining Allowable Height and Compliance Review:

Step 1: Review and apply applicable provisions of Section 503 for general building height determination.

Step 2: Review and apply applicable provisions of Section 504 regarding the determination of allowable building height in feet and allowable number of stories above grade plane.

Step 3: Determine allowable building height in feet as established in Table 504.3. Verify actual building height in feet does not exceed allowable height.

Allowable height in feet from Table 504.3:	60 feet	
Actual height:	32 feet	OK

Step 4: Determine allowable building height in stories above grade plane as established in Table 504.4. Verify actual number of stories above grade plane does not exceed allowable height.

Allowable height in stories above grade plane from Table 504.4:	2 stories	
Actual height in stories above grade plane:	2 stories	OK

Allowable height determination example

CHANGE SIGNIFICANCE: The allowable height of a building, both in feet and in number of stories above grade plane, is based upon the two fundamental aspects of building classification: occupancy classification and type of construction classification. Based upon these two criteria, Table 503 has historically established a basic allowable height limit. This limit can potentially be increased where the building is sprinklered. In order to increase the degree of user-friendliness in the process by which the allowable height provisions are determined, the previous table has now been reformatted as Tables 504.3 (allowable height in feet) and 504.4 (allowable number of stories above grade plane), and any applicable sprinkler increase has been incorporated directly into the new tables. The maximum allowable height values can now be obtained by simply applying the building's construction type, occupancy classification and sprinkler protection variables directly in the tables. No additional calculation is required.

Although the extent of the revision appears to be considerable, the technical aspects regarding allowable building height remain unchanged, as only the format and terminology were modified.

505.2.3, Exception 2

Mezzanine Openness

CHANGE TYPE: Modification

CHANGE SUMMARY: Direct access to at least one exit at the mezzanine level is no longer required for those enclosed mezzanines regulated by Exception 2 of Section 505.2.3.

2015 CODE: 505.2.3 Openness. A mezzanine shall be open and unobstructed to the room in which such mezzanine is located except for walls not more than 42 inches (1067 mm) in height, columns and posts.

Exceptions:

1. Mezzanines or portions thereof are not required to be open to the room in which the mezzanines are located, provided that the occupant load of the aggregate area of the enclosed space is not greater than 10.
2. A mezzanine having two or more exits or access to exits is not required to be open to the room in which the mezzanine is located~~ if at least one of the means of egress provides direct access to an exit from the mezzanine level~~.
3. Mezzanines or portions thereof are not required to be open to the room in which the mezzanines are located, provided that the aggregate floor area of the enclosed space is not greater than 10 percent of the mezzanine area.
4. In industrial facilities, mezzanines used for control equipment are permitted to be glazed on all sides.
5. In occupancies other than Groups H and I, that are no more than two stories above grade plane and equipped throughout with an automatic sprinkler system in accordance with Section 903.3.1.1, a mezzanine having two or more means of egress shall not be required to be open to the room in which the mezzanine is located.

505.2.3, Exception 2 continues

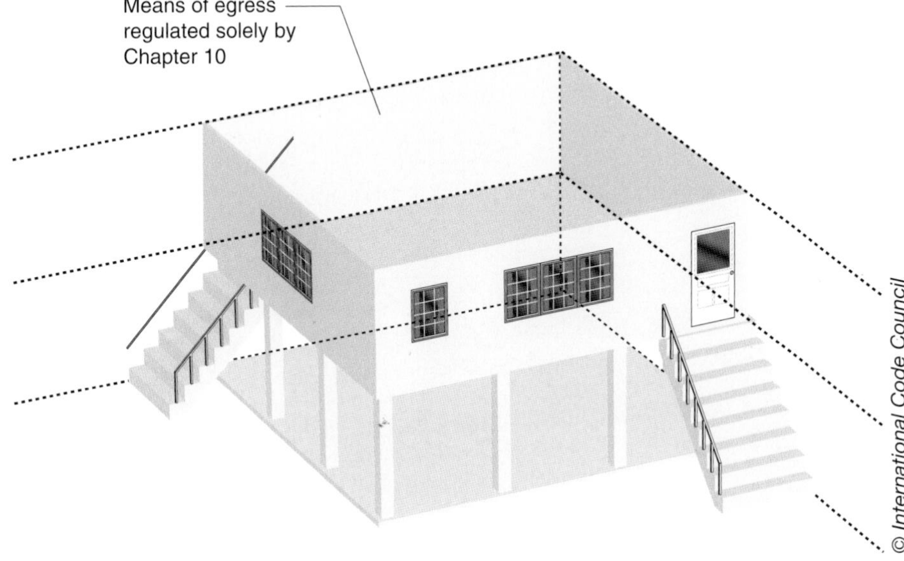
Means of egress regulated solely by Chapter 10
© International Code Council

Enclosed mezzanine with two means of egress

505.2.3, Exception 2 continued

CHANGE SIGNIFICANCE: As a general rule, a mezzanine must be open to the room or space in which it is located. An intermediate floor of limited size without enclosure has historically been viewed as having no significant safety hazard above that of the room or space below. The occupants of the mezzanine by means of sight, smell or hearing will be able to determine if there is some emergency or fire that takes place either on the mezzanine or in the room in which the mezzanine is located. This required openness, although important to the overall mezzanine concept, is not always mandated. Five varied exceptions to Section 505.2.3 permit partial or complete enclosure of the mezzanine level where specified conditions are met.

Exception 2 has previously allowed a mezzanine to be fully or partially enclosed where it is provided with at least two means of egress, but only if at least one of the means of egress provides direct access to an exit component. The exception has been modified by removing the mandate that direct access to at least one exit at the mezzanine level be provided. There has been some confusion as to the proper application of the term "direct access," resulting in varied applications of the provisions. Typically, it was expected that access be provided at the mezzanine level to an exit component, such as an interior exit stairway or exterior exit stairway, before vertical travel occurs. The change in the exception is deemed to be an acceptable reduction in the required egress protection afforded to occupants of an enclosed mezzanine.

It should be noted that this revision to Exception 2 effectively renders Exception 5 as unnecessary. Exception 2 as modified is applicable to all occupancy groups and there is no distinction as to the location within the building where the mezzanine is located. Therefore, it is less restrictive in its application than Exception 5 while providing the same allowance for mezzanine enclosure.

CHANGE TYPE: Clarification

Table 506.2

Building Area

CHANGE SUMMARY: In order to increase the degree of user-friendliness of the process by which the allowable building area provisions are determined, Table 503 has now been reformatted as new Table 506.2 (allowable area factor in square feet), and any applicable sprinkler increase has been incorporated directly into the new table.

2015 CODE: 506.2 Allowable Area Determination. The allowable area of a building shall be determined in accordance with the applicable provisions of Sections 506.2.1 through 506.2.4 and Section 506.3.

TABLE ~~503~~ 506.2[a,b] **Allowable ~~Building Heights and~~ Areas Factor (A_t = NS, S1, S13R or SM, as applicable) in Square Feet**

Occupancy Classification	See Footnotes	Type of Construction								
		Type I		Type II		Type III		Type IV	Type V	
		A	B	A	B	A	B	HT	A	B
A-1	NS	UL	UL	15,500	8,500	14,000	8,500	15,000	11,500	5,500
	S1	UL	UL	62,000	34,000	56,000	34,000	60,000	46,000	22,000
	SM	UL	UL	46,500	25,500	42,000	25,500	45,000	34,500	16,500
A-2	NS	UL	UL	15,500	9,500	14,000	9,500	15,000	11,500	6,000
	S1	UL	UL	62,000	38,000	56,000	38,000	60,000	46,000	24,000
	SM	UL	UL	46,500	28,500	42,000	28,500	45,000	34,500	18,000
A-3	NS	UL	UL	15,500	9,500	14,000	9,500	15,000	11,500	6,000
	S1	UL	UL	62,000	38,000	56,000	38,000	60,000	46,000	24,000
	SM	UL	UL	46,500	28,500	42,000	28,500	45,000	34,500	18,000

(Only a portion of Table 506.2 is shown above.)

CHANGE SIGNIFICANCE: The allowable area of a building is based upon the two fundamental aspects of building classification: occupancy classification and type of construction classification. Based upon these two criteria, Table 503 has historically established a basic allowable area limit. This limit can potentially be increased where the building is sprinklered, as well as where significant open space is provided at the building's perimeter. In order to increase the degree of user-friendliness in the process by which the allowable area provisions are determined, the previous table has now been reformatted as Table 506.2 (allowable area factor in square feet), and any applicable sprinkler increase has been

Table 506.2 continues

Table 506.2 continued

EXAMPLE:

GIVEN: A fully sprinklered Group A-2 restaurant in a building of Type VB construction. Building has two stories above grade plane with 8,500 square feet per story. A 25% increase is permited due to open frontage.

DETERMINE: If in compliance with maximum allowable building area.

2015 IBC Procedure for Determining Allowable Area and Compliance Review:

Step 1: Review and apply applicable provisions of Section 503 for general building area determination.

Step 2: Review and apply applicable provisions of Section 506 regarding the determination of allowable area.

Step 3: Determine allowable building area factor (A_t) as established in Table 506.2, based upon SM value (sprinklered, multi-story condition).

Allowable area factor in square feet from Table 506.2: A_t = 18,000 square feet

Step 4: Determine applicable allowable area frontage increase as established in Section 506.3.

Frontage increased based on example: $I_f = 0.25$

Step 5: Determine maximum building allowable area using Equation 5-2, $A_a = [A_t + (NS \times I_f)] \times S_a$

$[18{,}000 + (6{,}000 \times 0.25)] \times 2 = (18{,}000 + 1{,}500) \times 2 = 39{,}000$ square feet OK

Step 6: Determine maximum allowable area per story using Equation 5-2, with $S_a = 1$

$[18{,}000 + (6{,}000 \times 0.25)] \times 1 = (18{,}000 + 1{,}500) \times 1 = 19{,}500$ square feet OK

Allowable area determination example

incorporated directly into the new table. The maximum allowable building area can now be obtained by applying the building's construction type, occupancy classification and sprinkler protection variables directly in the table, along with any applicable increase for open frontage.

Although the extent of the revision appears to be considerable, the technical aspects regarding allowable building area remain unchanged, as only the format and terminology were modified.

507.1

Basements in Unlimited Area Buildings

CHANGE TYPE: Clarification

CHANGE SUMMARY: The allowance of a single-story basement in unlimited area buildings has now been clarified.

2015 CODE: **507.1 General.** The area of buildings of the occupancies and configurations specified in Sections 507.1 through 507.12 shall not be limited. Basements not more than one story below grade plane shall be permitted.

CHANGE SIGNIFICANCE: One of the key concepts in the application of the unlimited area building provisions of Section 507 is the limitation on building height. In order to provide the degree of fire department access necessary to address such buildings under fire conditions, a limit on vertical travel has always been an important element of the overall level of fire safety. Unlimited area buildings permitted by Section 507 cannot exceed two stories in height above grade plane, and in some cases a limitation of one story is established. The allowance for a single-story basement in such buildings, which has previously never been specifically addressed, is now clearly set forth in Section 507.1.

The allowances for various types of unlimited area buildings specifically all include conditions regulating the maximum number of stories that are permitted above the grade plane. In previous code editions, there has been no mention of how to address a story that is considered a basement. Although there was no specific prohibition of basements in unlimited area buildings, there has been much confusion as to the appropriateness of such a condition. In order to maintain consistency in relation to the level of exit discharge, basements are limited to one story below grade plane.

Unlimited area building

507.9

Group H-5 in Unlimited Area Buildings

CHANGE TYPE: Addition

CHANGE SUMMARY: Group H-5 buildings are now permitted to be unlimited in area under the special provisions of Section 507.

2015 CODE: **507.9 Unlimited Mixed Occupancy Buildings with Group H-5.** The area of a Group B, F, H-5, M or S building no more than two stories above grade plane shall not be limited where the building is equipped throughout with an automatic sprinkler system in accordance with Section 903.3.1.1, and is surrounded and adjoined by public ways or yards not less than 60 feet (18 288 mm) in width, provided all of the following criteria are met:

1. Buildings containing Group H-5 occupancy shall be of Type I or II construction.
2. Each area used for Group H-5 occupancy shall be separated from other occupancies as required in Sections 415.11 and 508.4.
3. Each area used for Group H-5 occupancy shall not exceed the maximum allowable area permitted for such occupancies in Section 503.1 including modifications of Section 506.

   Exception: Where the Group H-5 occupancy exceeds the maximum allowable area, the Group H-5 shall be subdivided into areas that are separated by 2-hour fire barriers.

CHANGE SIGNIFICANCE: The unlimited area building provisions of Section 507 are provided as an alternative approach to the allowable building area requirements of Sections 503 and 506. As a general rule, as buildings become larger, the permitted types of construction are reduced. At some point, the floor area becomes so large that the building must be of Type I construction. Section 507 utilizes a combination of building features and conditions to allow buildings of any size to be constructed of various construction types. This allowance is based upon four key considerations: limited building height, significant perimeter open space, sprinkler protection, and low- and moderate-hazard occupancy. As a result, buildings containing significant high-hazard Group H occupancies have historically been prohibited from benefitting from the unlimited area buildings provisions of Section 507. The prohibition has now been lifted for Group H-5 buildings no more than two stories in height above grade plane.

The Group H-5 occupancy category was created some time ago to standardize regulations for semiconductor manufacturing facilities. The Group H-5 category requires engineering and fire-safety controls that reduce the overall hazard of the occupancy to a level regarded to be equivalent to that of a moderate-hazard Group B occupancy. The mitigating provisions of Section 415.11 have effectively equalized the two occupancies in terms of relative hazard such that, in concert with the three additional criteria newly established in Section 507.9, the allowance for Group H-5 unlimited area buildings is deemed to be appropriate.

The conditions under which Group H-5 occupancies are permitted in unlimited area buildings are consistent with those for other moderate-hazard occupancies. Group H-5 unlimited area buildings are limited to a maximum of two stories above grade plane, must be sprinklered throughout, and are required to be surrounded by a minimum of 60 feet of open space. Although

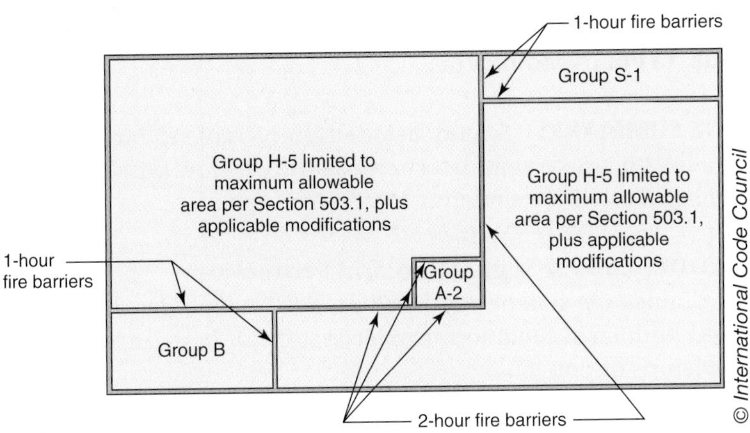

Group H-5 in unlimited area building example

EXAMPLE:

GIVEN: A mixed-occupancy semiconductor fabrication facility with related support occupancies of Groups A-2, B and S-1. The Type IIB building is fully sprinklered, surrounded with minimum 60-foot yards and public ways, and one story in height. The intent is to regulate the facility as an unlimited area building under the provisions of Section 507.9.

DETERMINE: The minimum required occupancy separations and maximum allowable area for the Group H-5 portions of the building.

SOLUTION: Per Table 508.4, a minimum 1-hour separation is required between the Group B occupancy and the Group H-5 occupancy. A minimum 1-hour separation is also required between the Group S-1 occupancy and the Group H-5 occupancy.

Per the Exception to Section 507.9, a minimum 2-hour separation is required between the Group H-5 occupancy areas in a manner that the maximum allowable area of 97,750 square feet for each of the Group H-5 areas is not exceeded as determined below.

Allowable area factor (A_t) for Group H-5 of Type IIB construction:	92,000 square feet
Frontage increase based on perimeter of each Group H-5 occupancy: (I_f = 23,000 x 0.25 = 5,750 square feet)	5,750 square feet
Maximum allowable area for each of the Group H-5 portions:	97,750 square feet

Note: The Group A-2 occupancy must comply with the accessory occupancy provisions of Section 508.2 as set forth in Section 507.1.1.

the provision will typically be applied where mixed-occupancy conditions exist, there is no prohibition of a single-occupancy Group H-5 unlimited area building. It is expected that laboratories, offices, conference rooms, storage areas and other uses with differing occupancy classifications will be present to support the multiple cleanrooms and other Group H-5 areas.

Three additional conditions have been established to directly address the presence of the Group H-5 activities. The unlimited area building must be of Type I or II noncombustible construction. The separated occupancy provisions of Section 508.4 are to be utilized, where applicable, along with any other fire separation requirements established in Section 415.11. And most uniquely, each area used for Group H-5 purposes is limited in floor area to the maximum allowable area permitted by Section 503, including applicable area increases. Where multiple Group H-5 areas are provided, they are required to be subdivided by minimum 2-hour fire barriers. Although an unlimited amount of Group H-5 floor area is permitted within a complying unlimited area building, it may be necessary to subdivide the floor area using minimum 2-hour separations due to the limits of Item 3.

Table 509

Fire Protection from Incidental Uses

CHANGE TYPE: Modification

CHANGE SUMMARY: A more detailed analysis of various support spaces within a healthcare or ambulatory care facility is now possible due to modifications to Table 509 regulating incidental uses.

2015 CODE: **509.4 Separation and Protection.** The incidental uses listed in Table 509 shall be separated from the remainder of the building or equipped with an automatic sprinkler system, or both, in accordance with the provisions of that table.

TABLE 509 Incidental Uses

Room or Area	Separation and/or Protection
Furnace room where any piece of equipment is over 400,000 Btu per hour input.	1 hour or provide automatic sprinkler system
Rooms with boilers where the largest piece of equipment is over 15 psi and 10 horsepower	1 hour or provide automatic sprinkler system
Refrigerant machinery room	1 hour or provide automatic sprinkler system
Hydrogen ~~cutoff~~ <u>fuel gas</u> rooms, not classified as Group H	1 hour in Group B, F, M, S and U occupancies; 2 hours in Group A, E, I and R occupancies.
Incinerator rooms	2 hours and provide automatic sprinkler system
Paint shops, not classified as Group H, located in occupancies other than Group F	2 hours; or 1 hour and provide automatic sprinkler system
In Group E occupancies, laboratories and vocational shops not classified as Group H	1 hour or provide automatic sprinkler system
In Group I-2 occupancies, laboratories not classified as Group H	1 hour ~~or~~ <u>and</u> provide automatic sprinkler system
<u>In ambulatory care facilities, laboratories not classified as Group H</u>	<u>1 hour or provide automatic sprinkler system</u>
Laundry rooms over 100 square feet	1 hour or provide automatic sprinkler system
<u>In Group I-2, laundry rooms over 100 square feet</u>	<u>1 hour</u>
Group I-3 cells <u>and Group I-2 patient rooms</u> equipped with padded surfaces	1 hour
<u>In Group I-2, physical plant maintenance shops.</u>	<u>1 hour</u>
In ambulatory care facilities or Group I-2 occupancies, waste and linen collection rooms <u>with containers that have an aggregate volume of 10 cubic feet or greater</u>	1 hour
<u>In other than ambulatory care facilities and Group I-2 occupancies,</u> waste and linen collection rooms over 100 square feet	1 hour or provide automatic sprinkler system
<u>In ambulatory care facilities or Group I-2 occupancies, storage rooms greater than 100 square feet</u>	<u>1 hour</u>
Stationary storage battery systems having a liquid electrolyte capacity of more than 50 gallons for flooded lead-acid, nickel cadmium or VRLA, or more than 1000 pounds for lithium-ion and lithium metal polymer used for facility standby power, emergency power or uninterruptable power supplies	1 hour in Group B, F, M, S and U occupancies; 2 hours in Group A, E, I and R occupancies.

For SI: 1 square foot = 0.0929 m^2, 1 pound per square inch (psi) = 6.9 kPa, 1 British thermal unit (Btu) per hour = 0.293 watts, 1 horsepower = 746 watts, 1 gallon = 3.785 L.

INCIDENTAL USES SPECIFIC TO AMBULATORY CARE FACILITIES		
Room or Area	**2015 IBC**	**2012 IBC**
Laboratories not classified as Group H occupancies	1-hour separation or provide automatic sprinkler system	Not considered as an incidental use
Waste and linen collection rooms	1-hour separation for rooms where containers have an aggregate volume of 10 cubic feet or more	1-hour regardless of amount of collection
Storage rooms more than 100 square feet in floor area	1-hour separation	Not considered as an incidental use

INCIDENTAL USES SPECIFIC TO GROUP I-2 OCCUPANCIES		
Room or Area	**2015 IBC**	**2012 IBC**
Laboratories not classified as Group H occupancies	1-hour separation *and* provide automatic sprinkler system	1-hour separation *or* provide automatic sprinkler system
Laundry rooms	1-hour separation where more than 100 square feet in floor area	1-hour separation *or* provide automatic sprinkler system where more than 100 square feet in floor area
Patient rooms equipped with padded surfaces	1-hour separation	Not considered as an incidental use
Physical plant maintenance shops	1-hour separation	Not considered as an incidental use
Waste and linen collection rooms	1-hour for rooms where containers have an aggregate volume of 10 cubic feet or more	1-hour regardless of amount of collection
Storage rooms more than 100 square feet in floor area	1-hour separation	Not considered as an incidental use

Incidental-use requirements specific to ambulatory care facilities and Group I-2 occupancies

CHANGE SIGNIFICANCE: There are occasionally one or more rooms or spaces in a building that pose risks not typically addressed by the provisions for the general occupancy group under which the building is classified. However, such rooms or spaces may functionally be an extension of the primary use. These types of spaces are considered in the IBC to be incidental uses and are specifically identified and addressed through the provisions of Section 509. Modifications to Table 509 now allow for a more detailed analysis of those spaces that are considered to have a higher hazard level than the remainder of a healthcare or ambulatory care facility, without the need to create a mixed-occupancy condition. The following incidental uses are either newly addressed in Table 509 or current provisions have been modified:

- Laboratories in Group I-2 occupancies where the laboratory is not classified as a Group H occupancy

Table 509 continues

Table 509 continued

- Laboratories in ambulatory care facilities where the laboratory is not classified as a Group H occupancy
- Laundry rooms greater than 100 square feet in floor area where located in Group I-2 occupancies
- Group I-2 patient rooms equipped with padded surfaces
- Physical plant maintenance shops in Group I-2 occupancies
- Waste and linen collection rooms in ambulatory care facilities and Group I-2 occupancies where the rooms have containers with an aggregate volume of 10 cubic feet or greater
- Storage rooms more than 100 square feet in floor area where located in ambulatory care facilities or Group I-2 occupancies

Laboratories in Group I-2 occupancies, other than those classified as Group H occupancies, now require a higher degree of compartmentalization in order to address the hazards involved. Previously, where sprinkler protection was provided, only smoke-resistant construction was required to surround the incidental-use laboratory. Because the IBC mandates automatic sprinkler systems in all Group I-2 occupancies, the only added degree of protection has been the construction of a smoke-resistant separation. Table 509 now mandates, in addition to the sprinkler protection, separation between the incidental-use laboratory and the remainder of the building through the use of 1-hour fire barriers, horizontal assemblies, or both. This increased degree of separation recognizes the potential for significant quantities of hazardous materials in non–Group H laboratories that are commonly found in hospitals and similar Group I-2 occupancies.

The regulation of laboratories as incidental uses is new to the 2015 IBC where such laboratories are located within ambulatory care facilities. Previously, only those laboratories in Group I-2 and E occupancies were considered as incidental uses. In all cases, any laboratory classified as Group H due to the type, state and amount of hazardous materials must be considered as a separate and unique occupancy rather than as an incidental use. Laboratories in ambulatory care facilities, although typically not rising to the level of a Group H classification, create an increased hazard, and as such must now be separated from other portions of the building with minimum 1-hour fire-resistance-rated fire barriers, horizontal assemblies or both, or, as an alternative, must utilize sprinkler protection and smoke-resistant construction to address the concern.

Laundry rooms over 100 square feet in floor area have previously been regulated in Group I-2 occupancies; however, the 1-hour fire-resistance-rated separation was an optional method of protection. It has previously been acceptable to use an automatic sprinkler system, along with smoke-resistant surrounding construction, as a means of addressing the hazards presented to the building due to the presence of the laundry room. This alternative approach is no longer permitted by Table 509; thus, a minimum 1-hour fire barrier, horizontal assembly, or both must now be used to isolate any laundry room over 100 square feet in floor area from the remainder of the building.

Group I-2 patient rooms now join Group I-3 cells as spaces requiring a minimum 1-hour fire-resistive separation from other building areas where such rooms have padded surfaces. Although the application of this additional requirement will be infrequent, it is important that the protection be provided in those rare cases where such rooms are provided in emergency facilities, inpatient psychiatric units and similar areas.

Physical plants and maintenance shops, although necessary support areas of a healthcare facility, have a distinctly different function from those typically encountered. The required fire-resistive separation of at least 1 hour has been mandated to ensure protection due to a variety of hazards—most importantly, the fire potential due to stored materials related to operation of the physical plant.

Waste and linen collection rooms in ambulatory care facilities and Group I-2 occupancies now only must be separated with fire-resistance-rated assemblies from other portions of the building where the container volume threshold is exceeded. The modification recognizes that many spaces within healthcare facilities, such as basic exam spaces, contain some level of waste containers and linen hampers. Most of these spaces do not rise to the hazard level as those where storage is the primary use. The provision requiring a minimum 1-hour fire barrier, horizontal assembly or both is intended to address those rooms and spaces where larger linen and waste receptacles are typically located, usually in a concealed environment. The aggregate container volume threshold of 10 cubic feet can much better address the potential hazard than regulating the concern by means of a floor-area threshold.

Large storage rooms are now required to be separated from other portions of a Group I-2 occupancy or ambulatory care facility by a minimum 1-hour fire barrier, horizontal assembly or combination of both. Concealed, typically unoccupied spaces can pose a significant fire hazard where the potential for combustible storage is probable, and as such, require some degree of fire separation from other portions of the building. The 1-hour enclosure requirement is only applicable where the floor area of the storage room exceeds 100 square feet.

510.2

Horizontal Building Separation

CHANGE TYPE: Modification

CHANGE SUMMARY: In the special provisions of Section 510.2 addressing pedestal buildings, there is no longer a limit of one story above grade plane for that portion of the structure that occurs below the 3-hour horizontal separation.

2015 CODE: 510.2 Horizontal Building Separation Allowance. A building shall be considered as separate and distinct buildings for the purpose of determining area limitations, continuity of fire walls, limitation of number of stories and type of construction where all of the following conditions are met:

1. The buildings are separated with a horizontal assembly having a fire-resistance rating of not less than 3 hours.

~~2. The building below the horizontal assembly is not greater than one story above grade plane.~~

~~3.~~ 2. The building below the horizontal assembly is of Type IA construction.

~~4.~~ 3. Shaft, stairway, ramp and escalator enclosures through the horizontal assembly shall have not less than a 2-hour fire-resistance rating with opening protectives in accordance with Section 716.5.

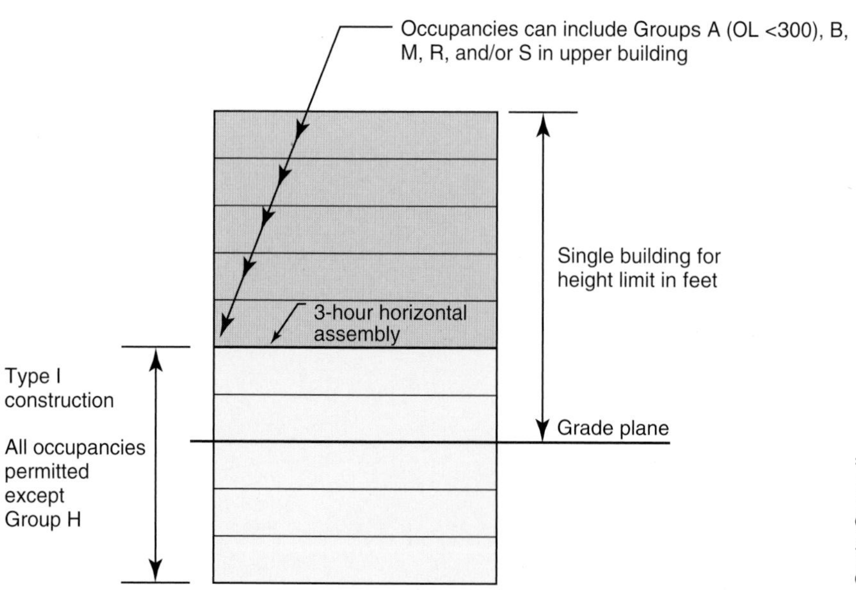

Horizontal building separation

602.4

Type IV Member Size Equivalencies

Structural composite lumber

© International Code Council

CHANGE TYPE: Addition

CHANGE SUMMARY: Equivalent size dimensions for structural composite lumber (SCL) in relationship to solid-sawn Type IV heavy-timber members have been introduced into Table 602.4.

2015 CODE: **602.4 Type IV.** Type IV construction (Heavy Timber, HT) is that type of construction in which the exterior walls are of noncombustible materials and the interior building elements are of solid or laminated wood without concealed spaces. The details of Type IV construction shall comply with the provisions of this section and Section 2304.11. Exterior walls complying with Section 602.4.1 or 602.4.2 shall also be permitted. ~~Fire-retardant-treated wood framing complying with Section 2303.2 shall be permitted within exterior wall assemblies with a 2-hour rating or less.~~ Minimum solid sawn nominal dimensions are required for structures built using Type IV construction (HT). For glued-laminated members and structural composite lumber (SCL) members, the equivalent net finished width and depths corresponding to the minimum nominal width and depths of solid sawn lumber are required as specified in Table 602.4. Cross-laminated timber (CLT) dimensions used in this section are actual dimensions.

602.4.1 Fire Retardant Treated Wood in Exterior Walls. Fire-retardant-treated wood framing complying with Section 2303.2 shall be permitted within exterior wall assemblies with a 2-hour rating or less.

TABLE 602.4 **Wood Member Size Equivalencies**

Minimum Nominal Solid Sawn Size		Minimum Glued-Laminated Net Size		Minimum Structural Composite Lumber Net Size	
Width, inch	Depth, inch	Width, inch	Depth, inch	Width, inch	Depth, inch
8	8	6¾	8¼	7	7½
6	10	5	10½	5¼	9½
6	8	5	8¼	5¼	7½
6	6	5	6	5¼	5½
4	6	3	6⅞	3½	5½

For SI: 1 inch = 25.4 mm

CHANGE SIGNIFICANCE: Type IV buildings are designated as heavy-timber buildings. Heavy-timber members have large cross sections to achieve the slow-burning characteristic that made them common construction elements during the 1800s in the heavy industrial areas of the Northeast. Under a continued application of heat during a fire incident, charring continues, but at an increasingly slower rate, as the charred wood insulates the inner portion of the wood member. There is quite often enough sound wood remaining during and after a fire to prevent sudden structural collapse. Although today's construction methods result

602.4 continues

602.4 continued

in very few structures built as Type IV construction, it is not uncommon for buildings to contain one or more heavy-timber members. In keeping with the concept of slow burning by means of wood members with large cross sections, Table 602.4 specifies minimum nominal dimensions for wood members considered as heavy timber.

For years, only solid-sawn wood members were addressed in Table 602.4 establishing the minimum size of heavy-timber members. Glued-laminated sizes were first introduced into the IBC in the 2006 edition. New to the 2015 table are minimum net size dimensions for structural composite lumber (SCL). SCL includes a number of engineered wood products used for structural purposes. Through the process of layering dried and graded wood veneers, strands, or flakes with exterior type adhesives, structural wood members of various types are created.

Net dimensions of typical SCL members are similar to the net dimensions of nominal solid-sawn members; however, the minimum width dimensions are slightly less than the solid-sawn member widths and slightly greater than the glued-laminated member net widths. In order to determine the appropriate net dimensions for SCL members that have now been incorporated into Table 602.4, the initial section properties of solid-sawn and glued-laminated members were compared with the initial section properties of SCL members. Then, utilizing the common net widths of SCL members, the minimum net depths were established.

602.4.2

Cross-Laminated Timber in Exterior Walls

CHANGE TYPE: Addition

CHANGE SUMMARY: Cross-laminated timber is now permitted within the exterior walls of Type IV buildings where protected by one of three specified materials.

2015 CODE: **602.4.2 Cross-Laminated Timber in Exterior Walls.** Cross-laminated timber complying with Section 2303.1.4 shall be permitted within exterior wall assemblies with a 2-hour rating or less, provided the exterior surface of the cross-laminated timber is protected by one the following:

1. Fire-retardant-treated wood sheathing complying with Section 2303.2 and not less than $^{15}/_{32}$ inch (12 mm) thick;
2. Gypsum board not less than ½ inch (12.7 mm) thick; or
3. A noncombustible material.

CHANGE SIGNIFICANCE: Cross-laminated timber (CLT) has been used extensively in Europe for the last two decades as large-section structural timber. This large-section, engineered wood product is now defined in the code, ANSI/APA PRG 320 is referenced in Section 2303.1.4, and the newly developed consensus manufacturing standard has been added as a referenced standard to Chapter 35. Under the Type IV provisions of Chapter 6, CLT is now permitted within exterior walls required to be of noncombustible construction, much in the same manner as fire-retardant-treated wood.

Exterior surface materials are regulated where CLT is used in the exterior walls of Type IV buildings. The CLT must be covered on the exterior side with minimum $^{15}/_{32}$-inch-thick fire-retardant-treated wood sheathing, minimum ½-inch-thick gypsum board, or noncombustible materials.

Photo courtesy of Nordic Wood Structures

Cross-laminated timber construction.

603.1, Item 26

Wall Construction of Freezers and Coolers

© International Code Council

Freezer in retail sales building

CHANGE TYPE: Addition

CHANGE SUMMARY: Walls of freezers and coolers located in buildings of Type I and II construction may now be constructed of wood materials provided three conditions are met.

2015 CODE: 603.1 Allowable materials. Combustible materials shall be permitted in buildings of Type I or II construction in the following applications and in accordance with Sections 603.1.1 through 603.1.3:

(Applications 1 through 25 remain unchanged.)

26. Wall construction of freezers and coolers of less than 1000 square feet (92.9 m^2), in size, lined on both sides with noncombustible materials and the building is protected throughout with an automatic sprinkler system in accordance with Section 903.3.1.1.

CHANGE SIGNIFICANCE: Although buildings of Type I and II construction are considered as noncombustible structures, an extensive listing of applications is set forth in Section 603.1, where limited amounts of combustibles are permitted to be used in a building's construction. It has been determined that the level of combustibles permitted by Section 603.1, as well as their control, does not adversely impact the fire-severity potential caused by the materials of construction. A wide variety of applications involving wood members are identified, including millwork, stages and platforms, blocking, furring strips and trim.

Walls of freezers and coolers located in buildings of Types I and II construction may now also be constructed of wood materials provided three conditions are met: (1) the freezer or cooler is less than 1000 square feet in floor area, (2) the freezer/cooler walls are lined on both sides with noncombustible materials, and (3) the building is fully sprinklered. It is reasoned that the hazard created by introducing a limited amount of combustible freezer/cooler wall construction is no more of a concern than that introduced by other allowances established in Section 601.1, such as the wood door and window frames permitted by Item 6 and the light-frame wood partition construction allowed by Item 11. Additional safeguards are provided through the finish limitations of Section 803 and the foam plastic requirements of Section 2603.

PART 3

Fire Protection

Chapters 7 through 9

- **Chapter 7** Fire and Smoke Protection Features
- **Chapter 8** Interior Finishes
  No changes addressed
- **Chapter 9** Fire Protection Systems

The fire protection provisions of the *International Building Code* (IBC) are found primarily in Chapters 7 through 9. There are two general categories of fire protection: active and passive. The fire and smoke resistance of building elements and systems in compliance with Chapter 7 provides for passive protection. Chapter 9 contains requirements for various active systems often utilized in the creation of a safe building environment, including automatic sprinkler systems, standpipe systems and fire alarm systems. To further address the rapid spread of fire, the provisions of Chapter 8 are intended to regulate interior-finish materials, such as wall and floor coverings. ■

717.1.1

Ducts Transitioning between Shafts

717.3, 717.5

Corridor Dampers

903.2.1.6

Sprinkler Systems—Assembly Occupancies

903.2.1.7

Multiple Fire Areas

903.2.8

Sprinkler Systems—Group R Occupancies

903.3.1.1.2

Exempt Locations for NFPA 13 Sprinklers

903.3.1.2.2

Open-Ended Corridors

903.3.8

Limited Area Sprinkler Systems

904.13

Domestic Cooking Systems in Group I-2 Condition 1

907.2.3

Fire Alarms—Group E Occupancies

907.2.9.3

Alarm Systems—Group R-2 College and University Buildings

907.2.11.3, 907.2.11.4

Smoke Alarms Near Cooking Appliances and Bathrooms

909.21.1

Elevator Hoistway Pressurization

910

Smoke and Heat Removal

915

Carbon Monoxide Detection

704.4

Protection of Secondary Members

CHANGE TYPE: Clarification

CHANGE SUMMARY: For structural fire protection purposes, the secondary member protection requirements have been reformatted and clarifies that the secondary members in a horizontal assembly can be protected by a ceiling membrane.

2015 CODE: 704.4 Protection of Secondary Members. Secondary members that are required to have a fire-resistance rating shall be protected by individual encasement protection. ~~by the membrane or ceiling of a horizontal assembly in accordance with Section 711, or by a combination of both.~~

704.4.1 Light-Frame Construction. ~~King~~ Studs and boundary elements that are integral elements in load-bearing walls of light-frame construction shall be permitted to have required fire-resistance ratings provided by the membrane protection provided for the load-bearing wall.

704.4.2 Horizontal Assemblies. Horizontal assemblies are permitted to be protected with a membrane or ceiling where the membrane or ceiling provides the required fire-resistance rating and are installed in accordance with Section 711.

CHANGE SIGNIFICANCE: Secondary members in buildings requiring structural fire resistance are addressed in Section 704.4. The requirements for horizontal assemblies have previously been included in the base provision of Section 704.4. Although not shown as an actual exception, the new provisions in Section 704.4.2 can be viewed as being the "specific" requirement per Section 102.1 and therefore take precedence over the requirement for individual encasement of the secondary member as Section 704.4 would imply is required.

The intent of the revision is unchanged from the 2012 code, which allows secondary members within a horizontal assembly to be protected by either individual encasement, by the ceiling membrane of a horizontal

704.4 continues

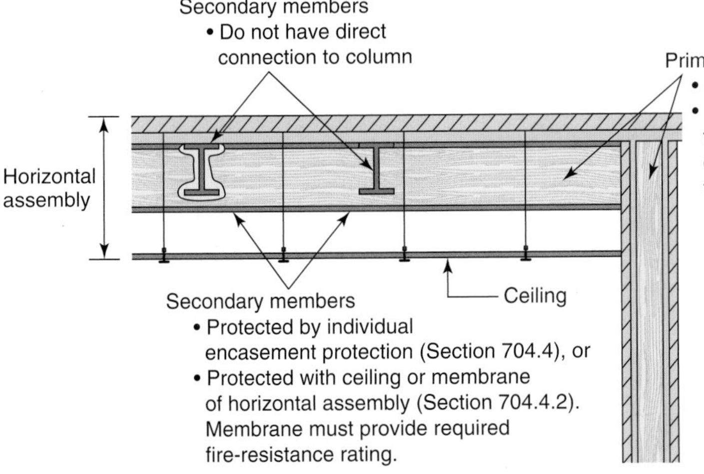

Protection of secondary members in horizontal assemblies

704.4 continued

assembly, or by a combination of the two methods. When horizontal assemblies are tested using the ASTM E 119 or UL 263 fire test, secondary members are evaluated as part of the overall assembly. The ceiling membrane of a floor/ceiling or roof/ceiling assembly typically does not provide the full amount of fire resistance that an assembly is required to have. Although the secondary members will therefore have less protection than what is provided on the unexposed floor or roof deck above, because the overall assembly did evaluate the exposure and impact on the secondary members, the ceiling membrane protection is adequate.

Section 704.4.1 has also been modified and now references all "studs" instead of just "king studs." King studs have essentially the same function, load ratio, and thermal properties as the other studs in the load-bearing wall. Therefore, there is no reason to make the distinction between types of studs or to treat king studs similar to columns and require individual encasement.

705.2

Projections at Exterior Walls

CHANGE TYPE: Modification

CHANGE SUMMARY: The minimum required separation between the leading edge of a projection and the line used to determine the fire separation distance has been modified in a manner that provides for a significant increase in the separation required.

2015 CODE: **705.2 Projections.** Cornices, eave overhangs, exterior balconies and similar projections extending beyond the exterior wall shall conform to the requirements of this section and Section 1406. Exterior

705.2 continues

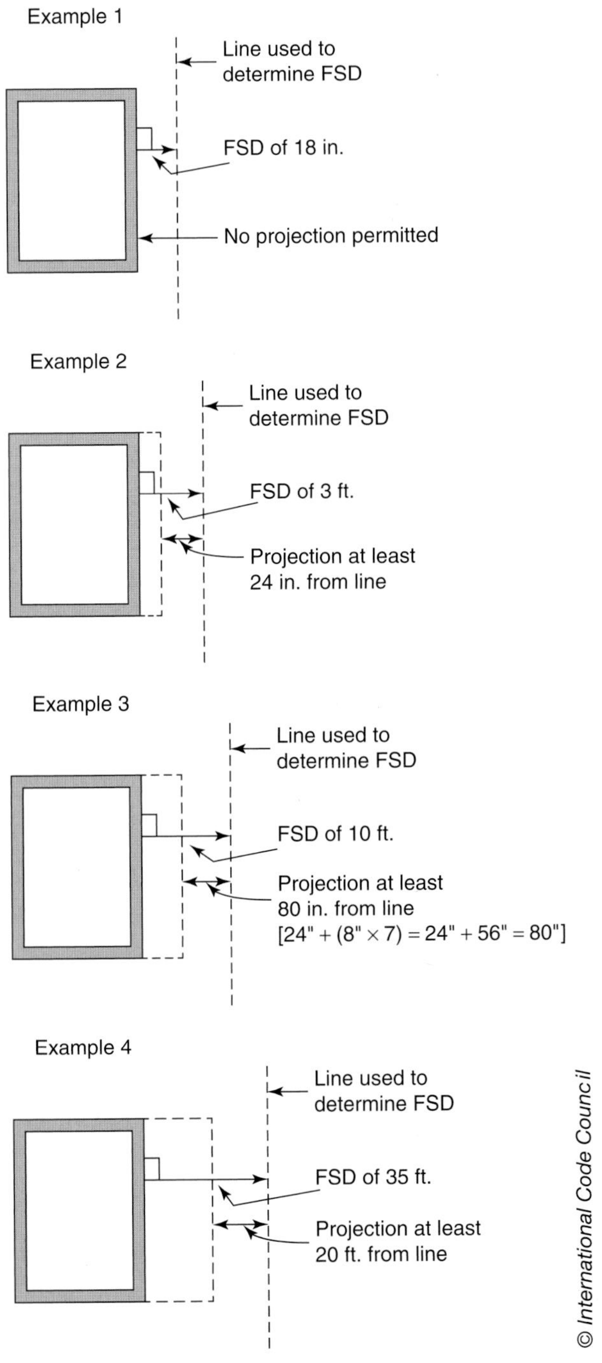

Minimum distance from projections

705.2 continued

egress balconies and exterior exit stairways and ramps shall also comply with Sections 1021 and 1027, respectively. Projections shall not extend any closer to the line used to determine the fire separation distance than shown in Table 705.2.

Exception: Buildings on the same lot and considered as portions of one building in accordance with Section 705.3 are not required to comply with this section for projections between the buildings.

TABLE 705.2 Minimum Distance of Projection

Fire Separation Distance (FSD)	Minimum Distance from Line Used to Determine FSD
0 feet to ~~less than~~ 2 feet	Projections not permitted
Greater than 2 feet to ~~less than 5 feet~~ 3 feet	24 inches
~~5 feet or~~ Greater than 3 feet to less than 30 feet	~~40 inches~~ 24 inches plus 8 inches for every foot of FSD beyond 3 feet or fraction thereof
30 feet or greater	20 feet

For SI: 1 foot = 304.8 mm; 1 inch = 25.4 mm.

CHANGE SIGNIFICANCE: Eave overhangs and similar projections located in close proximity to the lot line, or another building on the lot, create conditions where heated air from a fire is unable to escape vertically. The revised Table 705.2 continues the effort to simplify the projection requirements while addressing the potential hazards. Over the past three code editions, the projection provisions have been significantly modified in an effort to clarify the requirements and to ensure an adequate separation between the projections and adjacent property. The revisions within Table 705.2 continue to reflect that effort. In addition to providing a minimum clearance from the line used for measuring fire separation distance, the table now also factors in the location of the exterior wall. As the fire separation distance is increased, the minimum required distance between the projection and the line used to determine fire separation distance increases.

Applying the new requirements of Table 705.2, projections are not permitted to be closer than 24 inches to the lot line (or an assumed imaginary line between buildings on the same lot), and the minimum required separation is increased by 8 inches for every foot of additional fire separation distance beyond 3 feet. Providing a gradual increase in the required separation distance (8 inches per foot) eliminates an inconsistency in the 2012 code. The minimum separation distance is increased until the point that the exterior wall is located 30 feet or greater from the line used to determine fire separation distance. This 30-foot requirement coordinates with Tables 602 and 705.8, which first allow exterior walls with no fire-resistance rating and an unlimited amount of unprotected openings at the 30-foot distance.

The exception to Section 705.2 was also modified in a manner to limit its application. The exception can now only be applied to eliminate the minimum separation requirements between multiple buildings on the same site that use the provisions of Sections 705.3 and 503.1.2 to consider the buildings as being a single building. Where considered as being a portion of one building, it is reasonable that the code does not require a minimum separation between the projections. Previously, it was occasionally interpreted that the exception eliminated all projection requirements, including those between any building and any adjacent property.

705.2.3

Combustible Projections

CHANGE TYPE: Modification

CHANGE SUMMARY: The provisions regulating combustible projections adjacent to an interior lot line or other line used to determine the fire separation distance have been modified to provide a simple and consistent approach that is less restrictive than previously determined.

2015 CODE: 705.2.3 Combustible Projections. Combustible projections extending to within 5 feet (1524 mm) of the line used to determine the fire separation distance~~, or located where openings are not permitted, or where protection of some openings is required~~ shall be of not less than 1-hour fire-resistance-rated construction, Type IV construction, fire-retardant-treated wood or as required by Section 1406.3.

> **Exception:** Type VB construction shall be allowed for combustible projections in Group R-3 and U occupancies with a fire separation distance greater than or equal to 5 feet (1524 mm).

CHANGE SIGNIFICANCE: In addition to the required separation between a projection and the line used to determine the fire separation distance, such as an interior lot line, projections are further regulated in order to prevent a fire hazard from inappropriate use of combustible materials attached to exterior walls. Where located in close proximity to the fire separation distance line, combustible projections are prohibited unless of fire-retardant-treated wood, Type IV construction or 1-hour fire-resistance-rated construction.

Previously, three separate thresholds were established to identify the scope of the requirement. Where any one of the three thresholds was exceeded, the limitations on combustible projections would apply. As a result, the most restrictive condition was always applied. With the revision, only the 5-foot fire separation distance threshold is now the controlling point for where the additional protection requirements apply. Where a combustible projection has less than 5 feet of fire separation distance, then some increased level of protection is required.

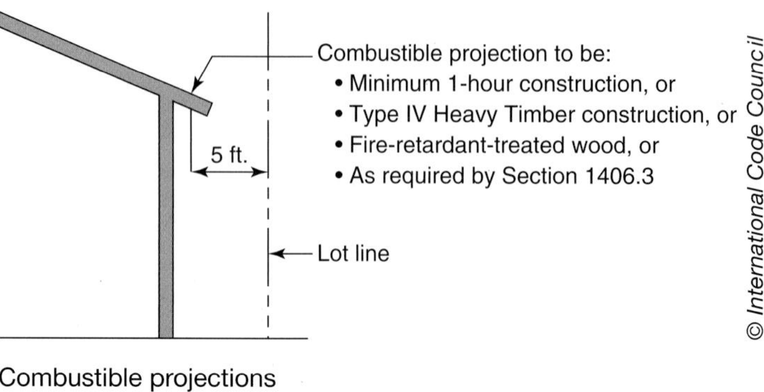

Combustible projections

705.3

Buildings on the Same Lot

CHANGE TYPE: Modification

CHANGE SUMMARY: Openings are permitted through adjacent exterior walls of a Group S-2 parking garage and a Group R-2 building on the same lot where such buildings are regulated as two buildings on the same lot and the fire separation distance is zero.

2015 CODE: 705.3 Buildings on the Same Lot. For the purposes of determining the required wall and opening protection, projections and roof-covering requirements, buildings on the same lot shall be assumed to have an imaginary line between them.

Where a new building is to be erected on the same lot as an existing building, the location of the assumed imaginary line with relation to the existing building shall be such that the exterior wall and opening protection of the existing building meet the criteria as set forth in Sections 705.5 and 705.8.

~~Exception~~ Exceptions:

1. Two or more buildings on the same lot shall either be regulated as separate buildings or shall be considered as portions of one building if the aggregate area of such buildings is within the limits specified in Chapter 5 for a single building. Where the buildings contain different occupancy groups or are of different types of construction, the area shall be that allowed for the most restrictive occupancy or construction.

705.3 continues

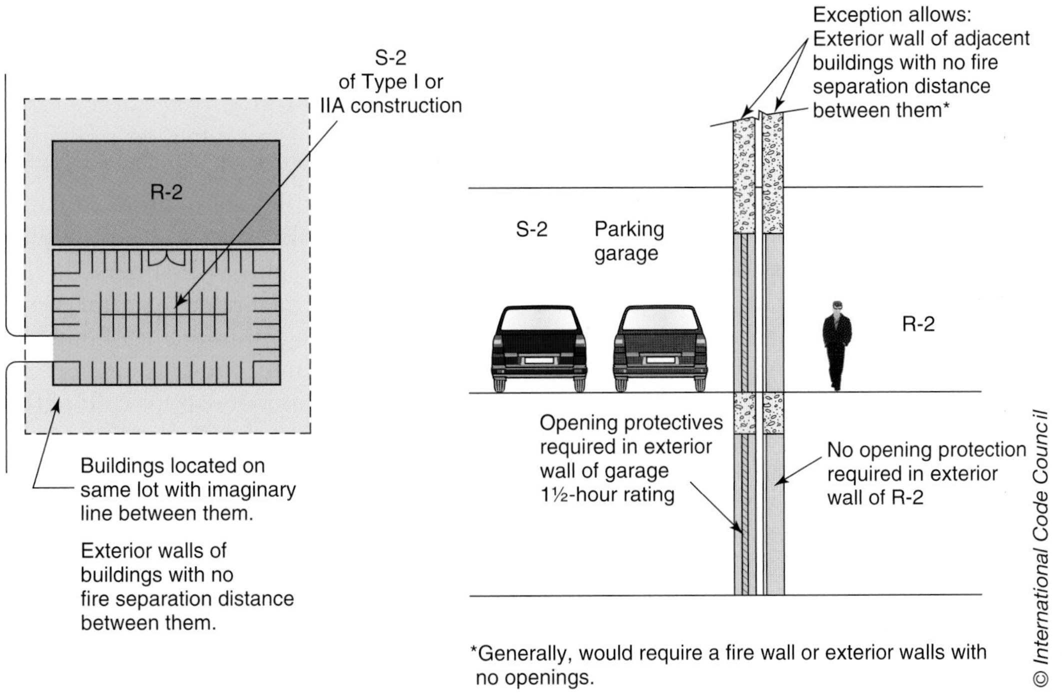

Separation of S-2 parking garage and R-2 building on the same lot

705.3 continued

2. Where an S-2 parking garage of Construction Type I or Type IIA is erected on the same lot as a Group R-2 building, and there is no fire separation distance between these buildings, then the adjoining exterior walls between the buildings are permitted to have occupant use openings in accordance with Section 706.8. However, opening protectives in such openings shall only be required in the exterior wall of the S-2 parking garage, not in the exterior wall openings in the R-2 building, and these opening protectives in the exterior wall of the S-2 parking garage shall be a minimum of 1½ hours fire protection rating.

CHANGE SIGNIFICANCE: When two or more buildings are located on the same lot, they are required to be regulated as either separate buildings or as a single building. A new exception specific to the combination of a Group R-2 occupancy building and an adjacent parking garage is applicable where evaluated as separate buildings. The opposing exterior walls of a Group S-2 parking garage of Type I or Type IIA construction and a Group R-2 building of any type of construction are regulated in a manner that allows for necessary openings in the walls. Previously, such a condition would have required a fire wall. Openings are now permitted in these adjoining exterior walls even though there is no fire separation distance, and the structures can be built independently and will not need to comply with the fire wall stability and continuity requirements of Section 706.

Generally, Section 705.8 and Table 705.8 prohibit any openings in an exterior wall that has a fire separation distance of less than 3 feet. Therefore, the typical solution where openings are desired between two opposing buildings with no fire separation distance would have been to construct a fire wall between the two adjacent structures so they could be viewed as separate buildings and have openings that allow access between the garage and the residential building. The new exception will now allow openings in the opposing exterior walls of the two buildings, and will only impose the requirement for opening protectives in the exterior wall of the Group S-2 parking garage. The fire protection rating for any door in the exterior wall of the garage would be a minimum of 1½ hours. This minimum would be consistent with the fire protection rating required by Table 716.5 for exterior walls with a 2- or 3-hour fire-resistance rating. As a specific requirement, this will supersede the rating that is required for a 1-hour fire-resistance-rated exterior wall. Although it may seem inconsistent to require a minimum 1½-hour fire protection rating to protect openings in a 1-hour fire-resistance-rated exterior wall, it is important to note that the adjoining exterior wall of the Group R-2 building will also have a fire-resistance rating that will improve the overall separation between the two structures.

Because this is an "exterior wall," code users should apply that portion of Table 716.5. Whether the exterior wall of the garage is a bearing or nonbearing wall regulated by Table 601 or 602, the minimum 1½-hour fire protection rating is appropriate. It is inappropriate to apply the fire wall or fire barrier provisions of Table 716.5 where this new exception is being used, as these are "exterior walls" between the two adjacent buildings.

Specifying that the opening protectives only need to be installed in the wall for the parking garage helps to resolve several issues that could arise if the opening protection was required in both walls. These issues or concerns include door swing and the details on how to install a listed door assembly in such a double-wall system. The fire history for parking garages shows that most garage fires are limited to a single vehicle fire; therefore, it is unlikely that the fire would spread to the Group R-2 or cause the parking garage to fail structurally. In addition, the sprinkler protection of the Group R-2 building helps to protect the openings. In the unlikely event a fire in the sprinklered Group R-2 is significant, the openings in the garage wall will still be protected.

705.6

Structural Element Bracing of Exterior Walls

CHANGE TYPE: Modification

CHANGE SUMMARY: Interior structural elements, such as floor or roof elements, that brace exterior walls are no longer required to be regulated for fire resistance due to the exterior wall's rating regardless of the building's proximity to a lot line.

2015 CODE: **705.6 Structural Stability.** ~~The wall~~ Exterior walls shall extend to the height required by Section 705.11. ~~and shall have sufficient structural stability such that it will remain in place for the duration of time indicated by the required fire-resistance rating. Where exterior walls have a minimum fire separation distance of not less than 30 feet (9144 mm), interior~~ Interior structural elements that brace the exterior wall but that are not located within the plane of the exterior wall shall have the minimum fire-resistance rating required in Table 601 for that structural element. Structural elements that brace the exterior wall but are located outside of the exterior wall or within the plane of the exterior wall shall have the minimum fire-resistance rating required in Tables 601 and 602 for the exterior wall.

CHANGE SIGNIFICANCE: It is not uncommon for elements of the floor and/or roof construction to provide bracing support for exterior walls. There has been confusion as to the exact type and performance of the protection of such bracing elements that are not located in the plane of the exterior wall. The deletion of the condition based on fire separation distance should result in more consistent application by focusing on the remaining provisions to establish the proper level of protection.

Perhaps the best way to explain the application of the revised provisions is through an example. Assume a building of Type VB construction where the floor and roof framing systems frame perpendicularly into an

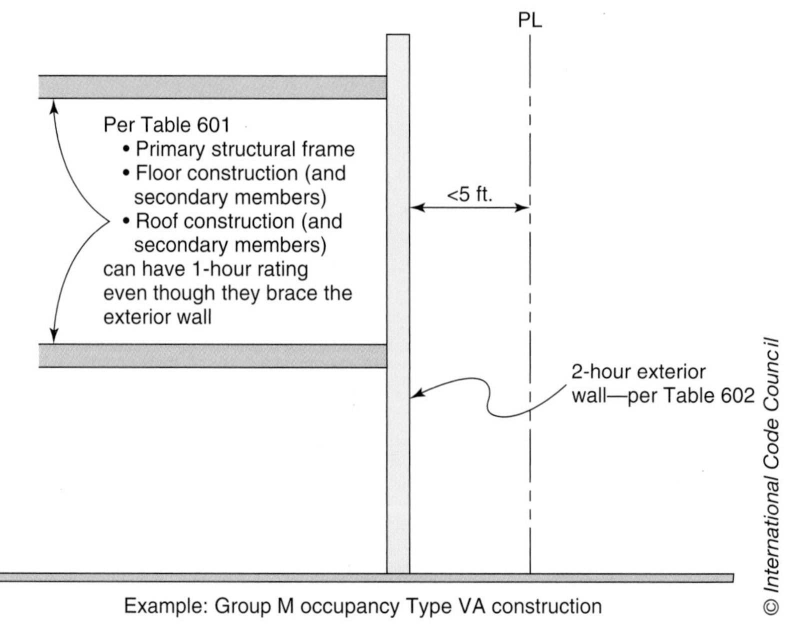

Interior structural elements bracing exterior wall

exterior wall that is required to have a 1-hour fire-resistance rating based solely on its location on the lot. In this situation, it was often questioned as to whether or not the horizontal assemblies (floor and roof members) needed to be provided with 1-hour fire-resistive protection simply because they serve as a means of bracing the exterior wall. It is now clear that those elements would not be required to have a fire-resistance rating, because floor and roof construction does not need to be rated in a Type VB building based on Table 601. On the other hand, if the bracing element is located "outside of the exterior wall or within the plane of the exterior wall," the protection requirement changes to where it must at least match the rating of the wall.

Deleting the previous text also provides clarification of the stability requirement to "remain in place for the duration of time indicated by the required fire-resistance rating." There has been a concern that while the ASTM E 119 fire test evaluates the capability of the wall to remain in place, previous code text required something above and beyond the standard fire-resistance rating of the exterior wall. There was a previous expectation that the wall needed to remain standing for that time period under whatever real-life fire conditions the wall may face. It has been clarified that the wall is to be evaluated under the general fire test requirements and that there is no additional structural protection or performance requirement imposed by the IBC.

705.8.5

Vertical Separation of Openings

CHANGE TYPE: Clarification

CHANGE SUMMARY: Where a fire-resistance-rated wall is provided to address the concern of a fire spreading out of the building and then exposing an upper-level opening, the exterior wall must be rated from both sides, regardless of the fire separation distance.

2015 CODE: **705.8.5 Vertical Separation of Openings.** Openings in exterior walls in adjacent stories shall be separated vertically to protect against fire spread on the exterior of the buildings where the openings are within 5 feet (1524 mm) of each other horizontally and the opening in the lower story is not a protected opening with a fire protection rating of not less than ¾ hour. Such openings shall be separated vertically at least 3 feet (914 mm) by spandrel girders, exterior walls or other similar assemblies that have a fire-resistance rating of at least 1 hour, rated for exposure to fire from both sides, or by flame barriers that extend horizontally at least 30 inches (762 mm) beyond the exterior wall. Flame barriers shall also have a fire-resistance rating of at least 1 hour. The unexposed surface temperature limitations specified in ASTM E 119 or UL 263 shall not apply to the flame barriers or vertical separation unless otherwise required by the provisions of this code.

Exceptions:

1. This section shall not apply to buildings that are three stories or less above grade plane.
2. This section shall not apply to buildings equipped throughout with an automatic sprinkler system in accordance with Sections 903.3.1.1 or 903.3.1.2.
3. Open parking garages.

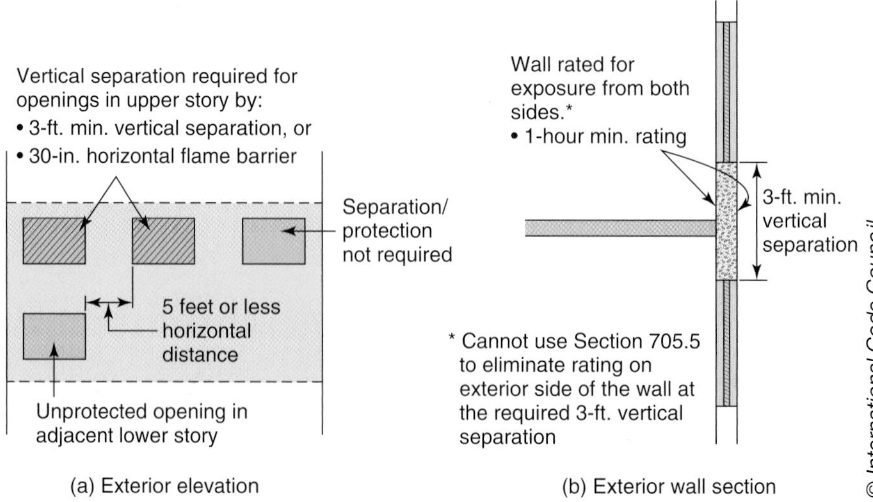

Vertical separation of openings in an exterior wall

CHANGE SIGNIFICANCE: In general, fire-resistance-rated exterior walls with a fire separation distance greater than 10 feet do not require testing from the exterior side. In the situations addressed by Section 705.8.5 where the concern is fire spreading from an exterior opening on a lower story to an exterior opening on the adjacent upper story, the exterior wall reductions allowed by Section 705.5 were deemed inappropriate. The concern is an exposure from a fire within the building spreading up to the adjacent story by rolling out of the lower-story opening and then impinging back into the building. Because the possibility of the fire exposure exists any time there are openings in the exterior walls in adjacent stories, the fire separation distance provisions are not appropriate in order to eliminate the exterior protection.

The Las Vegas Hilton Hotel fire of February 10, 1981 is one of the best examples of the importance of these provisions and one of the leading reasons that these requirements were originally placed into the code. In that fire, the flames from an elevator lobby fire spread up through the building by going out of the windows on a lower level and then into the windows on upper stories, ultimately causing the elevator lobbies above to also become involved.

A fire-resistance-rated exterior wall must now be provided with a fire-resistance rating based on exposure from both sides of the assembly when it is serving as the means of protection for vertical flame spread through the exterior openings.

706.2

Structural Stability of Fire Walls

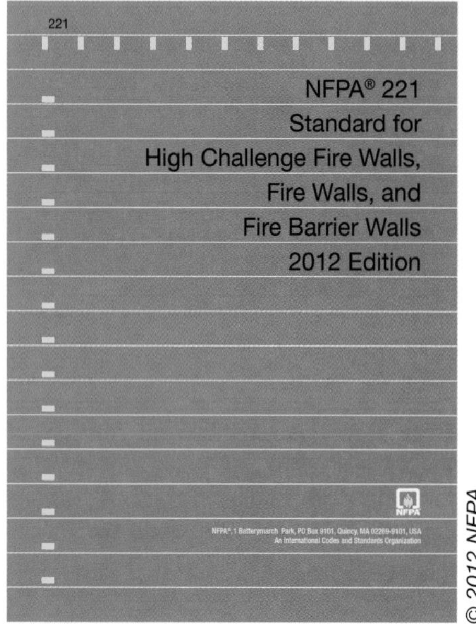

NFPA 221 Standard. Courtesy of NFPA.

CHANGE TYPE: Modification

CHANGE SUMMARY: The reference to NFPA 221 for fire wall design and construction has been expanded to permit the use of the "tied" and "cantilevered" options addressed in the standard.

2015 CODE: 706.2 Structural Stability. Fire walls shall ~~have sufficient structural stability under fire conditions~~ be designed and constructed to allow collapse of ~~construction~~ the structure on either side without collapse of the wall under fire conditions. ~~for the duration of time indicated by the required fire-resistance rating, or shall be~~ Fire walls designed and constructed ~~as double fire walls~~ in accordance with NFPA 221 shall be deemed to comply with this section.

CHANGE SIGNIFICANCE: Fire walls are fire-resistance-rated wall assemblies recognized by the IBC as a compliant means of dividing a structure into two or more buildings, allowing each of the buildings to be regulated separately. Specific provisions related to fire walls are established in Section 706. The IBC has previously only referenced NFPA 221, *Standard for High Challenge Fire Walls, Fire Walls, and Fire Barrier Walls*, for the design and construction of "double fire walls." The NFPA 221 standard contains three separate methods regarding fire wall construction that all result in structural behavior that is consistent with the guidance given in Section 706.2—that the wall remains standing if the structure on the fire-exposed side fails. The IBC will now also accept "tied" and "cantilevered" walls constructed using the NFPA 221 standard, in addition to the previously accepted "double fire walls." Because IBC Section 102.4 limits the use of referenced standards "to the prescribed extent of each such reference," the 2012 IBC restricted the use of the standard to only double fire walls while excluding the other methods. As it was determined that there is no reason to limit the code to the use of only one of the standard's methods, the revision will now allow designers the use of all three methods that are deemed to comply with the code's intended structural stability requirements.

In another revision, the expected performance of the fire wall has been clarified to address the concern that the stability requirement to "have sufficient structural stability under fire conditions . . . for the duration of time indicated by the required fire-resistance-rating" was imposing an additional real-time fire conditions performance requirement. Although the ASTM E 119 fire test evaluates the capability of the wall to remain in place and support any required load during the test process, it was felt that the provisions could be interpreted as looking for something above and beyond the standard fire-resistance rating of the wall, expecting the wall to remain standing for that time period under any real-life fire conditions it may face. The revision helps to clarify that the wall is to be evaluated under the general fire test requirements and that there is no additional structural protection or performance requirement imposed by the code to consider real-life fire conditions. The only extra structural performance requirement for a fire wall is that it must allow the collapse of the structure on either side without pulling the wall down.

709.4

Continuity of Smoke Barriers

709.4.1 is created to take care of the smoke barriers separating smoke compartments needing to go from outside wall to outside wall.
709.4.2 is now referenced for those situations where the barrier does not have to go from outside wall to outside wall (areas of refuge or elevator lobbies)

CHANGE TYPE: Clarification

CHANGE SUMMARY: The horizontal continuity of smoke barriers has been clarified for smoke barriers used to create smoke compartments, as well as for smoke barrier walls intended to create enclosures for elevator lobbies or areas of refuge.

2015 CODE: **709.4 Continuity.** Smoke barriers shall form an effective membrane continuous ~~from outside wall to outside wall and~~ from the top of the foundation or floor/ceiling assembly below to the underside of the floor or roof sheathing, deck or slab above, including continuity through concealed spaces, such as those found above suspended ceilings, and interstitial structural and mechanical spaces. The supporting construction shall be protected to afford the required fire-resistance rating of the wall or floor supported in buildings of other than Types IIB, IIIB or VB construction. Smoke-barrier walls used to separate smoke compartments shall comply with Section 709.4.1. Smoke-barrier walls used to enclose areas of refuge in accordance with Section 1009.6.4 or to enclose elevator lobbies in accordance with Section 405.4.3, 3007.6.2 or 3008.6.2 shall comply with Section 709.4.2.

Exceptions:

~~1.~~ Smoke-barrier walls are not required in interstitial spaces where such spaces are designed and constructed with ceilings or exterior walls that provide resistance to the passage of fire and smoke equivalent to that provided by the smoke-barrier walls.

709.4 continues

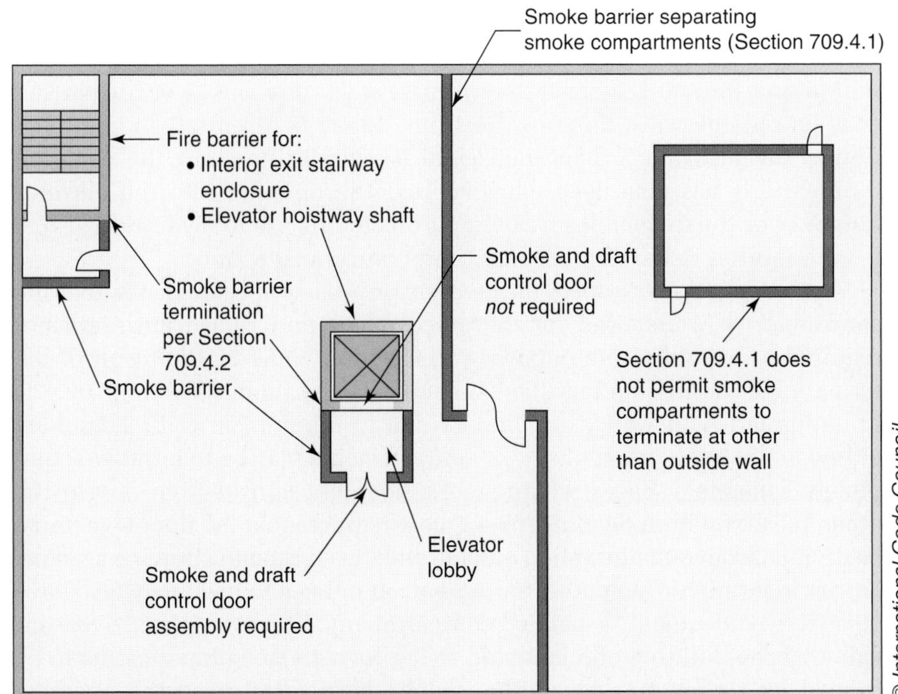

Smoke barrier continuity or termination

709.4 continued

2. ~~Smoke barriers used for elevator lobbies in accordance with Section 405.4.3, 3007.4.2 or 3008.11.2 are not required to extend from outside wall to outside wall.~~
3. ~~Smoke barriers used for areas of refuge in accordance with Section 1007.6.2 are not required to extend from outside wall to outside wall.~~

709.4.1 Smoke-Barrier Walls Separating Smoke Compartments. Smoke-barrier walls used to separate smoke compartments shall form an effective membrane continuous from outside wall to outside wall.

709.4.2 Smoke-Barrier Walls Enclosing Areas of Refuge or Elevator Lobbies. Smoke-barrier walls used to enclose areas of refuge in accordance with Section 1009.6.4, or to enclose elevator lobbies in accordance with Section 405.4.3, 3007.6.2 or 3008.6.2, shall form an effective membrane enclosure that terminates at a fire barrier wall having a level of fire protection rating not less than 1 hour, another smoke-barrier wall or an outside wall. A smoke and draft control door assembly as specified in Section 716.5.3.1 shall not be required at each elevator hoistway door opening or at each exit doorway between an area of refuge and the exit enclosure.

CHANGE SIGNIFICANCE: Smoke barriers are used selectively in the IBC to create refuge spaces that afford protection from both fire and smoke for building occupants unable to egress efficiently. A distinction has now been made between the condition where smoke-barrier walls are used to create smoke compartments (such as those in hospitals, ambulatory care facilities or jails) and those situations where the walls are used to create an enclosure for an elevator lobby or for an area of refuge. The base provision now references Section 709.4.1 where the smoke-barrier wall is used to create separate smoke compartments and therefore must be "continuous from outside wall to outside wall." On the other hand, the reference to Section 709.4.2 for the enclosure of areas of refuge or elevator lobbies will permit the smoke barrier to terminate at another smoke or fire barrier, or at an outside wall. The new Section 709.4.2 is essentially a combination of the previous Exceptions 2 and 3 and clarifies how the wall can terminate. It has also been clarified that the opening into the elevator hoistway or the exit enclosure being protected by the lobby or refuge area is not required to have smoke and draft control protection.

Although a major aspect of the revision was simply to clarify that the smoke-barrier walls used for elevator lobbies and refuge areas are not required to extend to the outside wall (Section 709.4.2), the result of the formatting is clearly forcing all other smoke compartments (Section 709.4.1) to be formed with walls that do continue to the outside wall. Therefore, although the basic intent of smoke compartments may be to form two completely separated spaces where smoke does not spread from one to the other, the text within Section 709.4.1 now requires that the floor level must be divided from outside wall to outside wall, even though allowing a smoke barrier to terminate at another smoke barrier or back to itself may also have created two adequately separated compartments. However, it seems reasonable that the building official could still allow smoke compartments to be formed by terminating at another smoke barrier by using the alternate method provisions of Section 104.11 and ensuring that the two compartments were effectively separated.

711, 712

Horizontal Assemblies and Vertical Openings

CHANGE TYPE: Modification

CHANGE SUMMARY: The reorganization of Sections 711 and 712 has been continued such that Section 711 now contains only the construction requirements for floor and roof assemblies, and Section 712 only contains the requirements related to the protection of vertical openings.

2015 CODE:

SECTION 711
FLOOR AND ROOF ASSEMBLIES ~~HORIZONTAL ASSEMBLIES~~

711.1 General. ~~Floor and roof assemblies required to have a fire-resistance rating~~ Horizontal assemblies shall comply with Section 711.2 ~~this section~~. Non-fire resistance-rated floor and roof assemblies shall comply with Section 711.3 ~~714.4.2~~.

711.2 Horizontal Assemblies. Horizontal assemblies shall comply with Sections 711.2.1 through 711.2.6.

711.2.1 ~~711.2~~ Materials. ~~The floor and roof~~ Assemblies shall be of materials permitted by the building type of construction.

711.2.2 ~~711.4~~ Continuity. Assemblies shall be continuous without vertical openings, ~~penetrations or joints~~ except as permitted by this section and Sections~~s~~ 712.~~2, 714.4, 715, 1009.3 and 1022.1~~. ~~Skylights and other penetrations through a fire-resistance-rated roof deck or slab are permitted to be unprotected, provided that the structural integrity of the fire-resistance-rated roof assembly is maintained. Unprotected skylights shall not be permitted in roof assemblies required to be fire-resistance rated in accordance with Section 705.8.6.~~

711, 712 continues

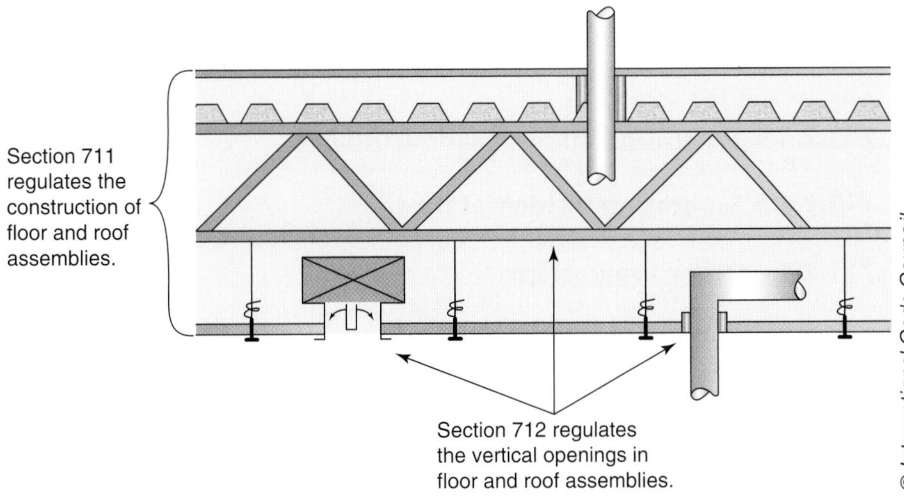
Section 711 regulates the construction of floor and roof assemblies.
Section 712 regulates the vertical openings in floor and roof assemblies.
© International Code Council

Horizontal assemblies and vertical openings

711, 712 continued

711.2.3 Supporting Construction. The supporting construction shall be protected to afford the required fire-resistance rating of the horizontal assembly supported.

> **Exception:** In buildings of Type IIB, IIIB or VB construction, the construction supporting the horizontal assembly is not required to be fire-resistance-rated at the following:
>
> 1. Horizontal assemblies at the separations of incidental uses as specified by Table 509 provided the required fire-resistance rating does not exceed 1 hour.
> 2. Horizontal assemblies at the separations of dwelling units and sleeping units as required by Section 420.3.
> 3. Horizontal assemblies at smoke barriers constructed in accordance with Section 709.

711.2.4 ~~711.3~~ Fire-Resistance Rating. The fire-resistance rating of ~~floor and roof~~ horizontal assemblies shall comply with Sections 711.2.4.1 through 711.2.4.6 but shall not be less than that required by the building type of construction.

711.2.4.1 Separating Mixed Occupancies.

711.2.4.2 Separating Fire Areas.

711.2.4.3 Dwelling Units and Sleeping Units. ~~Horizontal assemblies separating dwelling units in the same building and horizontal assemblies separating sleeping units in the same building shall be a minimum of 1-hour fire-resistance-rated construction.~~ Horizontal assemblies serving as dwelling or sleeping unit separations in accordance with Section 420.3 shall be not less than 1-hour fire-resistance-rated construction.

> **Exception:** Horizontal assemblies separating dwelling units and sleeping units shall be not less than ½-hour fire-resistance-rated construction ~~separations~~ in a building~~s~~ of Type IIB, IIIB and VB construction, ~~shall have fire-resistance ratings of not less than 1/2 hour in~~ where the building~~s~~ is equipped throughout with an automatic sprinkler system in accordance with Section 903.3.1.1.

711.2.4.4 Separating Smoke Compartments.

711.2.4.5 Separating Incidental Uses.

711.2.4.6 Other Separations.

711.2.5 ~~711.3.1~~ Ceiling Panels.

711.2.6 ~~711.3.3~~ Unusable Space.

711.3 Nonfire-Resistance-Rated Floor and Roof Assemblies. Nonfire-resistance-rated floor, floor/ceiling, roof and roof/ceiling assemblies shall comply with Sections 711.3.1 and 711.3.2.

711.3.1 Materials. Assemblies shall be of materials permitted by the building type of construction.

711.3.2 Continuity. Assemblies shall be continuous without vertical openings, except as permitted by Section 712.

~~**711.5 Penetrations.**~~ (Relocated to 712.1.4.)

~~**711.6 Joints.**~~ (Relocated to 712.1.5.)

~~**711.7 Ducts and Air Transfer Openings.**~~ (Relocated to 712.1.6.)

~~**711.8 Floor Fire Door Assemblies**~~ (Relocated to 712.1.13.1.)

~~**711.9 Smoke Barrier.**~~ (Relocated to 3006.)

SECTION 712
VERTICAL OPENINGS

712.1 General. ~~The provisions of this section shall apply to the~~ Each vertical opening ~~applications listed~~ shall comply in accordance with one of the protection methods in Sections 712.1.1 through 712.1.16~~8~~.

712.1.1 Shaft Enclosures.

712.1.2 Individual Dwelling Unit.

712.1.3 Escalator Openings.

712.1.4 Penetrations.

712.1.5 Joints. Joints shall be permitted where complying with Section 712.1.5.1 or 712.1.5.2, as applicable.

712.1.5.1 ~~711.6~~ Joints in or between Horizontal Assemblies.

712.1.5.2 ~~711.4.1~~ Joints in or between Nonfire-Resistance-Rated Floor Assemblies.

712.1.6 ~~712.1.5~~ Ducts and Air Transfer Openings. Penetrations by ducts and air transfer openings shall be protected in accordance with Section 717~~.6~~. Grease ducts shall be protected in accordance with the *International Mechanical Code.*

712.1.7 ~~712.1.6~~ Atriums.

712.1.8 ~~712.1.7~~ Masonry Chimney.

712.1.9 ~~712.1.8~~ Two-Story Openings. In other than Groups I-2 and I-3, a ~~floor~~ vertical opening that is not used as one of the applications listed in this section shall be permitted if the opening complies with all of the items below.

1. Does not connect more than two stories.
~~2. Does not contain a stairway or ramp required by Chapter 10~~.
~~3.~~ 2. Does not penetrate a horizontal assembly that separates fire areas or smoke barriers that separate smoke compartments.

711, 712 continues

711, 712 continued

~~4.~~ 3. Is not concealed within the construction of a wall or a floor/ceiling assembly.

~~5.~~ 4. Is not open to a corridor in Group I and R occupancies.

~~6.~~ 5. Is not open to a corridor on nonsprinklered floors.

~~7.~~ 6. Is separated from floor openings and air transfer openings serving other floors by construction conforming to required shaft enclosures.

712.1.10 ~~712.1.9~~ Parking Garages. Vertical openings in parking garages for automobile ramps, elevators and duct systems shall comply with Section 712.1.10.1, 712.1.10.2 or 712.1.10.3 as applicable.

712.1.10.1 Automobile Ramps.

712.1.10.2 ~~712.1.15~~ Elevators ~~in Parking Garages~~.

712.1.10.3 ~~712.1.16~~ Duct Systems ~~in Parking Garages~~.

712.1.11 ~~712.1.10~~ Mezzanine.

~~712.1.11 Joints.~~

712.1.12 ~~Unenclosed~~ Exit Access Stairways and Ramps. Vertical ~~floor~~ openings ~~created by unenclosed~~ containing exit access stairways or ramps in accordance with Section~~s 1009.2 and 1009.3~~ 1019 shall be permitted.

712.1.13 Openings. ~~Floor Fire Doors.~~ Vertical openings for floor fire doors and access doors shall be permitted where protected by Section 712.1.13.1 or 712.1.13.2. ~~floor fire doors in accordance with Section 711.8~~.

712.1.13.1 ~~711.8 Floor~~ Horizontal Fire Door Assemblies. ~~Floor~~ Horizontal fire door assemblies used to protect openings in fire-resistance-rated ~~floors~~ horizontal assemblies shall be tested in accordance with NFPA 288, and shall achieve a fire-resistance rating not less than the assembly being penetrated. ~~Floor~~ Horizontal fire door assemblies shall be labeled by an approved agency. The label shall be permanently affixed and shall specify the manufacturer, the test standard and the fire-resistance rating.

712.1.13.2 ~~711.3.2~~ Access Doors.

712.1.14 Group I-3.

~~712.1.17 Nonfire-Resistance-Rated Joints.~~

~~(Portion of 711.4)~~ 712.1.15 Skylights. Skylights and other penetrations through a fire-resistance-rated roof deck or slab are permitted to be unprotected, provided that the structural integrity of the fire-resistance-rated roof assembly is maintained. Unprotected skylights shall not be permitted in roof assemblies required to be fire-resistance rated in accordance with Section 705.8.6. The supporting construction shall be protected to afford the required fire-resistance rating of the horizontal assembly supported.

712.1.16 ~~712.1.18~~ Openings Otherwise Permitted. Vertical openings shall be permitted where allowed by other sections of this code.

CHANGE SIGNIFICANCE: Requirements related to the construction of horizontal assemblies and the protection of any vertical openings through them have been reformatted for clarification purposes. The revisions continue a clarification process that began with revisions between the 2009 and 2012 editions of the IBC with the establishment of separate scoping for each of the issues. In the 2012 edition of the IBC, Section 711 has a mixture of provisions related to the assembly construction and the protection of vertical openings through them. In the 2015 IBC, the provisions related to vertical openings (2012 Sections 711.5 through 711.9) have been relocated from Section 711 and placed into Section 712 or other applicable locations in the code. Section 712 then provides the starting point for protection requirements or direction to other provisions that deal with the protection of the openings in a horizontal assembly.

Some of the previous inconsistencies have been addressed, including those dealing with fire-resistance-rated assemblies and non-rated assemblies. Section 711 is now divided into separate subsections for rated and non-rated assemblies, and the scoping of Section 711.1 is revised to direct the code users to Section 711.3 for nonfire-resistance-rated floor and roof assemblies. This distinction is important because the term "horizontal assemblies" is defined within Chapter 2 as "a fire-resistance-rated floor or roof assembly."

Some of the more apparent revisions include:

- The previous continuity provisions of Section 711.4 have been moved to the beginning of the section and have been split into two subsections (711.2.2 and 711.2.3) to indicate that they deal with different subjects. Section 711.2.2 addresses the continuity of the fire-resistance protection and Section 711.2.3 deals with the supporting construction requirements. The skylight requirements from Section 711.4 in the 2012 IBC have been moved to Section 712 because they apply to a vertical opening versus the construction of the horizontal assembly.
- Section 711.2.4.3 has been modified so that it coordinates with the dwelling unit and sleeping unit separation requirements of Section 420, and regulates not only the separation between units but also between the units and other occupancies contiguous to the units.
- Previous Sections 711.5 through 711.8 have been moved to Section 712.
- Section 711.9 in the 2012 IBC has been relocated to the elevator lobby requirements, which were previously found in Section 713.14 of the 2012 IBC. They are now located in Section 3006 and are addressed in this text in the discussion of Chapter 30 changes.
- The scoping provisions of Section 712.1 have been modified to clearly indicate that each vertical opening must be addressed and protected using one of the options listed within Section 712 as opposed to simply stating that the requirements apply to the various applications.
- Item 2 from the "two-story opening" provisions of Section 712.1.8 in the 2012 IBC has been deleted. That item restricted an opening between adjacent levels from containing "a stairway or ramp required by Chapter 10." The limitation was removed from new Section 712.1.9 because it would cause conflict and confusion between Section 712 and Chapter 10. The deletion reinforces the intent that

711, 712 continues

711, 712 continued

the requirements in Chapter 10 (specifically, new Section 1019) are to be used for the opening protection provisions for exit access stairways. This intent is further emphasized by the provisions of Section 712.1.12 specifically stating that the openings for exit access stairways are to be protected in accordance with Section 1019.

- The vertical opening protection requirements related to parking garages have all been consolidated into a single location under Section 712.1.10. This format change brings the 2012 IBC provisions found in Sections 712.1.9, 712.1.15 and 712.1.16 together, where they are more easily located.
- Similar to the garage requirements, provisions related to opening protectives have been consolidated into new subsections within Section 712.1.13. This includes the floor fire door requirements previously found in Section 711.8 and ceiling access door requirements previously found in Section 711.3.2.
- Section 712.1.13.1 was revised to coordinate with changes made in the referenced NFPA 288 standard. The referenced edition of the standard has been modified so that it now includes assemblies that may be installed in a fire-resistance-rated roof.

714.4.2

Membrane Penetrations

CHANGE TYPE: Modification

CHANGE SUMMARY: Where the double top plates of a wall interrupt the ceiling membrane of a horizontal assembly, the wall must now be sheathed only with Type X gypsum wallboard. The wall will not require a fire-resistance rating unless needed due to some other code requirement.

2015 CODE: ~~714.4.1.2~~ 714.4.2 Membrane Penetrations. Penetrations of membranes that are part of a horizontal assembly shall comply with Section 714.4.1.1 or 714.4.1.2 ~~714.4.1.1.1 or 714.4.1.1.2~~. Where floor/ceiling assemblies are required to have a fire-resistance rating, recessed fixtures shall be installed such that the required fire resistance will not be reduced.

Exceptions:

1. (No change.)
2. (No change.)
3. (No change.)
4. (No change.)
5. (No change.)
6. (No change.)
7. The ceiling membrane of 1- and 2-hour fire-resistance-rated horizontal assemblies is permitted to be interrupted with the double wood top plate of a ~~fire-resistance-rated~~ wall assembly that is sheathed with Type X gypsum wallboard, provided that all penetrating items through the double top plates are protected in accordance with Section 714.4.1.1 or 714.4.1.2 and the ceiling membrane is tight to the top plates. ~~The fire-resistance rating of the wall shall not be less than the rating of the horizontal assembly.~~

714.4.2 continues

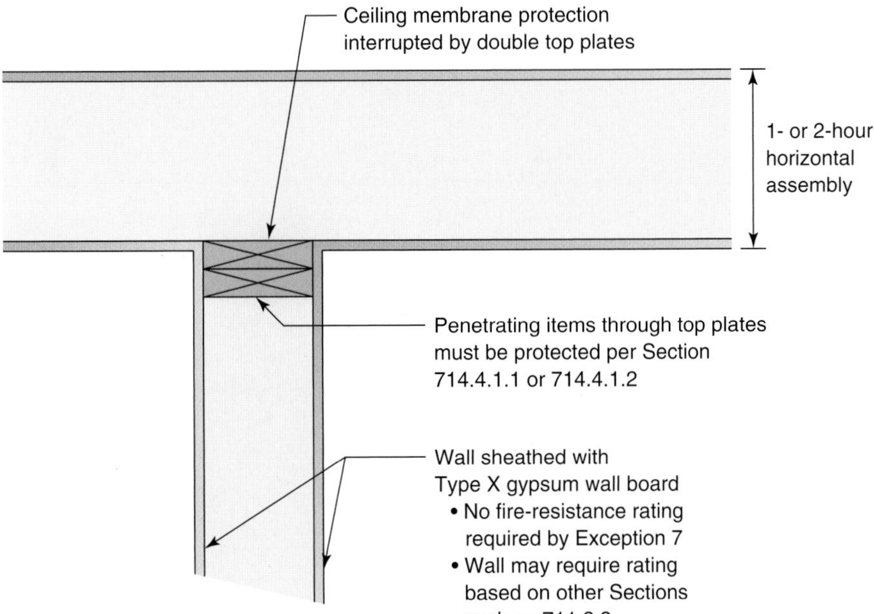
Ceiling membrane protection interrupted by double top plates
1- or 2-hour horizontal assembly
Penetrating items through top plates must be protected per Section 714.4.1.1 or 714.4.1.2
Wall sheathed with Type X gypsum wall board
• No fire-resistance rating required by Exception 7
• Wall may require rating based on other Sections such as 711.2.3

Ceiling membrane interrupted by wall top plates

714.4.2 continued

CHANGE SIGNIFICANCE: When Exception 7 was added to the 2012 IBC, it was the first time the code had specifically addressed the interruption of the ceiling membrane continuity at the intersection of a stud wall and a horizontal assembly. The exception required the wall to have a fire-resistance rating equivalent to the rating of the horizontal assembly. The new provisions simply require the wall to be sheathed with "Type X gypsum wallboard" and do not require any specific fire-resistance rating as a part of this exception.

Previously, the wall was required to have a fire-resistance rating equal to that of the horizontal assembly even for situations where the wall was a non-bearing wall. Under the revised exception, the rating of the wall, if required, is dependent upon other code provisions, such as Section 711.2.3, which regulates the supporting construction of horizontal assemblies, or where the wall is serving some other purpose such as a corridor or dwelling unit separation.

The normal methods of construction and the sequenced installation of gypsum wall and ceiling membranes are now intended to be an acceptable level of protection. The exception recognizes that the ceiling membrane of a fire-resistance-rated horizontal assembly will not be continuous over the top of the wall, but that the double top plate will ensure there is a minimum of 3 inches of solid wood at the point the ceiling membrane is interrupted. Provided any penetrations through the top plates are adequately protected (per Section 714.4.1.1 or 714.4.1.2), the ceiling membrane continuity will essentially be restored and capable of protecting the horizontal assembly. The ceiling membrane must be "tight to the top plates," and with the wall membrane typically being installed after the ceiling protection, this construction method will be acceptable.

717.1.1

Ducts Transitioning between Shafts

CHANGE TYPE: Clarification

CHANGE SUMMARY: Ducts are now expressly allowed to exit a shaft, transition horizontally, and then enter another shaft without continuous shaft construction.

2015 CODE: **717.1.1 Ducts and Air Transfer Openings.** Ducts transitioning horizontally between shafts shall not require a shaft enclosure provided that the duct penetration into each associated shaft is protected with dampers complying with this section.

CHANGE SIGNIFICANCE: Shafts are used to enclose vertical openings within a building in order to prevent the spread of smoke or fire from one story to the next. Although shafts are generally vertical, they are permitted by Section 713.2 to be constructed using fire barriers, horizontal assemblies or both. Section 717.1.1 now specifically indicates that it is permissible to have a duct come out of one shaft and then transition into a different shaft provided that dampers exist at the point where the duct penetrates each of the shafts.

The new duct and shaft enclosure provisions do not allow for the violation of any other code requirements, such as those in the *International Mechanical Code* (IMC), that prohibit the installation of dampers within the ductwork (such as in a clothes dryer exhaust system, IMC Section 504.2, or ducts serving a hazardous exhaust system, IMC 510.7.1) or that require the enclosure to be continuous to the outlet terminal (such as grease duct enclosures serving a Type I hood, IMC Section 506.3.11). However, in a typical HVAC duct system, the transition of a duct from one shaft to another without requiring the duct between the shafts to be within a fire-resistance-rated shaft enclosure is acceptable.

717.1.1 continues

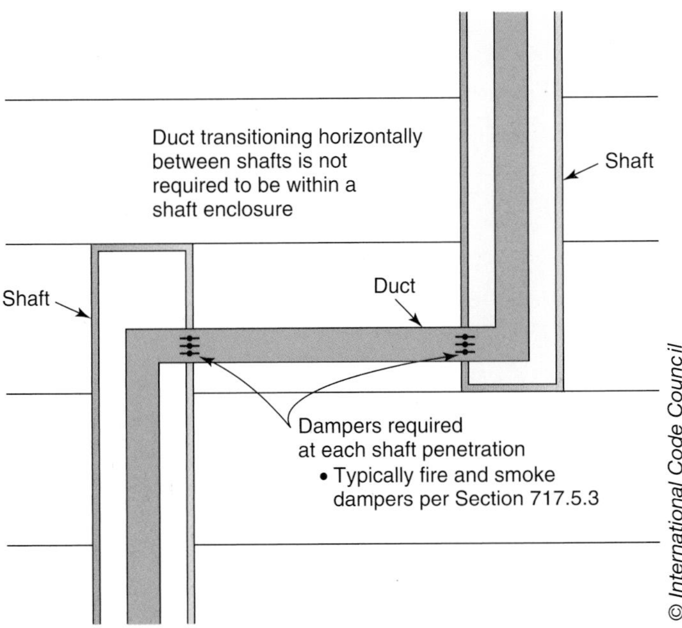
Duct transitioning horizontally between shafts is not required to be within a shaft enclosure
Shaft
Shaft
Duct
Dampers required at each shaft penetration
• Typically fire and smoke dampers per Section 717.5.3
© International Code Council

Duct transitioning between shafts

717.1.1 continued

Because the ducts will be protected with dampers in accordance with Section 717 at both shafts, the code's basic intent of providing an appropriate separation between stories can be maintained. In general, duct penetrations of shafts require both fire and smoke dampers. Therefore, protecting the duct as it leaves one shaft and again as it enters another shaft enclosure will minimize the spread of fire and smoke through the building and help maintain the separation between stories.

Application wise, there is no technical change. Although not as clearly stated, this option is permissible in the 2012 and earlier editions of the IBC. The previous editions allow this design option by simply following the code's general requirements. The option to protect the ductwork with a continuous horizontal shaft enclosure or to provide dampers at each duct penetration of the associated discontinuous shaft enclosures has always been available. Having the provisions clearly stated allows the designer to determine which approach will be taken to protect the ductwork. Because properly constructing and supporting a horizontal shaft enclosure can be difficult and expensive, this option of providing dampers at each duct penetration may be used more frequently now that the IBC specifically addresses it.

717.3, 717.5
Corridor Dampers

CHANGE TYPE: Clarification

CHANGE SUMMARY: Where a duct penetration occurs in the ceiling of a fire-resistance-rated corridor where the lid of the corridor is constructed using a corridor wall placed horizontally, a corridor damper is now specifically mandated.

2015 CODE:

SECTION 202
DEFINITIONS

CORRIDOR DAMPER. A listed device intended for use where air ducts penetrate or terminate at horizontal openings in the ceilings of fire-resistance-rated corridors, where the corridor ceiling is permitted to be constructed as required for the corridor walls.

DAMPER. See "Ceiling radiation damper," "Combination fire/smoke damper," "Corridor damper," "Fire damper" and "Smoke damper."

717.3.1 Damper Testing. Dampers shall be listed and labeled in accordance with the standards in this section.

717.3, 717.5 continues

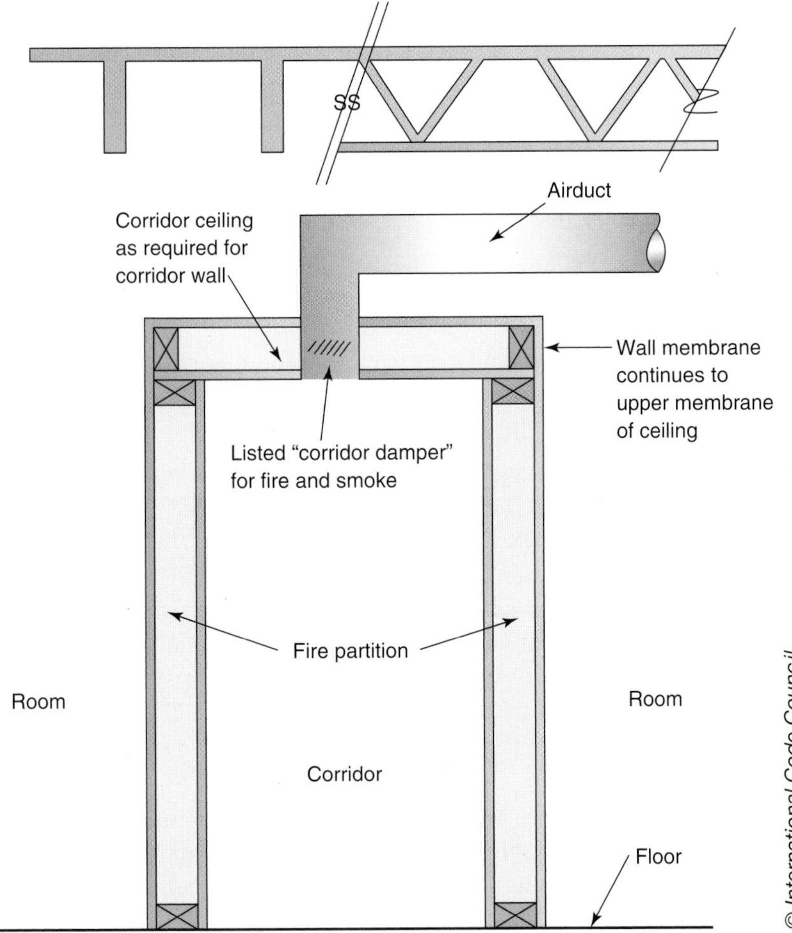

Corridor damper at "tunnel corridor"

717.3, 717.5 continued

1. Fire dampers shall comply with the requirements of UL 555. Only fire dampers and ceiling radiation dampers labeled for use in dynamic systems shall be installed in heating, ventilation and air-conditioning systems designed to operate with fans on during a fire.
2. Smoke dampers shall comply with the requirements of UL 555S.
3. Combination fire/smoke dampers shall comply with the requirements of both UL 555 and UL 555S.
4. Ceiling radiation dampers shall comply with the requirements of UL 555C or shall be tested as part of a fire-resistance-rated floor/ceiling or roof/ceiling assembly in accordance with ASTM E119 or UL 263.
5. Corridor dampers shall comply with requirements of both UL 555 and UL 555S. Corridor dampers shall demonstrate acceptable closure performance when subjected to 150 feet per minute (0.76 mps) velocity across the face of the damper during the UL 555 fire exposure test.

717.3.2.4 Corridor Damper Ratings. Corridor dampers shall have the following minimum ratings:

1. One-hour fire-resistance rating.
2. Class I or II leakage rating as specified in Section 717.3.2.2.

717.3.3.5 Corridor Damper Actuation. Corridor damper actuation shall be in accordance with Sections 717.3.3.1 and 717.3.3.2.

717.5 Where Required. Fire dampers, smoke dampers, ~~and~~ combination fire/smoke dampers, ceiling radiation dampers and corridor dampers shall be provided at the locations prescribed in Sections 717.5.1 through 717.5.7 and 717.6. Where an assembly is required to have both fire dampers and smoke dampers, combination fire/smoke dampers or a fire damper and a smoke damper shall be ~~required~~ provided.

717.5.4.1 Corridors. Duct and air transfer openings that penetrate corridors shall be protected with dampers as follows.

1. A corridor damper shall be provided where corridor ceilings, constructed as required for the corridor walls as permitted in Section 708.4, Exception 3, are penetrated.
2. A ceiling radiation damper shall be provided where the ceiling membrane of a fire-resistance-rated floor-ceiling or roof-ceiling assembly, constructed as permitted in Section 708.4, Exception 2, is penetrated.
3. A listed smoke damper designed to resist the passage of smoke shall be provided at each point a duct or air transfer opening penetrates a corridor enclosure required to have smoke and draft control doors in accordance with Section 716.5.3.

Exceptions:

1. Smoke dampers are not required where the building is equipped throughout with an approved smoke control system in accordance with Section 909, and smoke dampers are not necessary for the operation and control of the system.

2. Smoke dampers are not required in corridor penetrations where the duct is constructed of steel not less than 0.019 inch (0.48 mm) in thickness and there are no openings serving the corridor.

CHANGE SIGNIFICANCE: Where damper protection is required in a ceiling of a corridor constructed using what is often described as the "tunnel corridor construction" option permitted by Exception 3 of Section 708.4, the appropriate type of damper to be used is now identified as a "corridor damper." Where a hard ceiling is used to construct a corridor lid by essentially building a wall element and laying it horizontally over the top of the two sided wall fire partitions, the appropriate type of damper for this installation is a corridor damper. These dampers have been tested to both the UL 555 standard as a fire damper and the UL 555S standard as a smoke damper. Although the corridor dampers are tested to both standards, they are specifically listed as a "corridor damper" and not as a "combination fire/smoke damper."

Corridor dampers are tested to be installed in the hard ceiling of a corridor and are the appropriate damper to be used as opposed to a ceiling radiation damper, a combination fire/smoke damper or both fire and smoke dampers. Only the corridor damper is designed, tested and listed to be mounted in a wall that has been installed in a horizontal position as permitted by Section 708.4, Exception 3.

Section 717.3.1 assists the code user in recognizing the appropriate testing requirements for each of the various types of dampers that are acceptable under the IBC. Provisions have been reformatted such that each particular type of damper is now listed within a separate item. Corridor damper testing is addressed in new Item 5. In further reviewing Sections 717.3.1 and 717.3.2.4, a corridor damper both needs to be listed for a fire-resistance rating of 1 hour and have either a Class I or II leakage rating when tested at the elevated temperature rates using UL 555S. These dampers must also be capable of closing against the pressure caused by air flowing across the face of the damper at a minimum velocity of 150 feet per minute. What makes the corridor damper testing and listing different from the regular combination fire/smoke damper is that the testing is done in a wall assembly installed in the horizontal position.

The actuation of a corridor damper is regulated by the text found within new Section 717.3.3.5, which simply references the general actuation requirements of Section 717.3.3.1 for fire dampers and Section 717.3.3.2 for smoke dampers. Because the corridor damper is both a fire damper and smoke damper, it is appropriate that the actuation of the damper comply with each of these two sections. In the same manner, Section 717.5.4.1 also has been revised to help ensure that the appropriate type of damper is used to protect openings into a corridor. The general fire partition requirements of Section 717.5.4 will require or exempt the fire damper. Section 717.5.4.1 provides additional details for duct and air transfer openings into a corridor by addressing the need for a corridor damper, ceiling radiation damper or smoke damper depending on the corridor's method of construction and the damper's location. It is important to note that the exceptions within Section 717.5.4.1 are only applicable to the "smoke damper" required under Item 3, not to a corridor damper or ceiling radiation damper.

717.3, 717.5 continues

717.3, 717.5 continued

The revisions should be viewed as a clarification rather than a modification to the code requirements in the 2012 edition. Although corridor dampers are not specifically mentioned within the 2012 IBC or IMC, they were the only appropriate damper specifically designed, tested and listed for this particular installation. Using 2012 IBC Sections 717.5.4.1, 717.5.4 and/or 717.6.1, it might be implied that these types of duct or air transfer openings are to be protected by combination fire/smoke dampers or by fire dampers and smoke dampers. However, none of these methods is in accordance with the listing if it is installed in the described location.

903.2.1.6

Sprinkler Systems—Assembly Occupancies

CHANGE TYPE: Addition

CHANGE SUMMARY: An automatic sprinkler system is now required to be installed in a building when the roof is used for a Group A-2 assembly occupancy with an occupant load exceeding 100, as well as for other Group A occupancies where the occupant load exceeds 300.

2015 CODE: **903.2.1.6 Assembly Occupancies on Roofs.** Where an occupied roof has an assembly occupancy with an occupant load exceeding 100 for Group A-2 and 300 for other Group A occupancies, all floors between the occupied roof and the level of exit discharge shall be equipped with an automatic sprinkler system in accordance with Section 903.3.1.1 or 903.3.1.2.

> **Exception:** Open parking garages of Type I or Type II construction.

CHANGE SIGNIFICANCE: As a general provision, assembly occupancies require the installation of an automatic sprinkler system when the fire area containing the Group A use is "located on a floor other than a level of exit discharge." Where the assembly use is located on the roof of the building, the stories of the building that the occupants must pass through are also now required to be sprinklered. The new sprinkler requirement applies to buildings where the Group A-2 roof-top occupancy exceeds 100 occupants. For the other Group A occupancy classifications, the provisions are applicable where the occupant load exceeds 300.

Because an occupied roof does not meet the definition for a fire area, the provisions are addressed separately from the other Group A requirements. By requiring the building beneath the assembly occupancy to be sprinklered, the requirement is consistent with other provisions of Section 903.2.1 and protects the occupants from hazards elsewhere in the building. Whether building occupants are located on an upper story or on the roof, they are exposed to a similar hazard as they travel down through the building prior to reaching the level of exit discharge. It should be noted that this provision does not require the roof to be sprinklered or provided with any alternative fire-extinguishing system. The sprinkler protection is mandated only on the floors between the occupied roof and the level of exit discharge.

903.2.1.6 continues

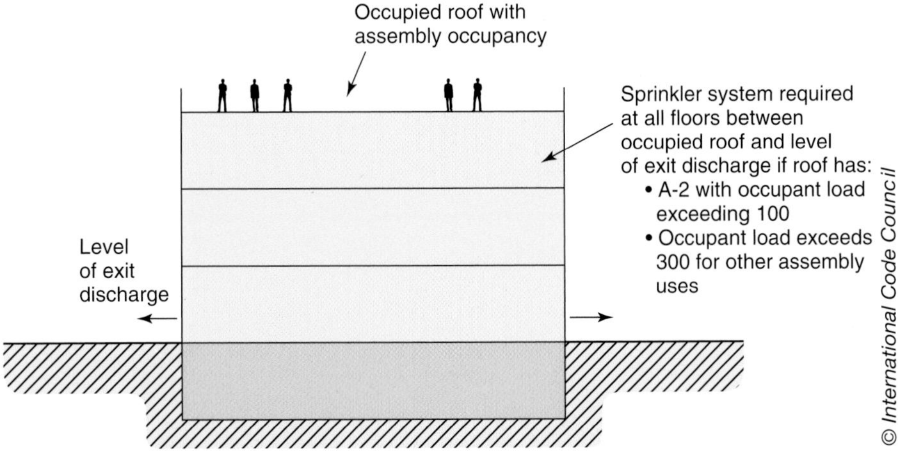

Assembly occupancy on roof

903.2.1.6 continued

Code users will need to be cautious of how they apply this requirement because it is located within Section 903.2.1 and only applicable to occupied roofs that are used for Group A occupancies. Although most occupied roofs that do have occupant loads reflected in the provisions are generally used for an assembly purpose, if the roof is classified as any other occupancy category, such as a retail space, then the requirements of Section 903.2.1.6 are not applicable. The provisions do allow an NFPA 13R sprinkler system to be provided where the assembly use is located on the roof of a hotel, apartment building or other Group R occupancy.

The application of this new provision should not be extended to other areas of the code. For example, the occupied roof is not considered as *building area, fire area* or a *story*. Therefore, even though an occupied roof is viewed as being an assembly occupancy, the limitations of Table 504.4 and many other code provisions would not apply.

The exception eliminates the sprinkler requirement in situations where the building beneath the occupied roof is an open parking garage of either Type I or II construction. Due to the very limited hazard these types of garages pose, allowing the roof-top occupants to exit through the non-sprinklered garage will not significantly affect their safety.

903.2.1.7

Multiple Fire Areas

CHANGE TYPE: Addition

CHANGE SUMMARY: Where small Group A fire areas share a common means of egress, the occupant load of the spaces must now be added together to determine if a sprinkler system is required.

2015 CODE: **903.2.1.7 Multiple Fire Areas.** An automatic sprinkler system shall be provided where multiple fire areas of Group A-1, A-2, A-3 or A-4 occupancies share exit or exit access components and the combined occupant load of these fire areas is 300 or more.

CHANGE SIGNIFICANCE: Through the creation of limited-size fire areas, it has historically been possible to eliminate the requirement for sprinklers in Group A occupancies. However, where the fire areas do not have individual egress systems but rather require the occupants to travel through a common or shared egress system serving 300 or more persons, the code now requires the building to be sprinklered in accordance with Section 903.2.1.

The option of compartmentalizing Group A occupancies into separate fire areas simply to avoid the sprinkler requirement is no longer available if the fire areas share a common egress system and the combined occupant load of the fire areas meets or exceeds the limitations typically found in the subsections of Section 903.2.1 for the individual Group A occupancies. It is important that code users notice all three of the triggers for this requirement, which are (1) multiple fire areas of Group A-1, A-2, A-3 or A-4 occupancies; (2) the fire areas share exit or exit access components; and (3) the combined occupant load of the fire areas is 300 or more. If any one of these conditions does not occur, then Section 903.2.1.7 is not applicable. For example, the new provision is not applicable (a) if there is only a

903.2.1.7 continues

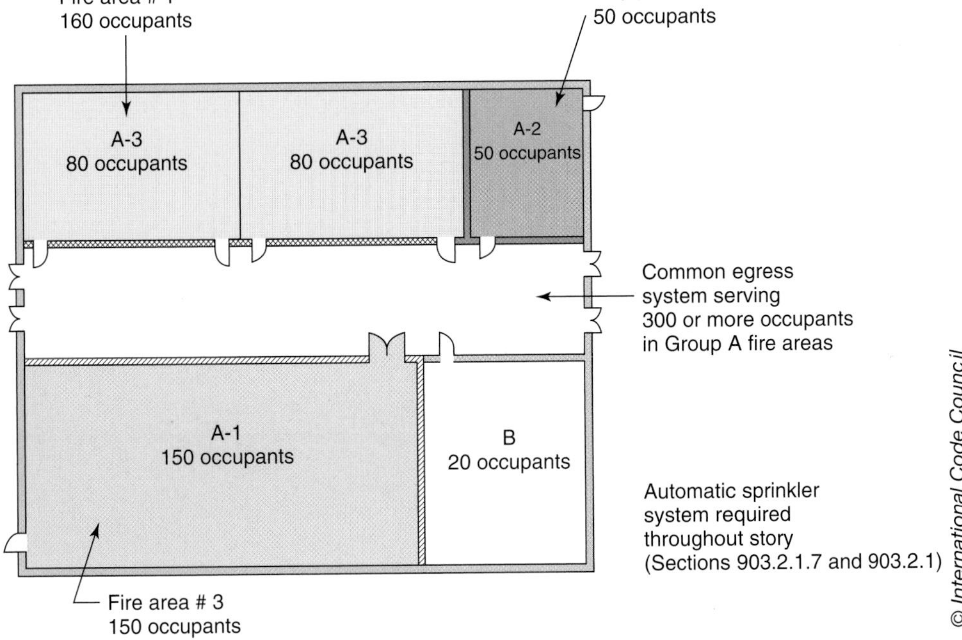

Group A fire areas with shared egress components

903.2.1.7 continued

single Group A fire area and the other fire areas are of different occupancy groups, or (b) if the Group A fire areas have separate or independent egress systems, or (c) if the aggregate occupant load of the assembly fire areas is less than 300. The concern is that any fire event near or within the common egress system has the same hazard potential whether the egress system serves 300 occupants from one fire area or from three separate fire areas having 100 occupants each.

903.2.8

Sprinkler Systems—Group R Occupancies

CHANGE TYPE: Modification

CHANGE SUMMARY: Sprinkler requirements for Group R-4 occupancies are now dependent on the capabilities of the occupants. In buildings where occupants require limited assistance when responding to an emergency condition, additional sprinkler protection is required for attic spaces.

2015 CODE: 903.2.8 Group R. An automatic sprinkler system installed in accordance with Section 903.3 shall be provided throughout all buildings with a Group R fire area.

903.2.8.1 Group R-3 ~~or R-4 Congregate Residences~~. An automatic sprinkler system installed in accordance with Section 903.3.1.3 shall be permitted in Group R-3 occupancies. ~~or R-4 congregate residences with 16 or fewer residents.~~

903.2.8.2 Group R-4 Condition 1. An automatic sprinkler system installed in accordance with 903.3.1.3 shall be permitted in Group R-4 Condition 1 occupancies.

903.2.8.3 Group R-4 Condition 2. An automatic sprinkler system installed in accordance with 903.3.1.2 shall be permitted in Group R-4 Condition 2. Attics shall be protected in accordance with Section 903.2.8.3.1 or 903.2.8.3.2.

903.2.8.3.1 Attics Used for Living Purposes, Storage or Fuel-Fired Equipment. Attics used for living purposes, storage or fuel-fired equipment shall be protected throughout with an automatic sprinkler system installed in accordance with Section 903.3.1.2.

903.2.8 continues

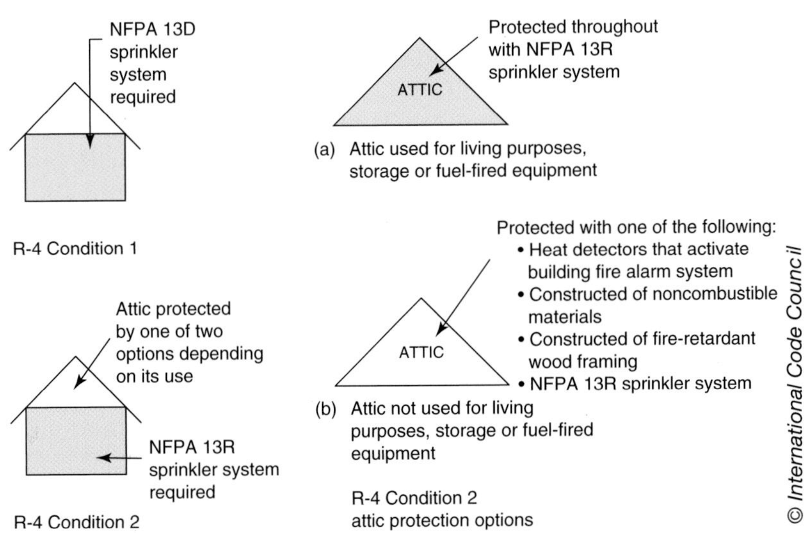

Group R-4 sprinkler requirements

903.2.8 continued

903.2.8.3.2 Attics Not Used for Living Purposes, Storage or Fuel-Fired Equipment. Attics not used for living purposes, storage or fuel-fired equipment shall be protected in accordance with one of the following:

1. Attics protected throughout by a heat detector system arranged to activate the building fire alarm system in accordance with Section 907.2.10.
2. Attics constructed of non-combustible materials.
3. Attics constructed of fire-retardant-treated wood framing complying with Section 2303.2.
4. The automatic fire sprinkler system shall be extended to provide protection throughout the attic space.

~~903.2.8.2~~ 903.2.8.4 Care Facilities. An *automatic sprinkler system* installed in accordance with Section 903.3.1.3 shall be permitted in care facilities with five or fewer individuals in a single-family dwelling.

CHANGE SIGNIFICANCE: The sprinkler requirements for Group R-4 occupancies have been modified based upon whether it is a "Condition 1" or "Condition 2" facility. Where the occupants are capable of responding on their own during emergencies (Condition 1), the requirements remain the same as they were in the 2012 edition. In those facilities, the installation of an NFPA 13D sprinkler system is permitted with no additional requirements. In facilities where the residents require limited assistance with evacuation (Condition 2), the installation of an NFPA 13R (Section 903.3.1.2) sprinkler system is permitted; however, the previously accepted NFPA 13D system is no longer acceptable.

Condition 2 facilities now also require any attic to be protected. The subsections of Section 903.2.8.3 establish the required attic protection depending upon how the attic is used. In situations where the attic is occupied, used for storage, or contains fuel-fired equipment, the sprinkler system must now be extended into the attic space. Where the attic is not used for living purposes or storage, and does not contain any fuel-fired equipment, Section 903.2.8.3.2 now provides four options to protect the attic, including the option of extending the sprinkler system into the attic space. The other options allow the installation of a heat detection system within the attic or regulate the materials that may be used to construct the attic. Providing a heat detection system within the attic will provide the residents with an early warning of an attic fire. By restricting the construction materials, the chance of an attic fire will be reduced. These options provide useful alternatives for group homes that are constructed in colder climates where freezing of the automatic sprinklers within an unheated attic is a concern.

The sprinkler requirements for Group R-3 occupancies and limited-size care facilities in single-family dwellings remain unchanged from the 2012 code. Both of these uses will continue to accept the installation of an NFPA 13D sprinkler system.

903.3.1.1.2

Exempt Locations for NFPA 13 Sprinklers

CHANGE TYPE: Modification

CHANGE SUMMARY: An exemption for sprinkler systems in small resident bathrooms has been introduced into the IBC because the provision was removed from the current edition of the referenced NFPA 13 standard.

2015 CODE: 903.3.1.1 NFPA 13 Sprinkler Systems. Where the provisions of this code require that a building or portion thereof be equipped throughout with an automatic sprinkler system in accordance with this section, sprinklers shall be installed throughout in accordance with NFPA 13 except as provided in Sections 903.3.1.1.1 and 903.3.1.1.2.

903.3.1.1.1 Exempt Locations. (*No changes.*)

903.3.1.1.2 Bathrooms. In Group R occupancies, other than Group R-4 occupancies, sprinklers shall not be required in bathrooms that do not exceed 55 square feet in area and are located within individual dwelling units or sleeping units, provided that walls and ceilings, including the walls and ceilings behind a shower enclosure or tub, are of noncombustible or limited-combustible materials with a 15-minute thermal barrier rating.

903.3.1.1.2 continues

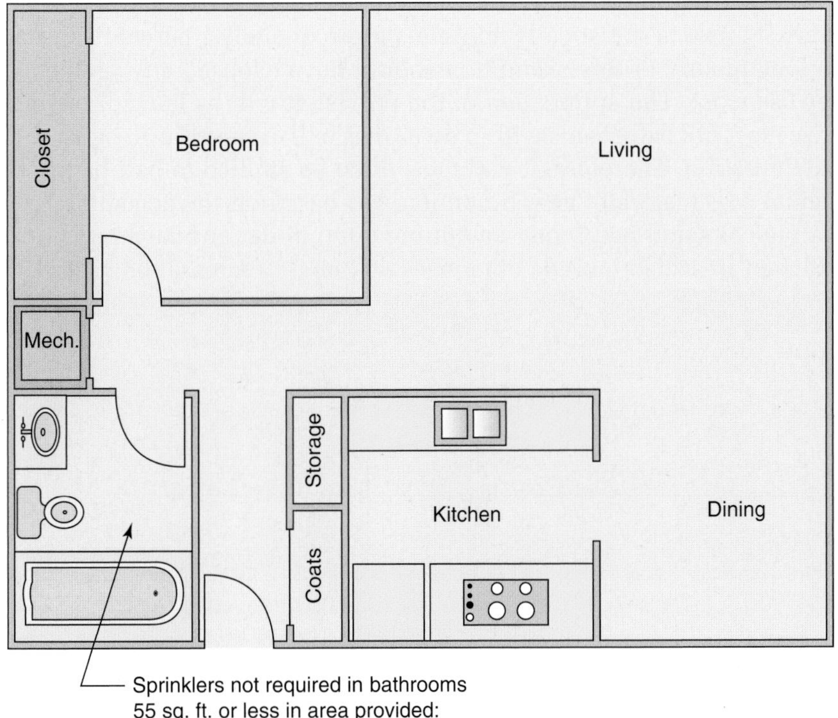
Closet
Bedroom
Living
Mech.
Storage
Kitchen
Dining
Coats
Sprinklers not required in bathrooms
55 sq. ft. or less in area provided:
• Located within individual unit
• Walls and ceilings of noncombustible or limited-combustible material
(including walls behind tub or shower enclosure)
Group R (other than R-4)
© International Code Council

Sprinkler exception for small bathrooms

903.3.1.1.2 continued

CHANGE SIGNIFICANCE: The IBC references NFPA 13, *Installation of Sprinkler Systems*, for the technical requirements regarding the installation of automatic sprinkler systems. Section 8.15.8.1.1 of the 2010 edition of the NFPA 13 standard establishes a sprinkler exemption for small bathrooms in most Group R occupancies. However, when the 2010 NFPA 13 standard was updated to the 2013 edition, this allowance was removed. Because the 2015 IBC adopts by reference the 2013 edition of NFPA 13, the exemption would not apply under the 2015 IBC unless inserted into the code itself. Therefore, the allowance has been included as IBC Section 903.3.1.1.2. Code users should remember that, based on Section 102.4.1, the IBC will be the controlling document in this situation and will take precedence over the standard. As a result, the change will not result in any difference of application between the 2012 and 2015 editions of the code.

It is important to recognize that this new provision is only applicable to NFPA 13 sprinkler systems because both the NFPA 13R and NFPA 13D systems continue to include this exception within the standards themselves. In addition to making this exemption consistent for all three types of sprinkler systems, including this provision within the IBC reinstates an exemption that has been applicable for these sprinkler systems since 1976. Given that there is more than 35 years of history behind this exemption, and that the NFPA standards committee itself had originally indicated there were no technical data provided to support the change, it was determined that the exemption should be maintained by placing it within the IBC.

Supporting information indicated that the history of apartment unit bathroom fires is statistically minimal (approximately 1 percent), and that the long history of this exemption should have exposed any problems if they did exist. The application of the provision will be fairly limited because most full bathrooms within dwellings will exceed the 55-square-foot size limitation. Therefore, it will most often be applied to half-baths. The original 55-square-foot area limitation was based on the accommodation of a typical small bathroom containing a tub, toilet and lavatory, but insufficient in size to include other uses that might increase the hazard.

903.3.1.2.2

Open-Ended Corridors

CHANGE TYPE: Clarification

CHANGE SUMMARY: Where an NFPA 13R sprinkler system is installed, the sprinkler protection must now be extended to any open-ended corridors and associated exterior stairways, clarifying that an open breezeway is considered as an interior portion of the building and not an exterior location for the application of the sprinkler requirements.

2015 CODE: **903.3.1.2.2 Open-Ended Corridors.** Sprinkler protection shall be provided in open-ended corridors and associated exterior stairways and ramps as specified in Section 1027.6, Exception 3.

CHANGE SIGNIFICANCE: It has been clarified that required sprinkler protection in residential buildings must extend to any "open-ended corridor" addressed under the provisions of Section 1027.6, Exception 3. Based on this specific requirement, the IBC takes precedence over the general allowance found within the NFPA 13R sprinkler standard to exclude sprinklers in open corridors and stairways.

As a general requirement, exterior stairways and ramps are required to be separated from the interior of the building by a fire-resistance-rated wall and a fire-protection-rated door assembly. The open-ended corridor provisions within Section 1027.6 allow the separation and door to be eliminated at the point where the exterior stairway or ramp connects to the corridor. Item 1 in the exception, which allows the separation/protection to be eliminated, states that "the building, including open-ended corridors, and stairways and ramps, shall be equipped throughout with an automatic sprinkler system in accordance with Section 903.3.1.1 or 903.3.1.2." It is intended that the sprinkler system be installed within the corridor and stairway to compensate for the elimination of the separation.

903.3.1.2.2 continues

Open-ended corridor

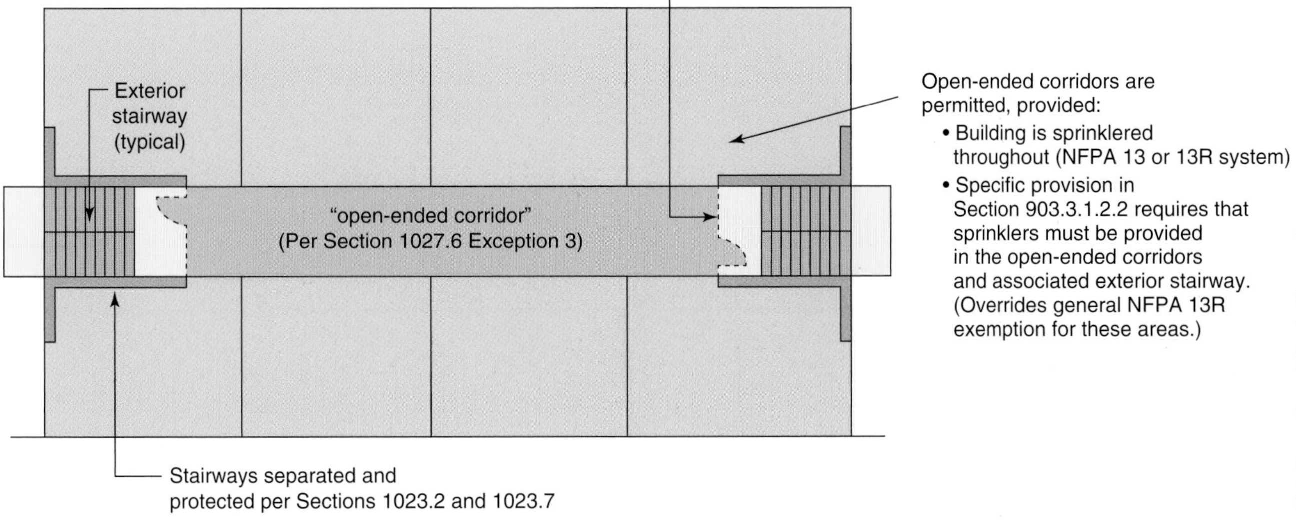

Wall and fire door eliminated by Section 1027.6 Exception 3

Exterior stairway (typical)

"open-ended corridor" (Per Section 1027.6 Exception 3)

Open-ended corridors are permitted, provided:
- Building is sprinklered throughout (NFPA 13 or 13R system)
- Specific provision in Section 903.3.1.2.2 requires that sprinklers must be provided in the open-ended corridors and associated exterior stairway. (Overrides general NFPA 13R exemption for these areas.)

Stairways separated and protected per Sections 1023.2 and 1023.7

Sprinklers required in an open-ended corridor

903.3.1.2.2 continued

Where a sprinkler system complying with Section 903.3.1.2 (an NFPA 13R system) is installed, the NFPA standard will typically allow the sprinklers to be excluded from the corridor and stairway areas based on Section 6.6.5 of the standard, which states: "Except as provided for in 6.6.5.1, sprinklers shall not be required in any porches, balconies, corridors, carports, porte cocheres, and stairs that are open and attached." Therefore, where the residential building is equipped with an NFPA 13R sprinkler system and the "open-ended corridor" requirements of Section 1027.6 are applied, this specific requirement in IBC Section 903.3.1.2.2 will require the sprinklers to be provided within the corridor and stairways in order to satisfy the code requirement that the building is "equipped throughout with an automatic sprinkler system in accordance with Section . . . 903.3.1.2."

903.3.8

Limited Area Sprinkler Systems

CHANGE TYPE: Modification

CHANGE SUMMARY: Additional restrictions have been placed on limited area sprinkler systems, including a reduction in the system size to a maximum of six sprinklers within a single fire area.

2015 CODE: **903.3 Installation Requirements.** Automatic sprinkler systems shall be designed and installed in accordance with Sections 903.3.1 through 903.3.8~~6~~.

903.3.5.1 Domestic Services. Where the domestic service provides the water supply for the automatic sprinkler system, the supply shall be in accordance with this section.

~~**903.3.5.1.1 Limited area Sprinkler Systems.** Limited area sprinkler systems serving fewer than 20 sprinklers on any single connection are permitted to be connected to the domestic service where a wet automatic standpipe is not available. Limited area sprinkler systems connected to domestic water supplies shall comply with each of the following requirements:~~

~~1. Valves shall not be installed between the domestic water riser control valve and the sprinklers.~~

   ~~**Exception:** An approved indicating control valve supervised in the open position in accordance with Section 903.4.~~

~~2. The domestic service shall be capable of supplying the simultaneous domestic demand and the sprinkler demand required to be hydraulically calculated by NFPA 13, NFPA 13D or NFPA 13R.~~

903.3.8 continues

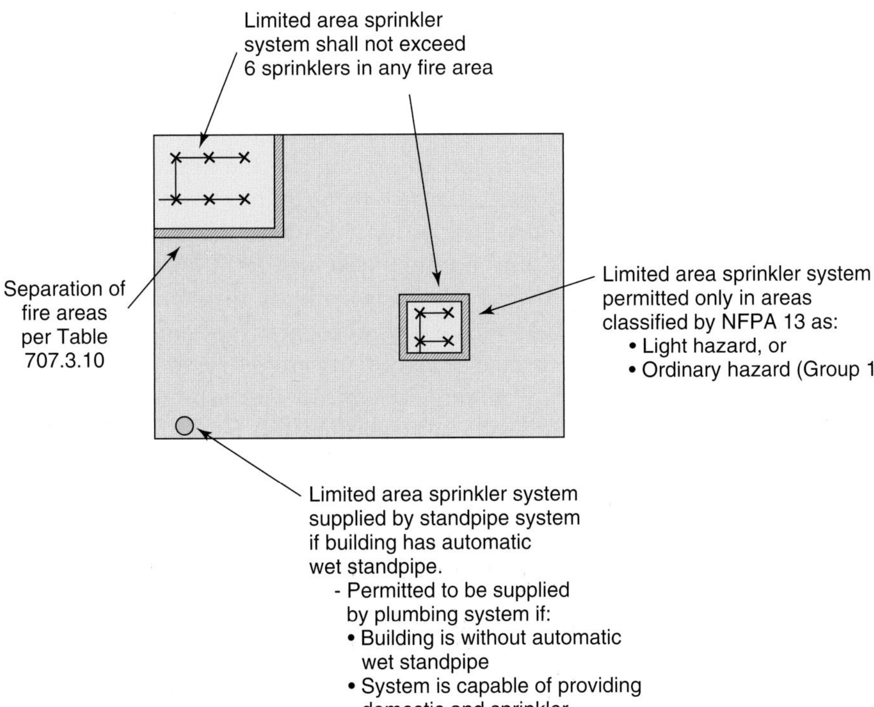
Limited area sprinkler system shall not exceed 6 sprinklers in any fire area
Separation of fire areas per Table 707.3.10
Limited area sprinkler system permitted only in areas classified by NFPA 13 as:
• Light hazard, or
• Ordinary hazard (Group 1)
Limited area sprinkler system supplied by standpipe system if building has automatic wet standpipe.
- Permitted to be supplied by plumbing system if:
• Building is without automatic wet standpipe
• System is capable of providing domestic and sprinkler water flow demand

Limited area sprinkler system

903.3.8 continued

~~903.3.5.1.2~~ 903.3.5.2 Residential Combination Services. A single combination water supply shall be allowed provided that the domestic demand is added to the sprinkler demand as required by NFPA 13R.

903.3.8 Limited Area Sprinkler Systems. Limited area sprinkler systems shall be in accordance with the standards listed in Section 903.3.1 except as provided in Sections 903.3.8.1 through 903.3.8.5.

903.3.8.1 Number of Sprinklers. Limited area sprinkler systems shall not exceed six sprinklers in any single fire area.

903.3.8.2 Occupancy Hazard Classification. Only areas classified by NFPA 13 as Light Hazard or Ordinary Hazard Group 1 shall be permitted to be protected by limited area sprinkler systems.

903.3.8.3 Piping Arrangement. Where a limited area sprinkler system is installed in a building with an automatic wet standpipe system, sprinklers shall be supplied by the standpipe system. Where a limited area sprinkler system is installed in a building without an automatic wet standpipe system, water shall be permitted to be supplied by the plumbing system provided that the plumbing system is capable of simultaneously supplying domestic and sprinkler demands.

903.3.8.4 Supervision. Control valves shall not be installed between the water supply and sprinklers unless the valves are of an approved indicating type that are supervised or secured in the open position.

903.3.8.5 Calculations. Hydraulic calculations in accordance with NFPA 13 shall be provided to demonstrate that the available water flow and pressure are adequate to supply all sprinklers installed in any single fire area with discharge densities corresponding to the hazard classification.

CHANGE SIGNIFICANCE: Specific additional restrictions have been placed on the design and use of limited area sprinkler systems. The most obvious restriction is the reduction to six on the maximum number of sprinklers permitted. Previously, up to 20 sprinklers were allowed on limited area sprinkler systems. In addition, although the 2012 IBC is not specific as to how multiple limited area systems are to be provided, the code now limits the system and the six sprinklers to a single fire area. Previously, it was permissible to have several separate systems protecting a large space. Because these limited area systems are permitted to be served from the general domestic plumbing system as opposed to a dedicated fire protection line, the new fire area limitation will help ensure that the demand on the water supply can be met and reduce or eliminate the potential need for multiple systems for any single event. Because limited area sprinkler systems do not afford the same level of protection that would be required for an NFPA 13 or 13R system, these new requirements place additional restrictions and controls on their design and use.

The area being protected by the limited area system is restricted to being classified as a "light hazard" or "ordinary hazard, Group 1" location based on the NFPA 13 classifications (Section 903.3.8.2). Under the NFPA standard, "light hazard" spaces are occupancies or areas where the quantity and/or combustibility of the contents is low and fires with relatively low rates of heat release are expected (NFPA 13 Section 5.2). According to the NFPA commentary, examples of uses that fall under this classification typically include animal shelters, churches, educational facilities, hospitals, nursing homes, libraries (except large stack rooms), museums, offices, residential uses, restaurant seating areas, theaters and auditoriums (excluding stages and prosceniums) and unused attics. NFPA describes "ordinary hazard (Group 1) occupancies" as occupancies where combustibility is low, quantity of combustibles is moderate, stockpiles of combustibles do not exceed 8 feet and fires with moderate rates of heat release are expected. Typical examples given by NFPA commentary include automobile parking and showrooms, bakeries, beverage manufacturing, canneries, dairy products manufacturing and processing, electronics plants, glass and glass products manufacturing, laundries and restaurant service areas.

Where limited area sprinkler systems are installed and supplied by the plumbing system, Sections 903.3.8.3, 903.3.8.4 and 903.3.8.5 each contain requirements to ensure their effectiveness and operation. The water supply demand requirements of both Sections 903.3.8.3 and 903.3.8.5 ensure an adequate supply to operate and discharge as the sprinklers are intended to. Section 903.3.8.4 limits the installation of valves that could be used to isolate or shut off the sprinkler supply lines. If the system is designed with control valves, then valves must be supervised or secured into the open position. These special valve requirements are applicable to any limited area sprinkler system, regardless of the number of sprinklers.

Although the monitoring of sprinkler systems is generally required by Section 903.4, an exception exempts limited area sprinkler systems that comply with Section 903.3.8. Therefore, if (a) there are no control valves between the water supply and sprinklers, or (b) the valves are of an approved indicating type and are secured in the open position, then the system will not require monitoring or alarms. Code users should also recognize that although the monitoring requirements for limited area systems were also previously established at 20 sprinklers, that requirement has also been revised to coordinate with the provisions of Section 903.3.8.

904.13

Domestic Cooking Systems in Group I-2 Condition 1

CHANGE TYPE: Addition

CHANGE SUMMARY: Requirements for domestic appliances installed within commercial facilities but used only for domestic cooking have been clarified, including provisions for an appropriate fire-extinguishing system for domestic cooking equipment in nursing homes, assisted living facilities and similar buildings.

2015 CODE: **904.13 Domestic Cooking Systems in Group I-2 Condition 1.** In Group I-2 Condition 1, occupancies where cooking facilities are installed in accordance with Section 407.2.6 of this code, the domestic cooking hood provided over the cooktop or range shall be equipped with an automatic fire-extinguishing system of a type recognized for protection of domestic cooking equipment. Preengineered automatic extinguishing systems shall be tested in accordance with UL 300A and listed and labeled for the intended application. The system shall be installed in accordance with this code, its listing and the manufacturer's instructions.

904.13.1 Manual System Operation and Interconnection. Manual actuation and system interconnection for the hood suppression system shall be installed in accordance with Sections 904.12.1 and 904.12.2.

904.13.2 Portable Fire Extinguishers for Domestic Cooking Equipment in Group I-2 Condition 1. A portable fire extinguisher complying with Section 906 shall be installed within 30 feet (9144 mm) distance of travel from domestic cooking appliances.

CHANGE SIGNIFICANCE: The new provisions in Section 904.13 relate directly to other code changes in IBC Section 407.2.6 and the IMC related to the installation and use of domestic cooking equipment in nursing homes, assisted living facilities and similar buildings. The domestic cooking appliance requirements of the IMC were modified to provide better guidance on hood and exhaust system requirements for domestic cooking equipment

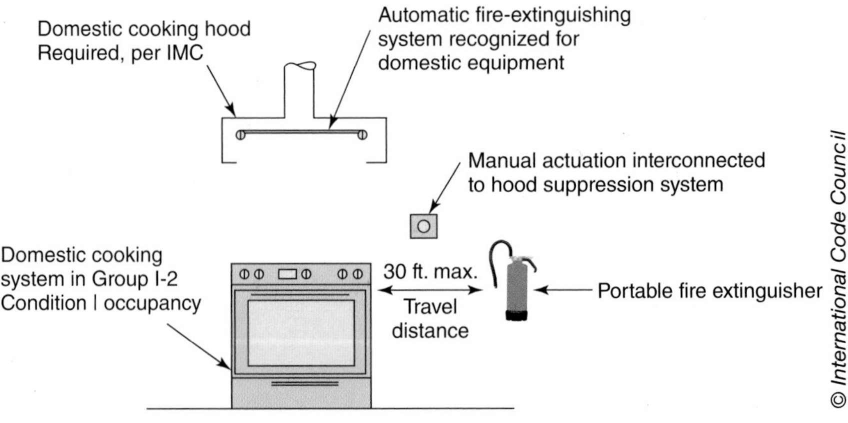

Protection of domestic cooking systems in Group I-2

installed within facilities such as small nursing homes and assisted living facilities where the facility may be set up in a more residential or group home arrangement as opposed to an institutional-type setting.

As many assisted living facilities and nursing homes are striving to produce a more residential atmosphere, domestic ranges are being provided either within the unit or in common use areas. Because the IMC requires domestic range hoods and exhaust systems where domestic cooking appliances are utilized for domestic purposes in other than Group R occupancies, this new IBC requirement mandates that an appropriate automatic fire-extinguishing system also be provided.

Section 904.13 requires that an extinguishing system that is "recognized for protection of domestic cooking equipment" must be utilized and provides a standard for the testing and listing of pre-engineered systems. These domestic hood and extinguishing systems differ from the systems that are required for commercial cooking appliances. The reference to the UL 300A standard will help to ensure that the appropriate protection is provided and that the standard is referenced from within the IBC.

As an added level of protection for these types of cooking facilities, Section 904.13.1 requires that a manual activation device be installed with the hood system in addition to the automatic activation required by the base requirement of Section 904.13. In addition, a portable fire extinguisher is required by Section 904.13.2 and must be installed within 30 feet of the cooking appliance. With all of these additional safety features required by the IMC and Section 904.13, the hazards associated with these limited cooking operations will be minimized.

907.2.3

Fire Alarms—Group E Occupancies

Alarm requirements vary based on school size

CHANGE TYPE: Modification

CHANGE SUMMARY: The threshold for alarm systems in Group E occupancies has been increased such that a manual fire alarm is required where the occupant load exceeds 50, and an emergency voice/alarm communication (EVAC) system must only be provided where the occupant load exceeds 100.

2015 CODE: 907.2.3 Group E. A manual fire alarm system that initiates the occupant notification signal utilizing an emergency voice/alarm communication system meeting the requirements of Section 907.5.2.2 and installed in accordance with Section 907.6 shall be installed in Group E occupancies. When automatic sprinkler systems or smoke detectors are installed, such systems or detectors shall be connected to the building fire alarm system.

Exceptions:

1. A manual fire alarm system is not required in Group E occupancies with an occupant load of ~~30~~ 50 or less.
2. Emergency voice/alarm communication systems meeting the requirements of Section 907.5.2.2 and installed in accordance with Section 907.6 shall not be required in Group E occupancies with occupant loads of 100 or less, provided that activation of the manual fire alarm system initiates an approved occupant notification signal in accordance with Section 907.5.

~~2.~~ 3. Manual fire alarm boxes are not required in Group E occupancies where all of the following apply:

~~2.1.~~ 3.1. Interior corridors are protected by smoke detectors.

~~2.2.~~ 3.2. Auditoriums, cafeterias, gymnasiums and similar areas are protected by heat detectors or other approved detection devices.

~~2.3.~~ 3.3. Shops and laboratories involving dusts or vapors are protected by heat detectors or other approved detection devices.

Group E Alarm Requirements

Occupant load	Type of system required
≤50	No Requirement (907.2.3 Exception 1)
51–100[a]	Manual Fire Alarm System (907.2.3 Exception 2) • Activates occupant notification system complying with Section 907.5
>100[a]	Emergency voice/alarm communication (EVAC) system (907.2.3) • See Section 907.5.2.2 for special occupant notification requirements

[a] Manual fire alarm boxes may be eliminated at specific locations and under specific conditions. (See Section 907.2.3 Exceptions 3 and 4.)

~~3.~~ 4. Manual fire alarm boxes shall not be required in Group E occupancies where all of the following apply:

- 4.1. The building is equipped throughout with an approved automatic sprinkler system installed in accordance with Section 903.3.1.1.
- 4.2. The emergency voice/alarm communication system will activate on sprinkler waterflow.
- 4.3. Manual activation is provided from a normally occupied location.

CHANGE SIGNIFICANCE: As a general requirement, all Group E occupancies must be provided with a manual fire alarm system that initiates the occupant notification signal utilizing an emergency voice/alarm communication (EVAC) system. Early notification is critical in the overall fire- and life-safety approach to educational facilities. Exception 1 in the 2012 IBC indicated that the manual alarm and EVAC systems were not required where the occupant load of the Group E occupancy did not exceed 30. The exceptions have been modified in the 2015 IBC to increase the thresholds at which both systems are now required.

Exception 1 has been modified so that small educational facilities with an occupant load of 50 or less will not require an alarm system. This revision generally restores the threshold that was established in the 2009 IBC and all previous editions. Although the 2012 IBC introduced a reduced alarm threshold of 30, a return to the previous level of 50 was deemed appropriate. It was established that there is no loss history to indicate that the 50-occupant load threshold was inadequate. For example, an occupant load threshold of 31 resulted in a facility as small as 620 square feet in floor area having an alarm system required. With such a small facility, the occupants would typically be in close visual or audible contact with all occupied spaces and with each other. Therefore, it seems it would be highly unlikely that the alarm system would be capable of providing any additional advanced warning.

Setting the occupant load trigger at 50 will generally coordinate with several other important code provisions, allow facilities to be slightly larger without an alarm system and coordinate with the historic requirements that have worked without major incidents.

Exception 2 has also raised the threshold for when the added features of an EVAC alarm system are required. Based upon this exception, the emergency voice/alarm communication aspects of the alarm system will not be required until the occupant load of the Group E occupancy exceeds 100 persons. Where a facility is large enough to have an occupant load exceeding 100, then it is likely that the occupants would not be within direct visual or audible contact of each other; at that point, the benefits and features provided by the EVAC system would be appropriate and useful during emergencies.

As an overview, an educational facility with an occupant load of 50 or less does not require any type of alarm system. A facility with 51 to 100 occupants requires a typical manual alarm system; those facilities with an occupant load of 101 or more would require the manual alarm system to include the added features of an EVAC system as set forth in Section 907.5.2.2.

907.2.9.3

Alarm Systems—Group R-2 College and University Buildings

CHANGE TYPE: Modification

CHANGE SUMMARY: The scope of the fire alarm provisions for Group R-2 college and university buildings has been revised to apply to facilities "operated by" the college or university whether owned by the school or not.

2015 CODE: 907.2.9.3 Group R-2 College and University Buildings. An automatic smoke detection system that activates the occupant notification system in accordance with Section 907.5 shall be installed in Group R-2 occupancies operated by a college ~~and~~ or university for student or staff housing ~~buildings~~ in the following locations:

1. Common spaces outside of dwelling units and sleeping units.
2. Laundry rooms, mechanical equipment rooms and storage rooms.
3. All interior corridors serving sleeping units or dwelling units.

Exception: An automatic smoke detection system is not required in buildings that do not have interior corridors serving sleeping units or dwelling units and where each sleeping unit or dwelling unit either has a means of egress door opening directly to an exterior exit access that leads directly to an exit or a means of egress door opening directly to an exit.

Required smoke alarms in dwelling units and sleeping units in Group R-2 occupancies operated by a college ~~and~~ or university for student or staff housing ~~buildings~~ shall be interconnected with the fire alarm system in accordance with NFPA 72.

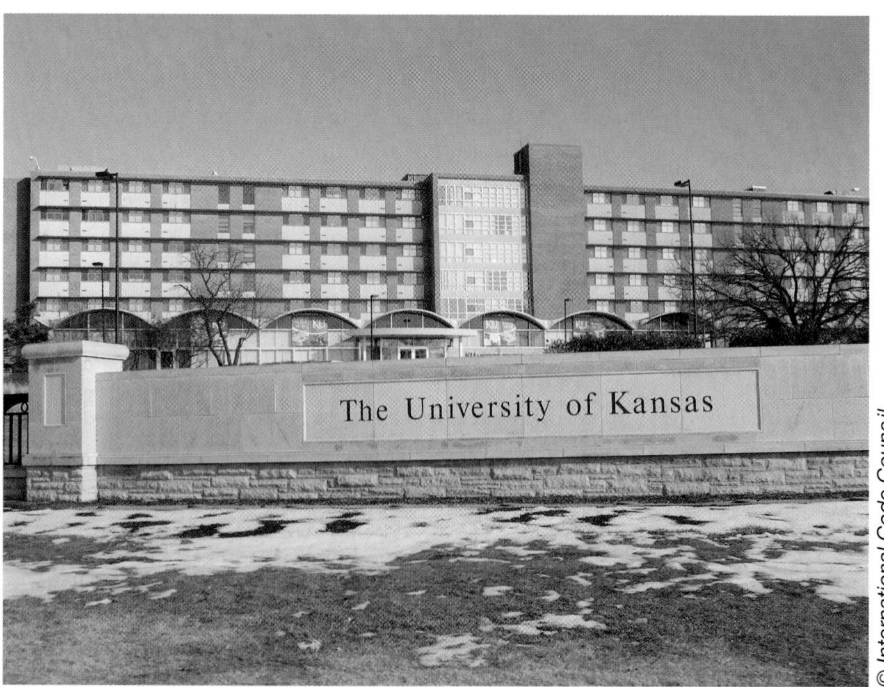

Group R-2 college and university building

Exception: ~~An automatic smoke detection system is not required in buildings that do not have interior corridors serving sleeping units or dwelling units and where each sleeping unit or dwelling unit either has a means of egress door opening directly to an exterior exit access that leads directly to an exit or a means of egress door opening directly to an exit.~~

CHANGE SIGNIFICANCE: Due to the special hazards found in college and university residence halls and similar residential facilities, the code has recently increased the fire alarm requirements to a higher level than required for other Group R-2 occupancies. The extent as to what specific buildings are scoped by the provisions has been unclear; therefore the code has been revised to better define what facilities are to be regulated. Previous requirements provided no guidance as to how to define buildings and facilities that are under the control, but not directly owned, by the college or university, or where the building is not located on campus property.

In the 2012 IBC, it is not clear as to how a private dormitory or an apartment building, directly across the street from a college campus and only rented to students, is to be regulated. As another example, a fraternity or sorority house that might be associated with the school but is privately owned by the local chapter or the organization's national office was previously difficult to evaluate. In some situations, the school may rent or control a building that is off-campus but is owned by a private individual or corporation, or is owned or controlled by a separate school-related entity such as a foundation, trust or alumni association. Under the new provisions, whether or not the college or university actually owns the building is not important. The primary issue is whether or not the school has operational control over the building that is used for student or staff housing.

Understanding the scoping of the new provisions is important because it imposes a requirement not applied to any other Group R-2 occupancy. The school-operated housing facilities must be provided with an automatic smoke detection system, and, perhaps more uniquely, are required to interconnect the smoke alarms located within dwelling and sleeping units to the building's fire alarm system. This interconnection of unit smoke alarms with the building's fire alarm system is typically not required, encouraged or permitted in a Group R-2 occupancy.

In a format change, the exception has also been relocated within the section so that it is placed directly after the paragraph it is applicable to. Because the exception eliminates the requirement for a smoke detection system, it is appropriate to have it follow the code text that requires the system. The relocation of the exception does not create any change of intent between the 2012 and 2015 code but simply places the exception in its proper place.

907.2.11.3, 907.2.11.4

Smoke Alarms Near Cooking Appliances and Bathrooms

CHANGE TYPE: Modification

CHANGE SUMMARY: Requirements from the NFPA 72 standard addressing the installation of smoke alarms near cooking appliances and bathrooms have been introduced to the IBC in order to provide direct guidance on the placement of smoke alarms.

2015 CODE: **907.2.11.3 Installation Near Cooking Appliances.** Smoke alarms shall not be installed in the following locations unless this would prevent placement of a smoke alarm in a location required by Section 907.2.11.1 or 907.2.11.2:

1. Ionization smoke alarms shall not be installed less than 20 feet (6096 mm) horizontally from a permanently installed cooking appliance.
2. Ionization smoke alarms with an alarm-silencing switch shall not be installed less than 10 feet (3048 mm) horizontally from a permanently installed cooking appliance.
3. Photoelectric smoke alarms shall not be installed less than 6 feet (1829 mm) horizontally from a permanently installed cooking appliance.

907.2.11.4 Installation Near Bathrooms. Smoke alarms shall be installed not less than 3 feet (914 mm) horizontally from the door or opening of a bathroom that contains a bathtub or shower unless this would prevent placement of a smoke alarm required by Section 907.2.11.1 or 907.2.11.2.

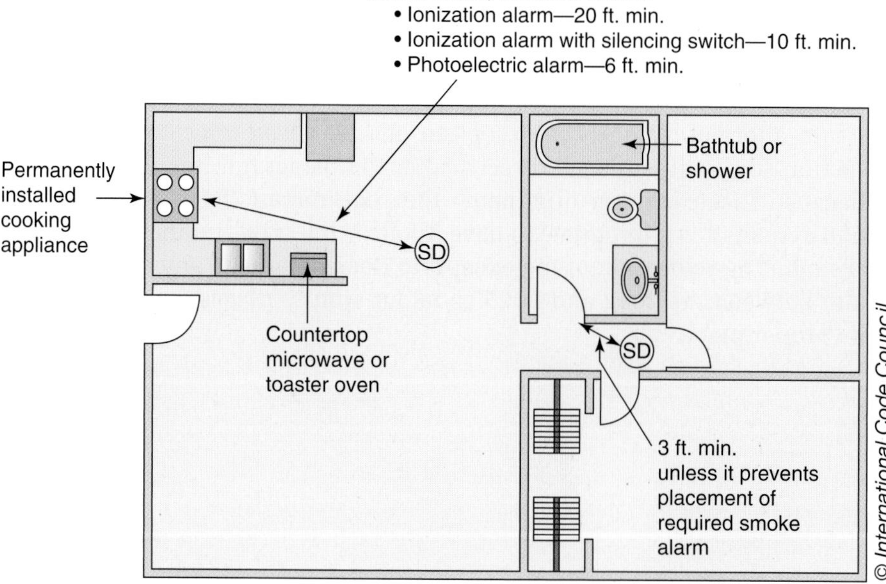

Location of smoke detector

CHANGE SIGNIFICANCE: Smoke alarms are an effective means of providing early detection, warning and life-safety functions, but only when they are properly installed and maintained. If the smoke alarms are installed incorrectly or are installed in locations where they are subject to inadvertent activation, the occupants are likely to either disconnect the alarms or become desensitized to their activation and ignore them when they do sound. Some of the most frequent causes of nuisance activation are steam or the effluent from cooking appliances, or the steam that is produced in a bathroom.

In order to provide inspectors and installers with more direct information on the proper placement of smoke alarms, Sections 907.2.11.3 and 907.2.11.4 have been added to the IBC. The provisions are consistent with, and contained within, the NFPA 72 standard; however, most inspectors may not have easy access to these specific provisions. Section 907.2.11.3 now provides guidance related to the installation of smoke alarms near cooking appliances depending on the type of smoke alarm used. Section 907.2.11.4 addresses the placement of smoke alarms near the door or other opening at a bathroom that contains a tub or shower. Including the installation limitations within the IBC is intended to reduce the number of nuisance alarms. Therefore, it is expected that occupants will not remove or disconnect the alarms and that the occupants' response when the alarms do sound will be improved.

When applying these provisions, there are two important factors that code users should recognize. The statement "unless this would prevent the placement of a smoke alarm in a location required by Section 907.2.11.1 or 907.2.11.2" is intended to provide guidance and direction on the proper location of the smoke alarms, but does not provide an exception or exemption that would allow an alarm required by another section to be eliminated. Therefore, the smoke alarms required in Groups R-1, R-2, R-3, R-4 and I-1 occupancies must be provided even if the only option is to install the device in the areas near the cooking appliance or bathroom that Sections 907.2.11.3 and 907.2.11.4 suggest they should not be located. In this situation, the requirements of Sections 907.2.11.1 and 907.2.11.2 override the provisions of Sections 907.2.11.3 and 907.2.11.4. The second important aspect of these new provisions is the recognition that guidance is provided on the proper location of smoke alarms but the provisions do not require the installation of any smoke alarm in unregulated occupancies. For example, if a shower area is located within a Group B office area, Section 907.2.11.4 does not require the installation of a smoke alarm outside of that space. Therefore Sections 907.2.11.3 and 907.2.11.4 should not be viewed as provisions that determine whether or not smoke alarms are installed, but simply as providing guidance and a best practice as to the location of the alarms required by other provisions of the code.

909.21.1

Elevator Hoistway Pressurization

CHANGE TYPE: Modification

CHANGE SUMMARY: Viable alternatives to the general elevator hoistway pressurization requirements are now available where pressurization is provided in lieu of an enclosed elevator lobby or an additional door.

2015 CODE: **909.21 Elevator Hoistway Pressurization Alternative.** Where elevator hoistway pressurization is provided in lieu of required enclosed elevator lobbies, the pressurization system shall comply with Sections 909.21.1 through 909.21.11.

909.21.1 Pressurization Requirements. Elevator hoistways shall be pressurized to maintain a minimum positive pressure of 0.10 inch of water (25 Pa) and a maximum positive pressure of 0.25 inch of water (67 Pa) with respect to adjacent occupied space on all floors. This pressure shall be measured at the midpoint of each hoistway door, with all elevator cars at the floor of recall and all hoistway doors on the floor of recall open and all other hoistway doors closed. The pressure differentials shall be measured between the hoistway and the adjacent elevator landing. The opening and closing of hoistway doors at each level must be demonstrated during this test. The supply air intake shall be from an outside, uncontaminated source located a minimum distance of 20 feet (6096 mm) from any air exhaust system or outlet.

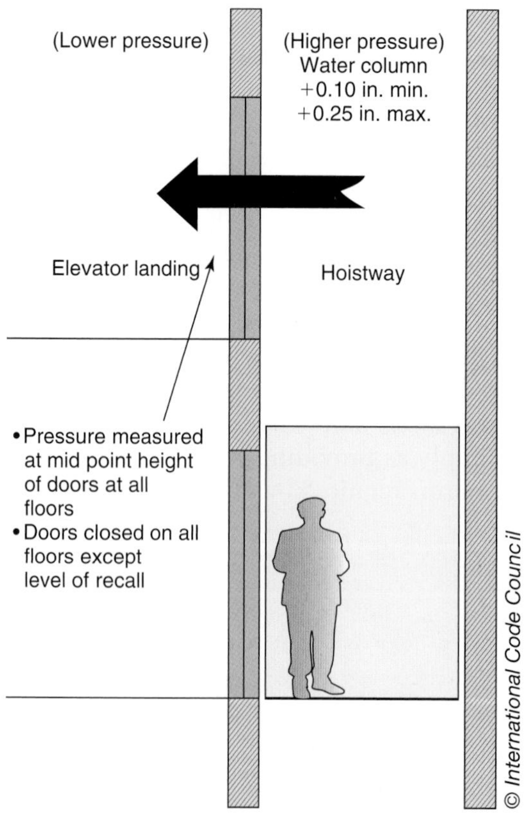

General requirement for hoistway pressurization

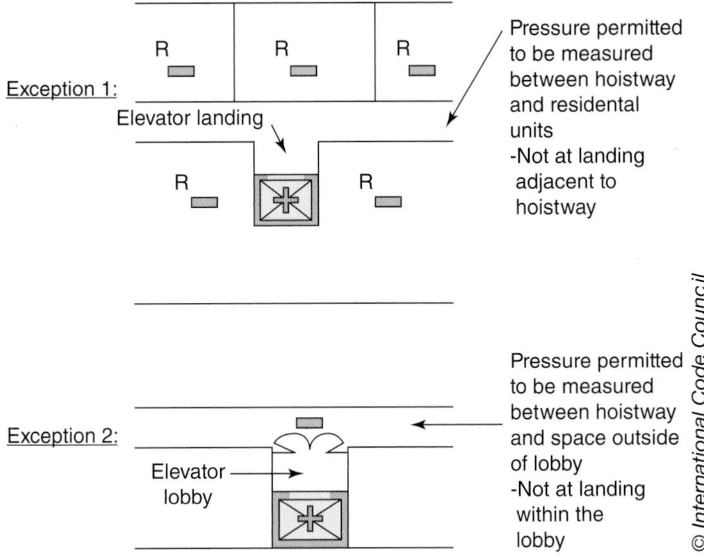

Exceptions 1 and 2 to pressurization requirement

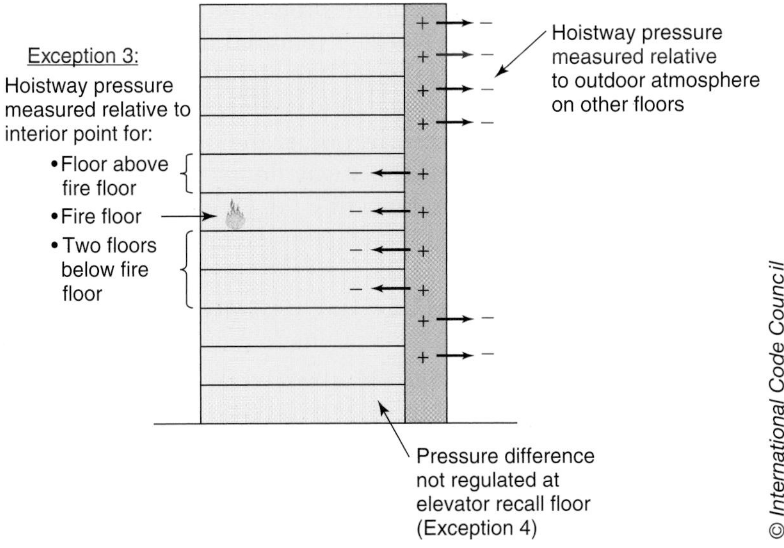

Exception 3 to pressurization requirement

Exceptions:

1. On floors containing only Group R occupancies, the pressure differential is permitted to be measured between the hoistway and a dwelling unit or sleeping unit.
2. Where an elevator opens into a lobby enclosed in accordance with Section 3007.6 or 3008.6 the pressure differential is permitted to be measured between the hoistway and the space immediately outside the door(s) from the floor to the enclosed lobby.

909.21.1 continues

909.21.1 continued

3. The pressure differential is permitted to be measured relative to the outdoor atmosphere on floors other than the following:
   3.1. The fire floor.
   3.2. The two floors immediately below the fire floor.
   3.3. The floor immediately above the fire floor.
4. The minimum positive pressure of 0.10 inch of water (25 Pa) and a maximum positive pressure of 0.25 inch of water (67 Pa) with respect to occupied floors is not required at the floor of recall with the doors open.

909.21.1.1 Use of Ventilation Systems. Ventilation systems, other than hoistway supply air systems, are permitted to be used to exhaust air from adjacent spaces on the fire floor, two floors immediately below and one floor immediately above the fire floor to the building's exterior where necessary to maintain positive pressure relationships as required in Section 909.21.1 during operation of the elevator shaft pressurization system.

CHANGE SIGNIFICANCE: Where elevator hoistway openings require protection as a means to address vertical smoke movement, complying pressurization is permitted as an appropriate protection option. Historically, the general provisions of Section 909.21.1 required the hoistway pressurization to be measured at all floors while the elevator was located at the recall level and the door there was open. It was shown in some studies that this approach resulted in overpressurization at the other levels and often interfered with the operation of the hoistway doors. A viable alternative, based on provisions developed and used by the City of Seattle, Washington, has now been provided to address what previously was considered as an unworkable system.

A specific method establishes how the hoistway pressurization is to be measured, with reliance on the exceptions to modify the general pressurization requirement based upon the building arrangement and floor under consideration. An exemption for pressurization is provided at the recall floor. The modified requirements are intended to still accomplish the code's goal of keeping smoke out of the hoistway while simplifying the process by clarifying the points where the pressure differential is measured. In addition, the design will be responsive to the most critical floors versus addressing all levels simultaneously.

Because residential buildings are highly compartmentalized and the fire is most likely to be within a unit, the first exception allows the pressure differential to be measured between the hoistway and the units as opposed to the elevator lobby. This will not only help keep smoke out of the hoistway but will also help reduce the smoke coming from the affected unit due to the positive pressure within the corridor outside of the affected unit. The second exception allows the pressure difference to be measured outside of the elevator lobby as opposed to within the lobby itself. Because the elevator lobby is constructed to provide smoke protection for the hoistway door opening, this exception allows the pressurization to occur in the hoistway and lobby, compared to the space outside of the lobby. If the smoke is kept out of the lobby, it will also be kept out of the hoistway.

Exception 3 allows the required minimum and maximum pressure difference to be measured internally, and only the four most critical floors are addressed—the floor of fire origin, the two floors immediately below, and one floor immediately above. All of the other floors are allowed to measure the pressure difference between the hoistway and the exterior of the building. The purpose of this allowance is to maintain a slightly positive pressure in the building when compared to the atmospheric pressure. This will serve to lower the neutral pressure plane in the building and therefore help reduce the driving force of the stack effect.

The fourth exception helps address how to deal with the pressure requirement at the level of elevator recall. As a part of the general elevator provisions, the code requires that both a primary and alternate recall floor be provided so the elevator does not return to the floor of fire origin as the designated level of recall. Smoke detectors are used in an elevator recall system to prevent the doors from opening when smoke is present. Because there is no smoke at the level of exit recall, there really is no reason for a pressure difference at that level to protect the hoistway. In addition, because the hoistway doors open when the elevator returns to the level of recall, there is no barrier between the hoistway and the adjacent elevator landing to allow for the pressurization to develop. If the hoistway is being pressurized at other levels, there will most likely be airflow out of the hoistway door opening that would function more like the airflow method of smoke control versus the pressurization method. This exception therefore exempts the level of recall from the pressure differential because that floor will not be the floor of fire origin and the smoke detectors would prevent the elevator from returning to that floor to open.

Code users should recognize that the four exceptions address different situations and that it would be permissible to use the various exceptions in conjunction with one another.

New Section 909.21.1.1 allows the use of the general building HVAC system to exhaust air to create or maintain the required pressure differential from Section 909.21.1. Because these general HVAC ventilation systems are being used as a part of the pressurization system, they would need to comply with all of the other requirements of Section 909.21 in order to be acceptable. It would be inappropriate to use the building's ventilation system if it was not held to the same standard as a dedicated pressurization system.

910

Smoke and Heat Removal

CHANGE TYPE: Modification

CHANGE SUMMARY: The format and technical requirements for smoke and heat removal systems have been revised, including a new allowance permitting a mechanical smoke removal system as an alternative to smoke and heat vents.

2015 CODE:

SECTION 910
SMOKE AND HEAT REMOVAL

910.1 General. Where required by this code, ~~or otherwise installed,~~ smoke and heat vents or mechanical smoke ~~exhaust~~ removal systems ~~and draft curtains~~ shall conform to the requirements of this section.

~~**Exceptions:**~~

~~1. Frozen food warehouses used solely for storage of Class I and II commodities where protected by an approved automatic sprinkler system.~~

~~2. Where areas of buildings are equipped with early suppression fast-response (ESFR) sprinklers, automatic smoke and heat vents shall not be required within these areas.~~

910.2 Where Required. Smoke and heat vents or a mechanical smoke removal system shall be installed ~~in the roofs of buildings or portions thereof occupied for the uses set forth in~~ as required by Sections 910.2.1 and 910.2.2.

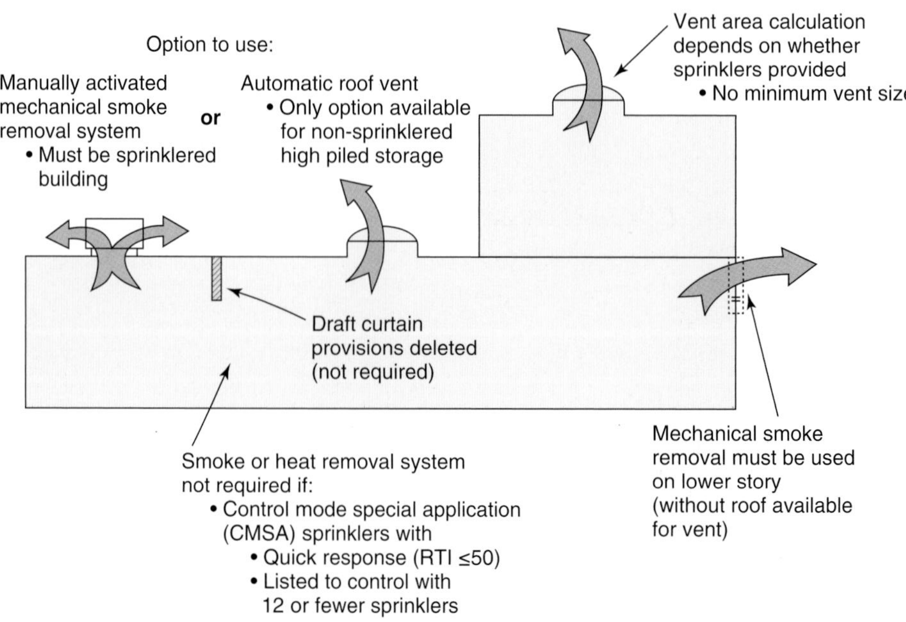
Option to use:
Manually activated mechanical smoke removal system
• Must be sprinklered building
or
Automatic roof vent
• Only option available for non-sprinklered high piled storage
Vent area calculation depends on whether sprinklers provided
• No minimum vent size
Draft curtain provisions deleted (not required)
Smoke or heat removal system not required if:
• Control mode special application (CMSA) sprinklers with
• Quick response (RTI ≤50)
• Listed to control with 12 or fewer sprinklers
Mechanical smoke removal must be used on lower story (without roof available for vent)
© International Code Council

Smoke and heat removal

Exceptions: ~~In occupied portions of a building where the upper surface of the story is not a roof assembly, mechanical smoke exhaust in accordance with Section 910.4 shall be an acceptable alternative.~~

1. Frozen food warehouses used solely for storage of Class I and II commodities where protected by an approved automatic sprinkler system.
2. Smoke and heat removal shall not be required in areas of buildings equipped with early suppression fast-response (ESFR) sprinklers.
3. Smoke and heat removal shall not be required in areas of buildings equipped with control mode special application sprinklers with a response time index of 50 $(m \cdot s)^{1/2}$ or less, which are listed to control a fire in the stored commodities with 12 or fewer sprinklers.

910.2.1 Group F-1 or S-1. Smoke and heat vents installed in accordance with Section 910.3 or a mechanical smoke removal system installed in accordance with Section 910.4 shall be installed in buildings and portions thereof used as a Group F-1 or S-1 occupancy having more than 50,000 square feet (4645 m^2) of undivided area. In occupied portions of a building equipped throughout with a sprinkler system in accordance with Section 903.3.1.1 where the upper surface of the story is not a roof assembly, a mechanical smoke removal system in accordance with Section 910.4 shall be installed.

Exception: Group S-1 aircraft repair hangars.

910.2.2 High-Piled Combustible Storage. Smoke and heat removal required by Table 3206.2 of the *International Fire Code* for buildings and portions thereof containing high-piled combustible ~~stock or rack~~ storage shall be installed in accordance with Section 910.3 in unsprinklered buildings. In buildings and portions thereof containing high-piled combustible storage equipped throughout with an automatic sprinkler system in accordance with Section 903.3.1.1 ~~in any occupancy group in accordance with Section 413 and the *International Fire Code*~~, a smoke and heat removal system shall be installed in accordance with Section 910.3 or 910.4. In occupied portions of a building equipped throughout with a sprinkler system in accordance with Section 903.3.1.1 where the upper surface of the story is not a roof assembly, a mechanical smoke removal system in accordance with Section 910.4 shall be installed.

910.3 Smoke and Heat Vents ~~Design and Installation~~. The design and installation of smoke and heat vents ~~and draft curtains~~ shall be ~~as specified~~ in accordance with Sections 910.3.1 through 910.3.3 ~~910.3.5.2 and Table 910.3~~.

910.3.1 ~~Design~~ Listing and Labeling. Smoke and heat vents shall be listed and labeled to indicate compliance with UL 793 or FM 4430.

~~910.3.2 Vent Operation.~~

~~910.3.3 Vent Dimensions.~~

910 continues

910 continued

~~910.3.4~~ 910.3.2 Smoke and Heat Vent Locations. *(No significant changes.)*

910.3.3 Smoke and Heat Vents Area. The required aggregate area of smoke and heat vents shall be calculated as follows:
For buildings equipped throughout with an automatic sprinkler system in accordance with Section 903.3.1.1:

$$A_{VR} = V/9000 \quad \textbf{(Equation 9-4)}$$

Where:
A_{VR} = the required aggregate vent area (ft^2)
V = volume (ft^3) of the area that requires smoke removal

For unsprinklered buildings:

$$A_{VR} = A_{FA}/50 \quad \textbf{(Equation 9-5)}$$

Where:
A_{VR} = the required aggregate vent area (ft^2)
A_{FA} = the area of the floor in the area that requires smoke removal

~~910.3.5 Draft Curtains.~~

910.4 Mechanical Smoke Removal Systems ~~Exhaust~~. ~~Where approved by the fire code official, engineered~~ Mechanical smoke removal systems ~~exhaust~~ shall be designed and installed in accordance with Sections 910.4.1 through 910.4.7 ~~an acceptable alternative to smoke and heat vents~~.

~~910.4.1 Location.~~

~~910.4.2 Size.~~

~~910.4.3 Operation.~~

910.4.1 Automatic Sprinklers Required. The building shall be equipped throughout with an approved automatic sprinkler system in accordance with Section 903.3.1.1.

910.4.2 Exhaust Fan Construction. Exhaust fans that are part of a mechanical smoke removal system shall be rated for operation at 221°F (105°C) Exhaust fan motors shall be located outside of the exhaust fan air stream.

910.4.3 System Design Criteria. The mechanical smoke removal system shall be sized to exhaust the building at a minimum rate of two air changes per hour based upon the volume of the building or portion thereof without contents. The capacity of each exhaust fan shall not exceed 30,000 cubic feet per minute (14.2 m^3/s).

910.4.3.1 Make-up Air. Make-up air openings shall be provided within 6 feet (1829 mm) of the floor level. Operation of make-up air openings shall be manual or automatic. The minimum gross area of make-up air inlets shall be 8 square feet per 1000 cubic feet per minute (0.74 m^2 per 0.4719 m^3/s) of smoke exhaust.

910.4.4 Activation. The mechanical smoke removal system shall be activated by manual controls only.

910.4.5 Manual Control Location. Manual controls shall be located so as to be accessible to the fire service from an exterior door of the building and be protected against interior fire exposure by not less than 1-hour fire barriers constructed in accordance with Section 707 or horizontal assemblies constructed in accordance with Section 711, or both.

~~910.4.4~~ 910.4.6 Control Wiring ~~and Control~~. Wiring for operation and control of mechanical smoke removal systems ~~exhaust fans~~ shall be connected ahead of the main disconnect in accordance with Section 701.12E of NFPA 70 and be protected against interior fire exposure to temperatures in excess of 1,000°F (538°C) for a period of not less than 15 minutes. ~~Controls shall be located so as to be immediately accessible to the fire service from the exterior of the building and protected against interior fire exposure by not less than 1-hour *fire barriers* constructed in accordance with Section 707 or *horizontal assemblies* constructed in accordance with Section 711, or both.~~

~~910.4.5 Supply Air.~~

~~910.4.6~~ 910.4.7 Controls ~~Interlocks~~. ~~On combination comfort air-handling/smoke removal systems or independent comfort air-handling systems, fans shall be controlled to shut down in accordance with the *approved* smoke control sequence.~~ Where building air-handling and mechanical smoke removal systems are combined or where independent building air-handling systems are provided, fans shall automatically shut down in accordance with the *International Mechanical Code*. The manual controls provided for the smoke removal system shall have the capability to override the automatic shutdown of fans that are part of the smoke removal system.

(*Because this code change affected substantial portions of Section 910, the entire code change text is too extensive to be included here. Refer to Code Changes F-195 and F-196 in the* 2015 IFC Code Changes Resource Collection *for the complete text and history of the code change.*)

CHANGE SIGNIFICANCE: The changes within Section 910 reflect the effort of the ICC Code Technology Committee's (CTC) Roof Vent Study Group (RVSG) to update the requirements mandating smoke and heat removal systems in industrial and storage buildings. The RVSG undertook this effort to determine an acceptable balance between active and passive fire protection and how the interaction of vents, building area, use, and sprinklers can impact the operation of the systems as well as how they impact fire-fighting operations and the overall fire- and life-safety objectives of the IBC. Much of the information used for the basis of updating these provisions did not exist when the requirements for roof vents were first included into the legacy codes in the 1970s and 1980s; therefore the provisions differ significantly from those in previous editions.

910 continues

910 continued

Because the primary purpose of smoke and heat removal is to assist fire-fighting operations after the control of a fire has been achieved, provisions will differ from what would be required for a life-safety system intended to protect the building's occupants. This is one of the primary reasons a manually activated mechanical smoke removal system is now permitted as an alternative to roof vents. Whereas the 2012 IBC only permitted a mechanical exhaust system where "approved by the fire code official," the code now allows the mechanical system as an acceptable alternative to smoke and heat vents and leaves that decision to the designer without requiring specific approval.

Some of the other primary revisions include:

- The scoping of Section 910.1 has been changed to limit the application to required systems. For non-required systems, the provisions of Section 910 are not mandatory. Previously non-required systems would have been included and regulated by the "or otherwise installed" phrasing.
- The scoping in Section 910.2 and its subsections will allow either automatic roof vents or a manually activated mechanical smoke removal system in industrial and storage buildings protected by a sprinkler system. Where high-piled storage is not protected by a sprinkler system, then a mechanical smoke removal system is not an acceptable alternative.
- A new exception allows the smoke removal requirements to be eliminated where a specific type of sprinkler system is used. No smoke or heat removal system (vents or mechanical) is required where the sprinkler system is designed using Control Mode Special Application (CMSA) sprinklers with both a quick response (response-time index of 50 or less) and if listed to control a fire with 12 or fewer sprinklers.
- Subsections in 910.2 specify that a mechanical smoke removal system must be used on lower stories that do not have the ability to vent through the roof above.
- Deletion of the requirements for draft curtains and minimum vent sizes occurred because research has demonstrated that they affect the sequence and operation of sprinklers and they may have an adverse effect on sprinkler operation.
- Section 910.3 addresses the requirements for the smoke and heat vents and provides two simplified calculations to determine required vent area, dependent upon whether or not the building is sprinklered. The vent area calculation results in 1 ft^2 of vent area for each 9,000 ft^3 of volume within a sprinklered building and 1 ft^2 of vent area for each 50 ft^2 of floor area in a non-sprinklered building. Almost all buildings will be regulated using the 1 ft^2 of vent per 9,000 ft^3 calculation because very few regulated buildings will not have a sprinkler system.
- Section 910.4 addresses the mechanical smoke removal system and includes a number of important provisions: limiting the system to only sprinklered buildings; a minimum amount of air changes and make-up air; and a restriction for the system to only be activated by manual controls. Because automatic operation of the smoke removal system could be detrimental to the operation of the sprinkler system, the requirement for manual activation ensures the fire department is in control of the system. Fire department control includes not

only the activation requirements of Section 910.4.4 but also the capability to override an automatic shutdown of the fans as addressed in Section 910.4.7.

- Section 910.4.6 allows a reduction in the protection requirements for the power supply and controls from what was previously found in Section 910.4.4. Because smoke removal in Section 910 is focused on fire-fighting and property protection as opposed to occupant life safety, a compromise on the reliability was made that is substantially different from the requirements found within Section 909.
- Section 910.5 establishes the maintenance requirements for both smoke and heat vents and for mechanical smoke removal systems.

For a more detailed section-by-section analysis and discussion related to the basis for the revisions, refer to the *2015 IFC Code Changes Resource Collection*, specifically the information related to code changes F195-13 and F196-13.

915

Carbon Monoxide Detection

Carbon monoxide detector

CHANGE TYPE: Modification

CHANGE SUMMARY: The carbon monoxide (CO) alarm provisions have been relocated, reformatted and revised; the scope has been modified to exclude Group I-3 occupancies while adding Group E occupancies.

2015 CODE: ~~**908.7 Carbon Monoxide Alarms.**~~

~~**908.7.1 Carbon Monoxide Detection Systems.**~~

SECTION 915
CARBON MONOXIDE DETECTION

915.1 Carbon Monoxide Detection. Carbon monoxide detection shall be installed in new buildings in accordance with Sections 915.1.1 through 915.6. Carbon monoxide detection shall be installed in existing buildings in accordance with Chapter 11 of the *International Fire Code*.

915.1.1 Where Required. Carbon monoxide detection shall be provided in Groups I-1, I-2, I-4 and R occupancies, and in classrooms in Group E occupancies in the locations specified in 915.2 where any of the conditions in Sections 915.1.2 through 915.1.6 exist.

915.1.2 Fuel-Burning Appliances and Fuel-Burning Fireplaces. Carbon monoxide detection shall be provided in dwelling units, sleeping units and classrooms that contain a fuel-burning appliance or a fuel-burning fireplace.

915.1.3 Forced Air Furnaces. Carbon monoxide detection shall be provided in dwelling units, sleeping units and classrooms served by a fuel-burning, forced air furnace.

> **Exception:** Carbon monoxide detection shall not be required in dwelling units, sleeping units and classrooms if carbon monoxide detection is provided in the first room or area served by each main duct leaving the furnace, and the carbon monoxide alarm signals are automatically transmitted to an approved location.

915.1.4 Fuel-Burning Appliances Outside of Dwelling Units, Sleeping Units and Classrooms. Carbon monoxide detection shall be provided in dwelling units, sleeping units and classrooms located in buildings that contain fuel-burning appliances or fuel-burning fireplaces.

> **Exceptions:**
>
> 1. Carbon monoxide detection shall not be required in dwelling units, sleeping units and classrooms if there are no communicating openings between the fuel-burning appliance or fuel-burning fireplace and the dwelling unit, sleeping unit or classroom.

2. Carbon monoxide detection shall not be required in dwelling units, sleeping units and classrooms where carbon monoxide detection is provided in one of the following locations:
    2.1 In an approved location between the fuel-burning appliance or fuel-burning fireplace and the dwelling unit, sleeping unit or classroom,
    2.2 On the ceiling of the room containing the fuel-burning appliance or fuel-burning fireplace.

915.1.5 Private Garages. Carbon monoxide detection shall be provided in dwelling units, sleeping units, and classrooms in buildings with attached private garages.

Exceptions:

1. Carbon monoxide detection shall not be required where there are no communicating openings between the private garage and the dwelling unit, sleeping unit or classroom.
2. Carbon monoxide detection shall not be required in dwelling units, sleeping units and classrooms located more than one story above or below a private garage.
3. Carbon monoxide detection shall not be required where the private garage connects to the building through an open-ended corridor.
4. Where carbon monoxide detection is provided in an approved location between openings to a private garage and dwelling units, sleeping units or classrooms, carbon monoxide detection shall not be required in the dwelling units, sleeping units or classrooms.

915 continues

CO detection required in classrooms

915 continued

915.1.6 Exempt Garages. For determining compliance with Section 915.1.5, an open parking garage, complying with Section 406.5, or an enclosed parking garage complying with Section 406.6 shall not be considered a private garage.

915.2 Locations. Where required by Section 915.1.1, carbon monoxide detection shall be installed in the locations specified in Sections 915.2.1 through 915.2.3.

915.2.1 Dwelling Units. Carbon monoxide detection shall be installed in dwelling units outside of each separate sleeping area in the immediate vicinity of the bedrooms. Where a fuel-burning appliance is located within a bedroom or its attached bathroom, carbon monoxide detection shall be installed within the bedroom.

915.2.2 Sleeping Units. Carbon monoxide detection shall be installed in sleeping units.

> **Exception:** Carbon monoxide detection shall be allowed to be installed outside of each separate sleeping area in the immediate vicinity of the sleeping unit where the sleeping unit or its attached bathroom do not contain a fuel burning appliance and is not served by a forced air furnace.

915.2.3 Group E Occupancies. Carbon monoxide detection shall be installed in classrooms in Group E occupancies. Carbon monoxide alarm signals shall be automatically transmitted to an on-site location that is staffed by school personnel.

> **Exception:** Carbon monoxide alarm signals shall not be required to be automatically transmitted to an on-site location that is staffed by school personnel in Group E occupancies with an occupant load of 30 or less.

915.3 Detection Equipment. Carbon monoxide detection required by Sections 915.1 through 915.2.3 shall be provided by carbon monoxide alarms complying with Section 915.4 or carbon monoxide detection systems complying with Section 915.5.

915.4 Carbon Monoxide Alarms. Carbon monoxide alarms shall comply with Sections 915.4.1 through 915.4.3.

915.4.1 Power Source. Carbon monoxide alarms shall receive their primary power from the building wiring where such wiring is served from a commercial source, and when primary power is interrupted, shall receive power from a battery. Wiring shall be permanent and without a disconnecting switch other than that required for overcurrent protection.

> **Exception:** Where installed in buildings without commercial power, battery-powered carbon monoxide alarms shall be an acceptable alternative.

915.4.2 Listings. Carbon monoxide alarms shall be listed in accordance with UL 2034.

915.4.3 Combination Alarms. Combination carbon monoxide/smoke alarms shall be an acceptable alternative to carbon monoxide alarms. Combination carbon monoxide/smoke alarms shall be listed in accordance with UL 2034 and UL 217.

915.5 Carbon Monoxide Detection Systems. Carbon monoxide detection systems shall be an acceptable alternative to carbon monoxide alarms and shall comply with Sections 915.5.1 through 915.5.3.

915.5.1 General. Carbon monoxide detection systems shall comply with NFPA 720. Carbon monoxide detectors shall be listed in accordance with UL 2075.

915.5.2 Locations. Carbon monoxide detectors shall be installed in the locations specified in Section 915.2. These locations supersede the locations specified in NFPA 720.

915.5.3 Combination Detectors. Combination carbon monoxide/smoke detectors installed in carbon monoxide detection systems shall be an acceptable alternative to carbon monoxide detectors, provided they are listed in accordance with UL 2075 and UL 268.

915.6 Maintenance. Carbon monoxide alarms and carbon monoxide detection systems shall be maintained in accordance with the *International Fire Code.*

(*Because this code change affected substantial portions of Chapter 9, the entire code change text is too extensive to be included here. Refer to Code Changes F-360, F-180 and F-182 in the* 2015 IFC Code Changes Resource Collection *for the complete text and history of the code change.*)

CHANGE SIGNIFICANCE: Because carbon monoxide (CO) alarm systems are a bit different from the other types of emergency alarm systems covered within Section 908, the provisions have been relocated to a new Section 915. Although both types of systems provide warnings of hazardous situations, the emergency alarms addressed in Section 908 tend to provide indication and warning of exposure to items more closely tied to the hazardous materials provisions of the code. The emergency alarms are more related to the possible exposure to toxic materials and warning occupants prior to reaching the "permissible exposure limit (PEL)," the "immediately dangerous to life and health (IDLH) limit," or the "threshold limit value—time weighted average (TLV-TWA)." Carbon monoxide detectors, on the other hand, are more like smoke alarms and serve to detect and warn the occupants of a potential problem that needs to be addressed, but not necessarily an emergency situation.

In a review of the revised scoping provisions, it can be seen that the Group I-3 occupancies are no longer included. However, Group E occupancies are now regulated, as well as the continued application to all Group R and Group I-1, I-2 and I-4 occupancies. The educational occupancies have been added because certain states and local jurisdictions have begun to develop or add their own requirements for these types of uses. By including the requirement within the model code, it will help to

915 continues

915 continued ensure consistent requirements and lead to the development and maintenance of consensus-based provisions through the ICC code-development process. It should be noted that the Group E requirements differ somewhat from those addressing dwelling units and sleeping units.

The reformatting of the provisions helps clarify the requirements by describing specific locations where CO alarms are, and are not, required in regard to fuel-burning appliances. Section 915.1.2 references fuel-burning fireplaces that previously could have been assumed to be regulated but the intent was not clear because the term "appliance" was used.

Regardless of a forced-air furnace's location, Section 915.1.3 now requires alarms in the dwelling units or sleeping units if they are "served" by the furnace. Other types of appliances are regulated by Section 915.1.4. The provisions and exceptions of Section 915.1.4 address situations where the dwelling unit or sleeping unit does not contain a fuel-burning appliance but could be affected by an appliance in a common area of the building. An example would be a multistory hotel that has a fireplace in the lobby, a forced-air heating system serving common areas, or a boiler in an equipment room. In these situations, having a few strategically located CO alarms in the common area or area of hazard will provide a reasonable level of protection and alarms will not be needed within the dwelling or sleeping units. Exception 1 covers situations where CO emanating from the appliance has no direct path to a dwelling or sleeping unit. An example would be a water heater in an equipment room that only has access from the exterior of the building and no direct openings into the dwelling or sleeping unit.

Exception 2 of Section 915.1.4 need not be applied if CO alarms are installed within the dwelling units or sleeping units as required by the base paragraph, or the exception can be used to provide the alarms at a point between the units and the hazard or directly in the room with the hazard source.

The garage provisions continue to exempt naturally ventilated open parking garages and mechanically ventilated enclosed parking garages, and have also expanded the provisions addressing private garages. The exceptions for private garages are similar to those from the 2012 IBC but also now include a new exception that addresses garages connected through an open-ended corridor or breezeway.

Section 915.2 lists the specific locations where the CO alarms are to be installed. It should be noted that these are specific requirements and are to be applied instead of the locations specified in NFPA 720. Previously the code referenced the NFPA standard; however, now specific locations have been added to the IBC to make it more user-friendly. Although these locations do not coordinate exactly with the NFPA 720 standard, they are to be applied based upon the fact that the code no longer references the standard. For Group E occupancies, the specific locations identified in Section 915.2.3 are to be addressed where required by Sections 915.1.2, 915.1.3 and 915.1.4.

The power requirements of Section 915.4.1 and the allowance for CO alarms to be combined with smoke alarms per Section 915.4.3 are also now specifically addressed. The combination alarm systems must comply with both the CO and smoke alarm provisions; therefore, a combination system is acceptable.

PART **4**

Means of Egress

Chapter 10

■ **Chapter 10** Means of Egress

The criteria set forth in Chapter 10 regulating the design of the means of egress are established as the primary method for protection of people in buildings. Both prescriptive and performance language is utilized in the chapter to provide for a basic approach in the determination of a safe exiting system for all occupancies. Chapter 10 addresses all portions of the egress system and includes design requirements as well as provisions regulating individual components. A zonal approach to egress provides a general basis for the chapter's format through regulation of the exit access, exit, and exit discharge portions of the means of egress. ■

CHAPTER 10
Means of Egress

1004.1.1
Cumulative Occupant Loads

TABLE 1004.1.2
Occupant Load Factors

Claudio

1006, 1007
Numbers of Exits and Exit Access Doorways

1007.1
Exit and Exit Access Doorway Configuration

1009.8
Two-Way Communication Systems

Noel

1010.1.9
Door Operations—Locking Systems

1011.15, 1011.16
Ladders

1014.8
Handrail Projections

Bryan

1016.2
Egress through Intervening Spaces

1017.2.2
Travel Distance Increase for Groups F-1 and S-1

Charlie

1018.3
Aisles in Groups B and M

1020.2
Corridor Width and Capacity

Karol

1023.3.1
Stairway Extension

1029.13.2.2.1
Stepped Aisle Construction Tolerances

Craig

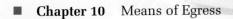

Also next time:
3006, 3102 through 3104 and 3302 from our amendments.

Chapter 10

Means of Egress

© International Code Council

© International Code Council

Means of egress

CHANGE TYPE: Modification

CHANGE SUMMARY: Provisions addressing the minimum required number of means of egress and their arrangement for rooms and space as well as stories have been reformatted and relocated.

2015 CODE: Section ~~1015~~ 1006 Numbers of Exits and Exit Access Doorways.

~~Section 1021 Number of Exits and Exit Configuration~~

Section 1007 Exit and Exit Access Doorway Configuration

Section ~~1007~~ 1009 Accessible Means of Egress

(Renumbering and formatting affect most remaining sections.)

CHANGE SIGNIFICANCE: Most of the egress provisions in 2012 IBC Sections 1015 and 1021 have been combined, reformatted, and relocated to the general provisions of Sections 1006 and 1007. Previously, both Sections 1015 and 1021 contained requirements addressing the number of means of egress and their arrangement. Section 1015 typically dealt with the required number and arrangement of the means of egress from an individual room or space, while Section 1021 addressed the requirements for the individual story or overall building. Because the relationship of and the distinction between these two sections was not always clear, it often led to confusion or unnecessary difficulty in applying the requirements.

Combining many of the provisions of Sections 1015 and 1021 with some other provisions into a single location should result in making the provisions easier to understand and apply, providing for a more uniform application of the code. The reformatting allows the requirements to be placed within technical context and clarifies whether or not the provisions apply to a room or a story. Provisions that were very similar have either been combined so that the distinction is clearer or eliminated, in order to reduce confusion. Section 1006 now addresses the number of means of egress that are required, with new Section 1007 addressing the arrangement and separation of means-of-egress routes.

Relocating the provisions of Sections 1015 and 1021 to the general provision portion of the chapter allows a natural sequential flow of requirements based on the occupant load. Section 1004 begins by providing the means to determine the occupant load, and Section 1005 establishes the capacity requirements for the egress path based on that occupant load. At that point, the two new sections follow, with Section 1006 establishing the number of means of egress required and Section 1007 addressing how those multiple egress paths must be arranged.

1004.1.1

Cumulative Occupant Loads

CHANGE TYPE: Modification

CHANGE SUMMARY: The determination of the cumulative design occupant load for intervening spaces, adjacent levels and adjacent stories has been clarified.

2015 CODE: **1004.1.1 Cumulative Occupant Loads.** Where the path of egress travel includes intervening rooms, areas or spaces, cumulative occupant loads shall be determined in accordance with this section.

1004.1.1.1 Intervening Spaces or Accessory Areas. Where occupants egress from one or more rooms, areas or spaces through ~~another~~ others, the design *occupant load* shall be the combined occupant load of interconnected accessory or intervening spaces. Design of egress path capacity shall be based on the cumulative portion of occupant loads of all rooms, areas or spaces to that point along the path of egress travel.

1004.1.1.2 Adjacent Levels for Mezzanines. That portion of the *occupant load* of a *mezzanine* ~~or story~~ with required egress through a room, area or space on an adjacent level shall be added to the *occupant load* of that room, area or space.

1004.1.1.3 Adjacent Stories. Other than for the egress components designed for convergence in accordance with Section 1005.6, the occupant load from separate stories shall not be added.

CHANGE SIGNIFICANCE: Efforts have been made to clarify how the occupant load of a space that passes through another space is viewed when determining both the number of means of egress and also the capacity (width) of the egress system. It has now been emphasized that rooms that share an egress path must be reviewed based on the aggregate occupant load in order to establish many of the minimum egress

1004.1.1 continues

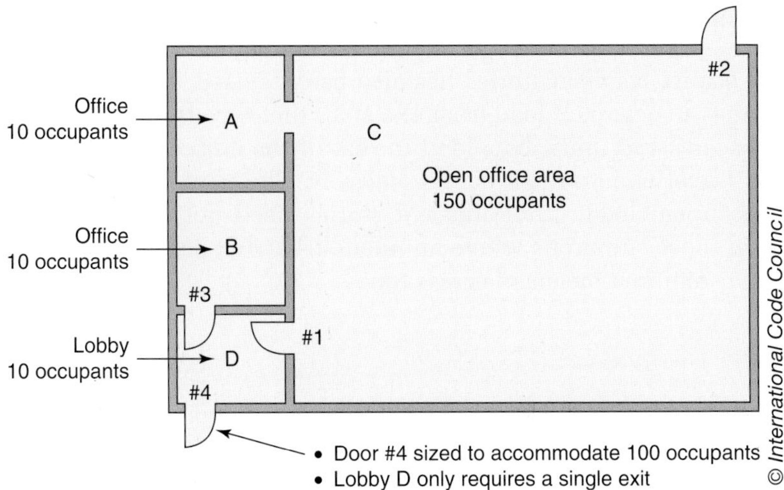

Cumulative occupant loads for intervening spaces

1004.1.1 continued

requirements. Each path of egress travel must be designed so the capacity of that path is capable of serving the accumulated occupant load that travels along that portion of the path.

The first sentence of Section 1004.1.1.1 indicates that where occupants egress from one space through another, the "design occupant load" is determined to be the combined or aggregate of the various interconnected or intervening spaces. This accumulated occupant load is to be used to establish many of the minimum requirements, such as the number of exits or exit access doorways that must be provided from the overall space, whether the doors must swing in the direction of egress travel, and the minimum component width of 36 inches or 44 inches for stairs and corridors. The second sentence indicates that it is only the egress capacity/width that is based on the accumulated occupants along that path of travel; the accumulation of occupants is not to be applied to items such as the number of means of egress.

The purpose of these changes is to reinforce the concept that the occupant load is assigned to each occupied area individually. Where there are intervening rooms, each area must be considered both individually and in the aggregate with the other interconnected occupied portions of the exit access to determine the number of means of egress and width of the exit access. Portions of the occupant load are accumulated along the egress path to determine the capacity of individual egress elements along those paths. However, once occupants from one area make a choice and travel along one of several independent paths of egress travel, their occupant load is not added to some other area to determine how many paths of travel are required from that different area.

Section 1004.1.1.2 recognizes that mezzanines may have independent egress similar to what is typical for a story. If the mezzanine occupants do not egress through the room or area it is a part of, then the occupant load is not added to the main room. If all of the occupants of a mezzanine must egress down through the main room, then their occupant load must be added to the main room or area. Where persons on the mezzanine have an option of egress paths, such as one independent exit and one through the room below, the occupant load may be divided among the available paths and the portion of the occupants exiting through the room below must be added to the occupant load of that space.

The method in which occupant accumulation is addressed where travel occurs between stories has also been revised. The 2012 IBC indicates that an occupant load from one story that travels through the area of an adjacent story must be added to that of the adjacent story where the egress travel is on an exit access stairway. The new provisions indicate that occupant loads from adjacent stories need not be added together, even in those situations where an unenclosed exit access stairway is utilized for required means of egress travel.

Table 1004.1.2

Occupant Load Factors

© iStockphoto/Juanmonino

Mercantile occupancy

CHANGE TYPE: Modification

CHANGE SUMMARY: The mercantile occupant load factor has been revised such that a single factor is now applicable regardless of the story on which the mercantile use is located.

2015 CODE:

TABLE 1004.1.2 Maximum Floor Area Allowances per Occupant

Function of Space	Occupant Load Factor[a]
Mercantile	60 gross
~~Areas on other floors~~	~~60 gross~~
~~Basement and grade floor areas~~	~~30 gross~~
Storage, stock, shipping areas	300 gross

For SI: 1 square foot = 0.0929 m^2.
a. Floor area in square feet per occupant.
(Remaining portions of table not shown are unchanged.)

CHANGE SIGNIFICANCE: Historically, the IBC has provided two occupant load factors for retail spaces. The occupant load factor applied to grade floors and basements has been considerably smaller than the occupant load factor applied to other floor levels. All retail spaces, regardless of floor level, are now regulated for occupant load by a single factor of 60 square feet per occupant.

When the previous factors were placed into the code, they were based on multistory single-operator buildings, such as mid-rise and high-rise department stores. Many of such stores had lower-density uses, such as furniture or housewares, on the upper levels while the spaces on the entry levels were used for higher-density sales areas. Below grade levels were often high-occupant-load spaces, often referred to as "bargain basements." Most large retail facilities are no longer constructed in taller multistory facilities but rather tend to use larger floor areas within a single story. With the current trend in retail display and merchandising,

Table 1004.1.2 continues

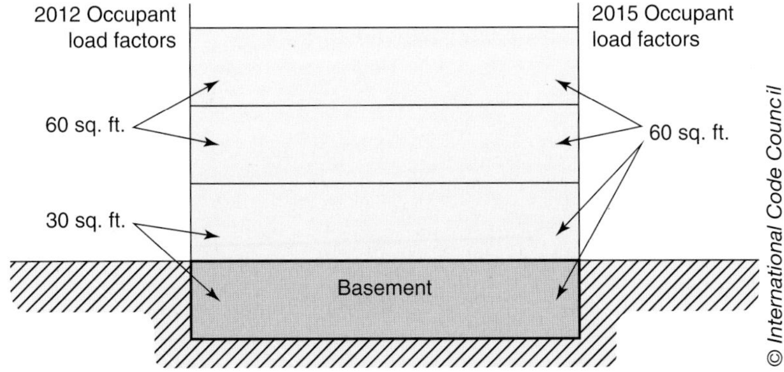

© International Code Council

Occupant load factor—mercantile

Table 1004.1.2 continued

it will now be easier to apply the requirements because all floor levels now apply the same occupant load factor. The use of the factor of 60 square foot per occupant (based on the gross area) is consistent with what was previously utilized for retail sale areas on floor levels other than basements and grade-level stories. This factor was felt to be a more reasonable number given today's retail environment and that much of the floor area is covered with display cases and counters.

1006, 1007

Numbers of Exits and Exit Access Doorways

CHANGE TYPE: Modification

CHANGE SUMMARY: The means of egress requirements for rooms and spaces, along with those for stories, have been consolidated in Chapter 10.

2015 CODE: ~~**1014.3 Common Path of Egress Travel.** The common path of egress travel shall not exceed the common path of egress travel distances in Table 1014.3.~~

1006, 1007 continues

~~**TABLE 1014.3 Common Path of Egress Travel**~~

	~~Without Sprinkler System (feet)~~		
	~~Occupant Load~~		
~~Occupancy~~	~~OL ≤ 30~~	~~OL > 30~~	~~With Sprinkler System (feet)~~
~~B, S[d]~~	~~100~~	~~75~~	~~100[a]~~
~~U~~	~~100~~	~~75~~	~~75[a]~~
~~F~~	~~75~~	~~75~~	~~100[a]~~
~~H-1, H-2, H-3~~	~~Not Permitted~~	~~Not Permitted~~	~~25[a,g]~~
~~R-2~~	~~75~~	~~75~~	~~125[b]~~
~~R-3[e]~~	~~75~~	~~75~~	~~125[b]~~
~~I-3~~	~~100~~	~~100~~	~~100[a]~~
~~All others[c,f]~~	~~75~~	~~75~~	~~75[a]~~

~~For SI: 1 foot = 304.8 mm.~~

~~a. Buildings equipped throughout with an automatic sprinkler system in accordance with Section 903.3.1.1.~~

~~b. Buildings equipped throughout with an automatic sprinkler system in accordance with Section 903.3.1.1 or 903.3.1.2. See Section 903 for occupancies where automatic sprinkler systems are permitted in accordance with Section 903.3.1.2.~~

~~c. For a room or space used for assembly purposes having fixed seating, see Section 1028.8.~~

~~d. The length of a common path of egress travel in a Group S-2 open parking garage shall not be more than 100 feet (30 480 mm).~~

~~e. The length of a common path of egress travel in a Group R-3 occupancy located in a mixed occupancy building.~~

~~f. For the distance limitations in Group I-2, see Section 407.4.~~

~~**TABLE 1015.1 Spaces with One Exit or Exit Access Doorway**~~

~~Occupancy~~	~~Maximum Occupant Load~~
~~A, B, E, F, M, U~~	~~49~~
~~H-1, H-2, H-3~~	~~3~~
~~H-4, H-5, I-1, I-2, I-3, I-4, R~~	~~10~~
~~S~~	~~29~~

1006, 1007 continued

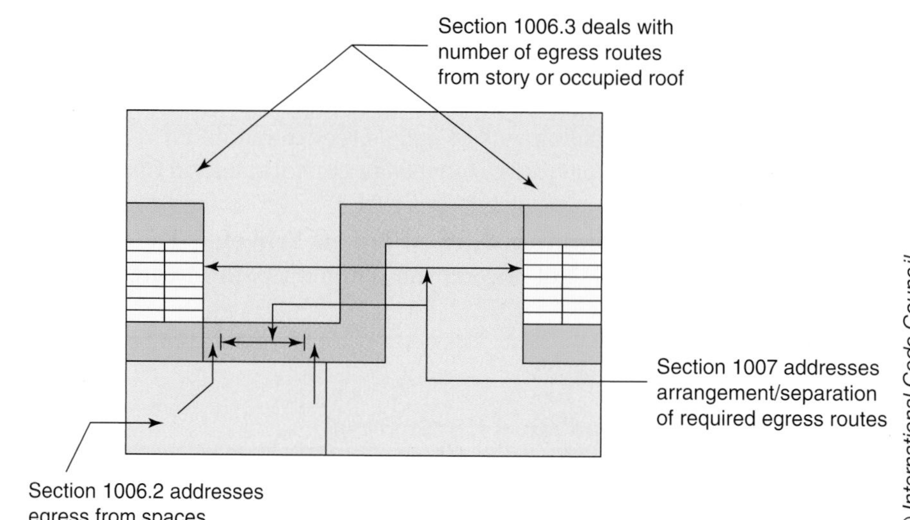
Section 1006.3 deals with number of egress routes from story or occupied roof
Section 1007 addresses arrangement/separation of required egress routes
Section 1006.2 addresses egress from spaces
© International Code Council

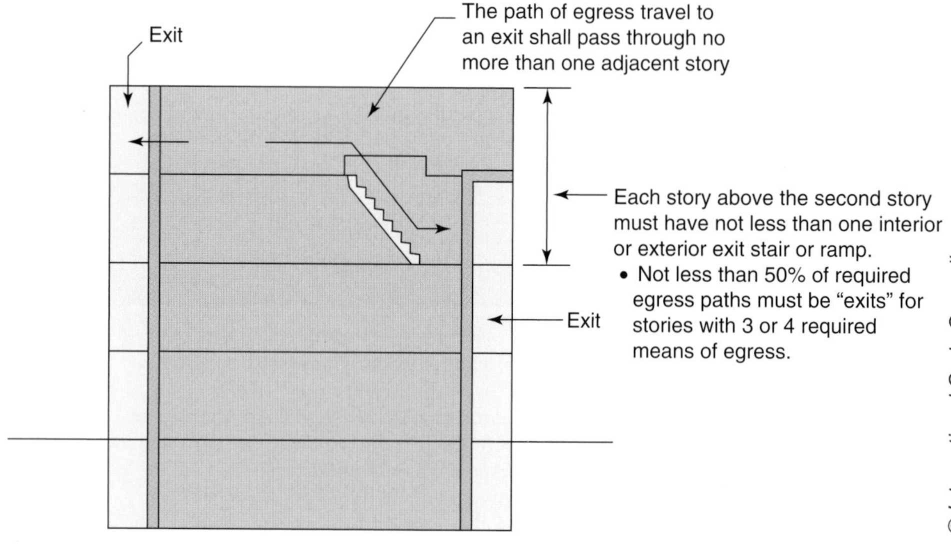
Exit
The path of egress travel to an exit shall pass through no more than one adjacent story
Each story above the second story must have not less than one interior or exterior exit stair or ramp.
• Not less than 50% of required egress paths must be "exits" for stories with 3 or 4 required means of egress.
Exit
© International Code Council

Number and configuration of means of egress

SECTION 1006
NUMBERS OF EXITS AND EXIT ACCESS DOORWAYS

1006.1 General. The number of exits or exit access doorways required within the means of egress system shall comply with the provisions of Section 1006.2 for spaces, including mezzanines, and Section 1006.3 for stories.

1006.2 Egress from Spaces. Rooms, areas or spaces, including mezzanines, within a story or basement shall be provided with the number of exits or access to exits in accordance with this section.

1006.2.1 Egress Based on Occupant Load and Common Path of Egress Travel Distance. Two exits or exit access doorways from any space shall be provided where the design occupant load or the common path of egress travel distance exceeds the values listed in Table 1006.2.1.

Exceptions:

1. In Group R-2 and R-3 occupancies, one means of egress is permitted within and from individual dwelling units with a maximum occupant load of 20 where the dwelling unit is equipped throughout with an automatic sprinkler system in accordance with Section 903.3.1.1 or 903.3.1.2 and the common path of egress travel does not exceed 125 feet (38 100 mm).
2. Care suites in Group I-2 occupancies complying with Section 407.4.

1006.3 ~~1021.3.1 Access to Exits at Adjacent Levels.~~ Egress from Stories or Occupied Roofs. The means of egress system serving any story or occupied roof shall be provided with the number of exits or access to exits based on the aggregate occupant load served in accordance

1006, 1007 continues

TABLE 1006.2.1 Spaces with One Exit or Exit Access Doorway

Occupancy	Maximum Occupant Load of Space	Maximum Common Path of Egress Travel Distance (feet)		
		Without Sprinkler System (feet)		With Sprinkler System (feet)
		Occupant Load		
		OL ≤ 30	OL > 30	
A[c], E, M	49	75	75	75[a]
B	49	100	75	100[a]
F	49	75	75	100[a]
H-1, H-2, H-3	3	NP	NP	25[b]
H-4, H-5	10	NP	NP	75[b]
I-1, I-2[d], I-4	10	NP	NP	75[a]
I-3	10	NP~~100~~	NP~~100~~	100[a]
R-1	10	NP~~75~~	NP~~75~~	75[a]
R-2	10	NP~~75~~	NP~~75~~	125[a]
R-3[e]	10	NP~~75~~	NP~~75~~	125[a]
R-4[e]	10	75	75	125[a]
S[f]	29	100	75	100[a]
U	49	100	75	75[a]

For SI: 1 foot = 304.8 mm.

NP = Not Permitted

a. Buildings equipped throughout with an automatic sprinkler system in accordance with Section 903.3.1.1 or 903.3.1.2. See Section 903 for occupancies where automatic sprinkler systems are permitted in accordance with Section 903.3.1.2.

b. Group H occupancies equipped throughout with an *automatic sprinkler system* in accordance with Section 903.2.5.

c. For a room or space used for assembly purposes having fixed seating, see Section ~~1028.8~~ 1029.8.

d. For the travel distance limitations in Group I-2, see Section 407.4.

e. The length of common path of egress travel distance in a Group R-3 occupancy located in a mixed occupancy building or within a Group R-3 or R-4 congregate living facility.

f. The length of common path of egress travel distance in a Group S-2 open parking garage shall be not more than 100 feet.

1006, 1007 continued

with this section. ~~Access to exits at other levels shall be by stairways or ramps. Where access to exits occurs from adjacent building levels, the horizontal and vertical exit access travel distance to the closest exit shall not exceed that specified in Section 1016.1. Access to exits at other levels shall be from an adjacent story.~~ The path of egress travel to an *exit* shall not pass through more than one adjacent story.

Each story above the second story of a building shall have not less than one interior or exterior exit stairway, or interior or exterior exit ramp. Where three or more exits or access to exits are required, not less than 50 percent of the required exits shall be interior or exterior exit stairways or ramps.

> **Exceptions:** ~~Landing platforms or roof areas for helistops that are less than 60 feet (18 288 mm) long, or less than 2,000 square feet (186 m²) in area, shall be permitted to access the second exit by a fire escape, alternating tread device or ladder leading to the story or level below.~~
>
> **1.** Interior exit stairways and interior exit ramps are not required in open parking garages where the means of egress serves only the open parking garage.
>
> **2.** Interior exit stairways and interior exit ramps are not required in outdoor facilities where all portions of the means of egress are essentially open to the outside.

~~SECTION 1021~~
~~NUMBER OF EXITS AND EXIT CONFIGURATION~~

(*Relocated 1021.1 to 1021.4 to new 1006.*)

SECTION 1007
EXIT AND EXIT ACCESS DOORWAY CONFIGURATION

(*Because this code change affected substantial portions of Chapter 10, the entire code change text is too extensive to be included here. Refer to Code Changes E-1, E-7, E-111, E-127, E-132 and E-136 in the* 2015 IBC Code Changes Resource Collection *for the complete text and history of the code change.*)

CHANGE SIGNIFICANCE: As a part of the extensive reformatting of Chapter 10, Section 1006 now provides information related to the required number of means of egress. Section 1006.2 now addresses the egress from spaces, Section 1006.3 deals with the number of egress routes from stories or occupied roofs, and Section 1007 addresses the arrangement or separation of those required means of egress.

A primary technical aspect to the reformatting effort includes relocating provisions in Sections 1014, 1015 and 1021 dealing with the number and arrangement of required means of egress to Sections 1006 and 1007. Perhaps the most apparent revision is seen in the new Table 1006.2.1. The format for this new table is similar to what was previously found in Table 1021.2(2) and combines the occupant load requirements from 2012 IBC Table 1015.1 along with the provisions for common path of egress travel from Section 1014.3 into a single location. Putting these two variables (design occupant load and occupant remoteness) into a single table makes the determination of the required minimum number of means of egress from a space easier and leads to more consistent application of the code.

The definition for "common path of egress travel" has also been revised to more closely align with the definition for "exit access travel distance," but notes that it terminates at an earlier point (where two egress routes are available). The definition was also revised to eliminate the reference to paths that merge. Because the common path is measured to the point where "occupants have separate and distinct access" to two exits or exit access doorways, paths that do merge no longer provide the two distinct paths and therefore continue to be considered within the measurement of the common path of egress travel.

Perhaps the most significant technical revision on this subject is found in Tables 1006.3.2(1) and 1006.3.2(2), where the last column has been changed from limiting the exit access travel distance to now limiting the common path of egress travel. Although the intention based on the table's title heading is to limit stories that have only a single means of egress, the fact that the distance specified can be measured to the point where distinct and separate means of egress are available instead of to the door of an actual exit is a substantial change. For example, consider a small second floor with only one means-of-egress stairway that is located above a larger first-floor area that has two exterior exit doors. Previously the occupants on the second floor were required to reach the protection of an exit within the specified travel distance shown in the table. Under the new provision, because it is now limiting the "common path of egress travel" rather than the "travel distance," the occupants' travel is measured across the second floor and down an unenclosed exit access stairway until that point on the first floor where the occupants first have "separate and distinct access to two exits."

Code users should also note a clarification that occurs in Section 1006.3 where the path-of-egress travel from one story through an adjacent level is limited to "one adjacent story." It has always been the intent of the IBC to limit an occupant's egress travel on an exit access stairway to travel only between adjacent stories. In the 2012 IBC, this intent is expressed in Section 1021.3.1, where it states "access to exits at other levels shall be from an adjacent story." The new provisions will help clarify the requirement and prohibit occupant travel for more than one story via an exit access stairway or exit access ramp in order to reach an exit that is located on another story.

Section 1007 addressing the separation or arrangement of the means of egress now provides specific guidance regarding how the distance between means-of-egress doors, exit access stairways or exit access ramps is to be measured. Previously, the code did not provide specific measuring points for determining the exit or exit access separation and therefore it occasionally led to confusion or differences in application. With the new Section 1007.1.1.1, the measurements should be more consistent and less subjective. The most commonly applied provision will be the measurement method for doorways, where the code now indicates that the separation distance is to be measured to any point along the width of the doorway.

1007.1

Exit and Exit Access Doorway Configuration

CHANGE TYPE: Modification

CHANGE SUMMARY: Specific information is now provided regarding the point where exit separation is to be measured. In addition, where three or more means of egress are required, performance language has been included to ensure the egress paths are adequately separated.

2015 CODE: **1007.1.1.1 Measurement Point.** The separation distance required in Section 1007.1.1 shall be measured in accordance with the following:

1. The separation distance to exit or exit access doorways shall be measured to any point along the width of the doorway.
2. The separation distance to exit access stairways shall be measured to the closest riser.
3. The separation distance to exit access ramps shall be measured to the start of the ramp run.

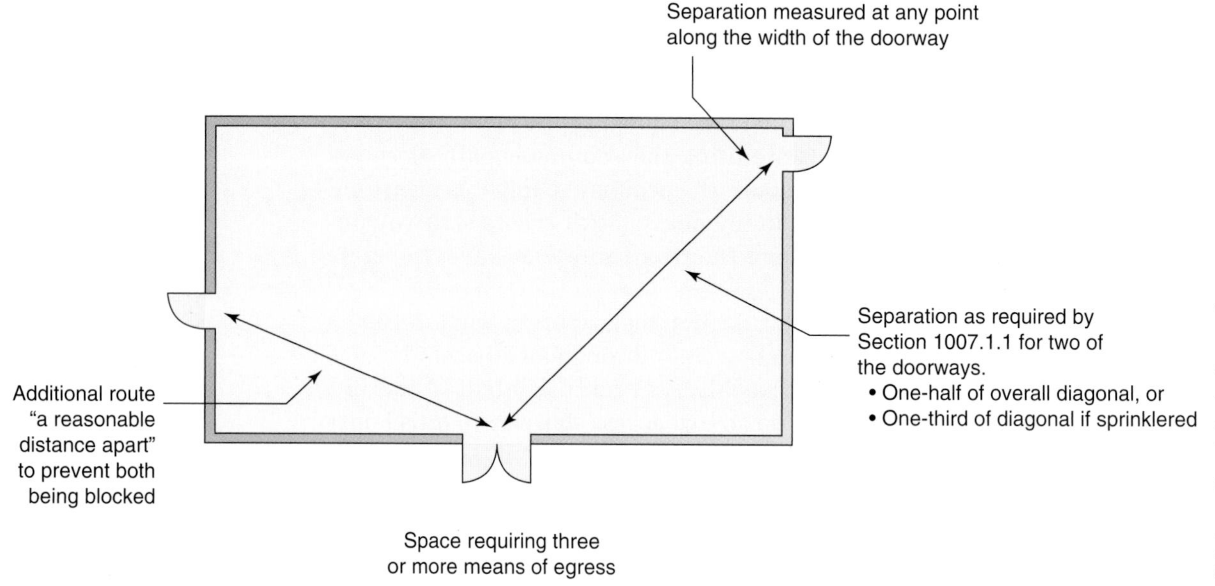
Separation measured at any point along the width of the doorway
Separation as required by Section 1007.1.1 for two of the doorways.
• One-half of overall diagonal, or
• One-third of diagonal if sprinklered
Additional route "a reasonable distance apart" to prevent both being blocked
Space requiring three or more means of egress
© International Code Council

Configuration for three or more doorways

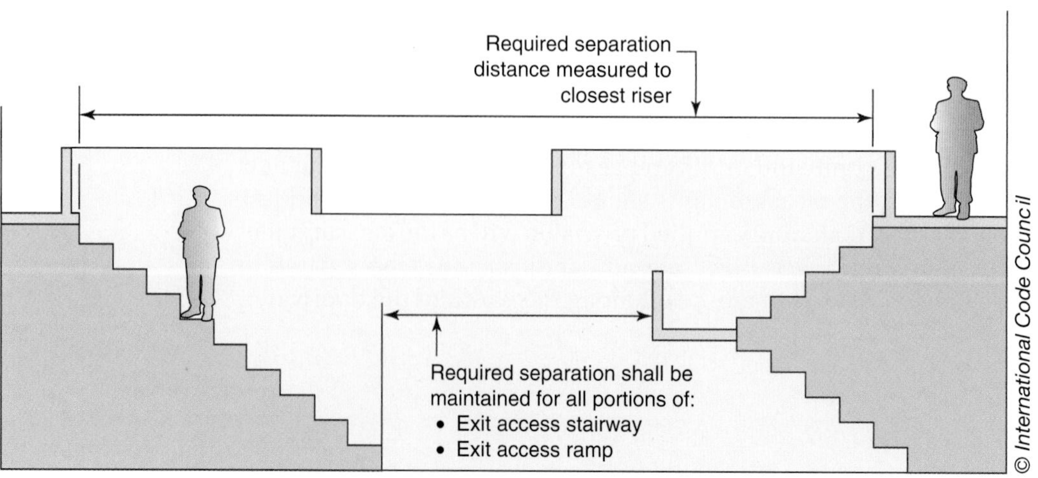
Required separation distance measured to closest riser
Required separation shall be maintained for all portions of:
• Exit access stairway
• Exit access ramp
© International Code Council

Measurement point for exit access stairways

1007.1.2 ~~1015.2.2~~ Three or More Exits or Exit Access Doorways. Where access to three or more exits is required, not less than two exit or exit access doorways shall be arranged in accordance with the provisions of Section 1007.1.1 ~~1015.2.1~~. Additional required exit or exit access doorways shall be arranged a reasonable distance apart so that if one becomes blocked, the others will be available.

1007.1.3 Remoteness of Exit Access Stairways or Ramps. Where two exit access stairways or ramps provide the required means of egress to exits at another story, the required separation distance shall be maintained for all portions of such exit access stairways or ramps.

1007.1.3.1 Three or More Exit Access Stairways or Ramps. Where more than two exit access stairways or ramps provide the required means of egress, not less than two shall be arranged in accordance with Section 1007.1.3.

CHANGE SIGNIFICANCE: Where two or more means of egress are required, the method for measuring the separation or remoteness of the means of egress has been specified. Three measurement methods have been established to clearly indicate how to measure between doors, stairways and ramps. The clarification will provide clarity and consistency in application. As an example, previously where an exit access stairway was provided, the point where the path of travel entered the stairway was by definition an "exit access doorway." There was often a question as to where the separation distance was to be measured; at the stair or at a point remote from it, perhaps the length of a required landing away. Each item is now specifically addressed and is clear as to where the measurement is made.

Coordination with new Sections 1007.1.3 and 1007.1.3.1 is now also provided such that the minimum separation distances between exit access stairways and ramps are maintained for the entire length of travel on the stairway or ramp. This provision will prohibit stairway and ramp runs that meet the separation distance requirement at the first riser or beginning slope, from converging toward another stair or ramp such that the separation is reduced below the minimum distance as the occupant goes either up or down the stairway or ramp. It is reasonable to expect the egress separation distance to be maintained in order to ensure a fire cannot affect all of the egress paths, whether it is at the beginning of the stair or ramp or at any point until the egress travel is completed. Code users will also notice that additional changes were made within Section 1007.1.1 to clarify that the separation provisions apply to the doors, stairs and ramps and not simply the exits or "exit access doorways."

Where three or more means of egress are required, the 2015 IBC now includes performance language in Section 1007.1.2 in order to address the location of the third and/or fourth path of travel. The new provision was previously found in the 2000 and 2003 editions of the IBC, but was removed from the 2006 edition because it was felt to be too subjective. However, without the performance language, there has been nothing to describe where the third and/or fourth element is to be located. The building official now has direction of intent and purpose, which is that a single fire is unlikely to take out multiple means of egress due to their being adequately separated and located remotely within the building. Although not specifically stated, it seems reasonable that the prescriptive guidance also applies to stairway and ramp conditions.

1009.8

Two-Way Communication Systems

A two-way communication system may serve multiple elevators

CHANGE TYPE: Clarification

CHANGE SUMMARY: It has been clarified that a two-way communication system may serve multiple elevators and that the systems are not required at service elevators, freight elevators or private residence elevators.

2015 CODE: 1009.8 ~~1007.8~~ Two-Way Communication. A two-way communication system complying with Sections 1009.8.1 and 1009.8.2 shall be provided at the landing serving each elevator ~~landing~~ or bank of elevators on each accessible floor that is one or more stories above or below the story of exit discharge ~~complying with Sections 1007.8.1 and 1007.8.2~~.

Exceptions:

1. Two-way communication systems are not required at the landing serving each elevator ~~landing~~ or bank of elevators where the two-way communication system is provided within areas of refuge in accordance with Section ~~1007.6.3~~ 1009.6.5.
2. Two-way communication systems are not required on floors provided with ramps conforming to the provisions of Section ~~1010~~ 1012.
3. Two-way communication systems are not required at the landings serving only service elevators that are not designated as part of the accessible means of egress or serve as part of the required accessible route into a facility.
4. Two-way communication systems are not required at the landings serving only freight elevators.
5. Two-way communication systems are not required at the landing serving a private residence elevator.

CHANGE SIGNIFICANCE: It has now been clarified which elevator landings are required to be provided with a two-way communication system. Perhaps the most helpful aspects of the revisions are the exceptions that address service elevators not used for an accessible route or as part of the accessible means of egress, freight elevators, and private residence elevators. Previously, the two-way communication system requirement was often applied inconsistently. Some jurisdictions applied the provision to only those elevator landings serving passenger elevators, whereas others applied it to each individual elevator and type of elevator, including freight elevators located in back-of-house areas.

The revised provisions also clarify that the communication system is not required at each elevator but rather only at the landing serving the elevators, allowing for a single communication system serving a number of elevators that open at the same landing. For example, if an elevator lobby has perhaps six or eight elevators opening into it, a single two-way communication system within the lobby will satisfy the requirement that a system is "provided at the landing serving each elevator or bank of elevators." The intent is that whether there is a single elevator or multiple elevators opening onto the landing, a single two-way communication system is acceptable.

New exceptions address conditions where an elevator landing would not require the installation of a two-way communication system. Exception 4 exempts landings that only serve freight elevators. Because freight elevators are not intended for passenger use, there is no reason for the communication system at the landing serving them. If the landing serves both freight and passenger elevators, then the exception is not applicable and the two-way communication system is required. Because private residence elevators are only allowed within individual dwelling units, Exception 5 eliminates the communication system requirements from the landings where this type of elevator is installed. Within an individual dwelling unit, a two-way communication system is not necessary or practical.

Exception 3 contains a number of items to be considered before eliminating the communication system. If an elevator is only intended as a service elevator, then the communication system may not be necessary at the landing. This exception could be used for many back-of-house service elevator installations, provided the landing is only for the service elevator and not for other passenger elevators, and the elevator is not designated as a part of the accessible means of egress or used as the accessible route to the floor.

It is important to focus on the fact the communication system is required "at the landing," rather than emphasize the reference to a "bank" of elevators. When addressing elevators, a "bank" of elevators includes all of the cars that respond to a single call button. The intent of the provision is to have a communication system at each landing or lobby area serving elevators regardless of the number of elevators or the number of "banks" or control systems involved. As an example, if an elevator lobby has three elevators that only serve upper-floor levels and three elevators that only serve lower levels, the lobby is typically served by two separate "banks" of elevators because each call system only has three elevators that respond to it. In this situation, it is reasonable that a single two-way communication system be installed in the lobby even though it is served by two "banks" of elevators. Because the landing serves both banks of elevators, a single communication system will still allow any occupants at that level to notify someone of their location and the complying communication system would be "provided at the landing" serving each bank of elevators, as the code requires.

1010.1.9

Door Operations—Locking Systems

© International Code Council

Electromagnetic lock

CHANGE TYPE: Modification

CHANGE SUMMARY: Numerous revisions throughout the locking provisions now help clarify requirements and their application through the use of consistent terminology.

2015 CODE: 1010.1.9 ~~1008.1.9~~ Door Operations. Except as specifically permitted by this section, egress doors shall be readily openable from the egress side without the use of a key or special knowledge or effort.

1010.1.9.3 ~~1008.1.9.3~~ Locks and Latches. Locks and latches shall be permitted to prevent operation of doors where any of the following exists:

1. (No change.)
2. In buildings in occupancy Group A having an occupant load of 300 or less, Groups B, F, M and S, and in places of religious worship, the main ~~exterior~~ door or doors are permitted to be equipped with key-operated locking devices from the egress side provided:
    - 2.1. The locking device is readily distinguishable as locked;
    - 2.2. A readily visible durable sign is posted on the egress side on or adjacent to the door stating: THIS DOOR TO REMAIN UNLOCKED WHEN ~~BUILDING~~ THIS SPACE IS OCCUPIED. The sign shall be in letters 1 inch (25 mm) high on a contrasting background; and
    - 2.3. The use of the key-operated locking device is revocable by the building official for due cause.
3. (No change.)
4. (No change.)
5. (No change.)

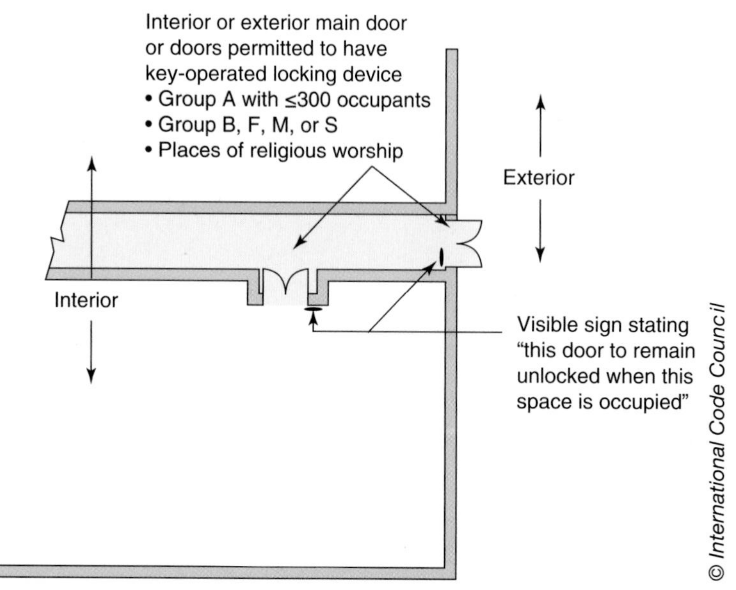

Key-locking hardware permitted on interior doors

1010.1.9.6 ~~1008.1.9.6 Special~~ Controlled Egress ~~Locking Arrangements in~~ Doors in Groups I-1 and ~~Group~~ I-2. ~~Approved, special egress locks~~ Electric locking systems, including electro-mechanical locking systems and electromagnetic locking systems, shall be permitted to be locked in the means of egress in ~~a~~ Group I-1 or I-2 ~~occupancy~~ occupancies where the clinical needs of persons receiving care require their containment. ~~such locking. Special egress locks~~ Controlled egress doors shall be permitted in such occupancies where the building is equipped throughout with an automatic sprinkler system in accordance with Section 903.3.1.1 or an approved automatic-smoke or heat detection system installed in accordance with Section 907, provided that the doors are installed and operate in accordance with all of the following.

(*Items 1 through 8 and two exceptions are not shown.*)

1010.1.9.7 ~~1008.1.9.7~~ Delayed Egress ~~Locks~~.

1010.1.9.8 ~~1008.1.9.8 Access Controlled~~ Sensor Release of Electrically Locked Egress Doors. The electric locks on sensor released ~~The entrance~~ doors located in a means of egress in buildings with an occupancy in Groups ~~-~~A, B, E, I-1, I-2, I-4, M, R-1 or R-2 and entrance doors to tenant spaces in occupancies in Groups ~~-~~A, B, E, I-1, I-2, I-4, M, R-1 or R-2 are permitted ~~to be equipped with an approved entrance and egress access control system, listed in accordance with UL 294, which shall be~~ where installed and operated in accordance with all of the following criteria:

(*Items 1 through 6 are not shown.*)

1010.1.9.9 ~~1008.1.9.9~~ Electromagnetically Locked Egress Doors. Doors in the means of egress in buildings with an occupancy in Group A, B, E, I-1, I-2, I-4, M, R-1 or R-2 and doors to tenant spaces in Group A, B, E, I-1, I-2, I-4, M, R-1 or R-2 shall be permitted to be ~~electromagnetically~~ locked with an electromagnetic locking system where ~~if~~ equipped with ~~listed~~ hardware that incorporates a built-in switch and ~~meet the requirements below~~, where installed and operated in accordance with all of the following.

(*Items 1 through 6 are not shown.*)

(*Because this code change affected substantial portions of Section 1010.1.9, the entire code change text is too extensive to be included here. Refer to Code Changes E-62, E-63, E-66, E-67, E-69, E-70, E-72, E-73, E-74, E-77, E-78, E-80, E-81, E-82, E-9 and E-2 in the* 2015 IBC Code Changes Resource Collection *for the complete text and history of the code change.*)

CHANGE SIGNIFICANCE: A substantial number of changes occurred in the door-locking provisions that, when looked at separately, may not seem significant, but because of the large number, they should be carefully reviewed. Many of the revisions now coordinate the code terminology with that used in the lock industry as well as in UL 294, *Access Control System Units.* The use of consistent terminology will help clarify acceptable methods and provide designers, manufacturers, inspectors and installers a degree of consistency in application.

1010.1.9 continues

1010.1.9 continued

The following is a brief overview of some of the revisions and their intent:

Section 1010.1.9.3: The provision allowing the main exit door to be locked with key-locking hardware when a sign is posted adjacent to the door has been modified to indicate that it is not limited to exterior doors but can also be used on the main egress door from an interior space within a building. The provision can now be applied to a means-of-egress door from a lecture hall within a college classroom building, or to means-of-egress doors serving a movie theater or restaurant located in a covered mall building. The text on the required sign has also been revised so that it now addresses the space served by that egress door and not the building.

Section 1010.1.9.6: The title of the section, and the terminology within it, has been revised to be more specific and consistent with what is used in the industry. The provisions have been expanded to include Group I-1 occupancies, as well as the previously accepted Group I-2 occupancies, because the concern with elopement of the patients is also a concern for the Group I-1 occupancies. In addition, a new exception eliminates requirements 1 through 4 of the provisions for areas such as hospital nurseries or obstetric areas where abduction of a child is a concern.

Section 1010.1.9.7: Revisions in this section include a number of terminology changes for consistency with the UL 294 standard. Item 4 has been modified such that the person attempting to egress must make a physical effort to open the hardware for "not more than three seconds." Previously, the limit was one second, which is not enough time for a person to realize that his or her action is what is causing the alarm and to back away prior to the system unlocking of the door.

The limitation of not passing through more than one delayed egress door has been relocated to the list of numbered requirements. This revision was made to include an exception allowing Groups I-2 and I-3 to pass through a maximum of two doors provided the overall delay does not exceed 30 seconds. Having multiple doors can help with preventing resident elopement and yet the overall delay does not exceed the previously accepted time period. An example of where the two-door arrangement may be helpful is a multistory facility where both the door from the story and the door from the building could be controlled. A maximum 30-second delay in these highly regulated, fully sprinklered facilities is viewed as being reasonable.

The signage requirements within Item 6 have been modified to recognize that the doors could open opposite the direction of travel and swing into the space. Therefore, language must be provided for the sign that is dependent upon the direction of the door swing. Previously, the wording for the sign only addressed doors swinging in the direction of egress travel. An exception has also been added to allow the signs to be eliminated where the patients require restraint or containment to prevent elopement. Based on the staff training within these facilities, the need to protect the residents by preventing elopement, and the fact these systems are required to be interconnected with the sprinkler or fire detection system and unlock upon loss of power, it was determined that eliminating the sign in these facilities is reasonable.

Section 1010.1.9.8: The title and language within this section has been revised to emphasize that it is the sensor release of a locked egress door, and not the access to the space, that is being controlled. Group I-1 and I-4 occupancies have been added to the list of acceptable uses due to the concern for security and safety regarding patient elopement or an occupant leaving a day-care facility. Because these types of locked egress doors allow for some control while at the same time allowing for free egress during an emergency, it is appropriate to include the Group I-1 and I-4 occupants in order for these needs to be addressed. In the numbered list of requirements, a new provision was added as Item 6 that will require these systems to comply with UL 294. The previous Item 6 requiring the doors serving a Group A, B, E, or M occupancy to be unlocked when the building was open to the public has been eliminated. If the doors comply with all of the other required items, they are ensured to be available for egress.

Section 1010.1.9.9: The requirements for electromagnetically locked egress doors have been revised to include Group I-1, I-2 and I-4 occupancies for security and safety reasons. The important distinction here is that Group I-2 occupancies have been added to this section, but were also previously permitted in Section 1010.1.9.8. Compliance with UL 294 has also been added to help ensure consistency between the various locking provisions.

Section 1010.1.10: An exception has been added to the panic hardware requirements for Group A and E occupancies, allowing electromagnetically locked systems to be installed provided they comply with Section 1010.1.9.9.

1011.15, 1011.16

Ladders

CHANGE TYPE: Addition

CHANGE SUMMARY: Locations where ladders can be used for access have now been identified and permanent ladders must follow the construction requirements from the *International Mechanical Code* (IMC).

2015 CODE: **1011.15** ~~**1009.14**~~ **Ships Ladders.** Ship ladders are permitted to be used in Group I-3 as a component of a means of egress to and from control rooms or elevated facility observation stations not more than 250 square feet (23 m^2) with not more than 3 occupants and for access to unoccupied roofs. The minimum clear width at and below the handrails shall be 20 inches (508 mm).

1011.15.1 Handrails of Ship Ladders. Handrails shall be provided on both sides of ship ladders.

1011.15.2 Treads of Ship Ladders. Ship ladders shall have a minimum tread depth of 5 inches (127 mm). The tread shall be projected such that the total of the tread depth plus the nosing projection is not less than 8½ inches (216 mm). The maximum riser height shall be 9½ inches (241 mm).

~~Handrails shall be provided on both sides of ship ladders. The minimum clear width at and below the handrails shall be 20 inches (508 mm).~~

© Jupiterimages/Photos.com/Thinkstock

Ladder access is permitted to limited spaces

1011.16 Ladders. Permanent ladders shall not serve as a part of the means of egress from occupied spaces within a building. Permanent ladders shall be permitted to provide access to the following areas:

1. Spaces frequented only by personnel for maintenance, repair or monitoring of equipment.
2. Nonoccupiable spaces accessed only by catwalks, crawl spaces, freight elevators or very narrow passageways.
3. Raised areas used primarily for purposes of security, life safety or fire safety including, but not limited to, observation galleries, prison guard towers, fire towers or lifeguard stands.
4. Elevated levels in Group U not open to the general public.
5. Non-occupied roofs that are not required to have stairway access in accordance with Section 1011.12.1.
6. Ladders shall be constructed in accordance with Section 306.5 of the *International Mechanical Code.*

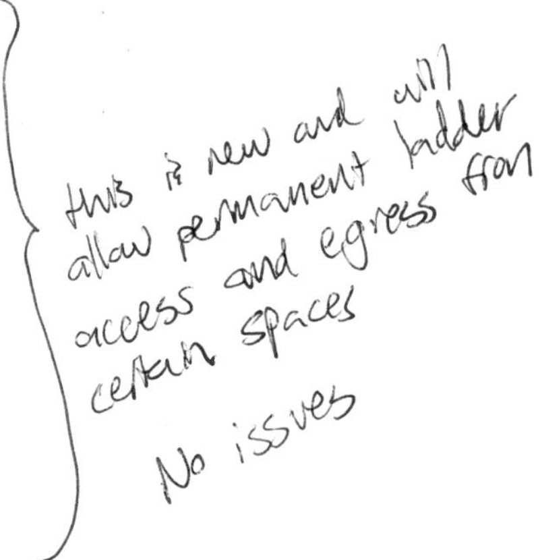
this is new and will
allow permanent ladder
access and egress from
certain spaces
No issues

CHANGE SIGNIFICANCE: Two main revisions addressing ladders have been made within Section 1011. The first is simply the reformatting of the "stairway" provisions to better address the various situations where ladders, ship ladders and alternating tread devices are allowed as an alternative to the typically required stairway. The second is the addition of Section 1011.16, which specifically addresses permanent ladders that may be used to provide access to certain spaces, along with the details of how the ladders are to be constructed.

The IBC has not previously been clear on whether or not a means of egress is required from certain support spaces such as catwalks above ceilings, mechanical equipment areas, service pits and so forth. These types of spaces are often found within buildings and occasionally face different requirements depending on whether or not they are viewed as being for limited service access or for some other purpose. The new provisions now will provide a listing of five locations where the IBC will permit a permanent ladder to be installed for both access and therefore egress from these infrequently occupied or limited-access spaces. In addition, the code references the provisions of IMC Section 306.5, which apply where mechanical equipment is located in an elevated space or on a roof. New Section 1011.16 helps delineate when ladders can be used to access certain areas that are not typically considered to be occupied, and thus do not need to be served by a means of egress.

Code users do need to remember that there may be additional provisions elsewhere in the code that allow for the use of ladders, such as those in Chapter 4 for technical production areas (Section 410.6.3.4, Item 6). However, such areas are also addressed by either Item 1 or 2 of the new provisions.

By referencing the provisions of IMC 306.5, the IBC also now provides the details and construction requirements for these permanent ladders. This will help to ensure that the ladders are safe and useable, while providing consistency for both the designer and building official. Although some jurisdictions have previously relied on the IMC requirements, there are no provisions in previous editions of the IBC that address situations where the ladders are not serving mechanical equipment.

1014.8

Handrail Projections

CHANGE TYPE: Clarification

CHANGE SUMMARY: Guidance is now provided regarding potential obstructions in the required egress width of a stairway where a pair of intermediate handrails is installed.

2015 CODE: **~~1012.8~~ 1014.8 Projections.** On ramps and on ramped aisles that are part of an accessible route, the clear width between handrails shall be 36 inches (914 mm) minimum. Projections into the required width of aisles, stairways and ramps at each side shall not exceed 4½ inches (114 mm) at or below the handrail height. Projections into the required width shall not be limited above the minimum headroom height required in Section ~~1009.5~~ 1011.3. Projections due to intermediate handrails shall not constitute a reduction in the egress width. Where a pair of intermediate handrails is provided within the stairway width without a walking surface between the pair of intermediate handrails and the distance between the pair of intermediate handrails is greater than 6 inches (152 mm), the available egress width shall be reduced by the distance between the closest edges of each such intermediate pair of handrails that is greater than 6 inches (152 mm).

CHANGE SIGNIFICANCE: Where an intermediate handrail is located on a stair or ramp, the IBC has previously indicated that the intermediate handrails do not constitute a reduction in the egress width. Although this approach is understandable when there is only a single intermediate

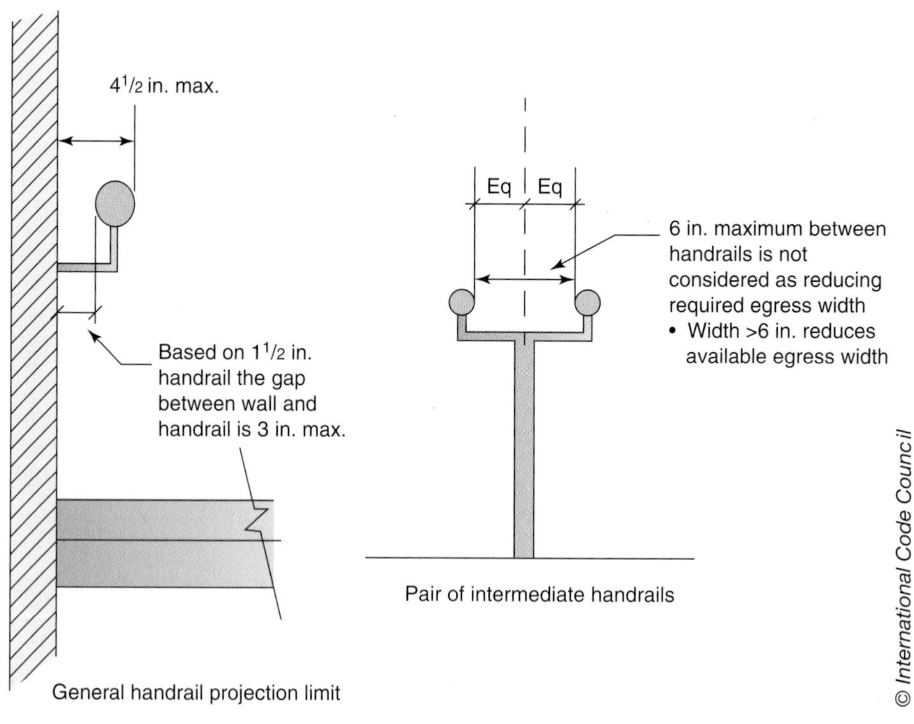
4½ in. max.
Based on 1½ in. handrail the gap between wall and handrail is 3 in. max.
General handrail projection limit
Eq
Eq
6 in. maximum between handrails is not considered as reducing required egress width
• Width >6 in. reduces available egress width
Pair of intermediate handrails
© International Code Council

Handrail projection at stairways

handrail, it becomes confusing when the intermediate handrail is actually constructed using double railings so that the users on each side have their own handrail to grasp as they egress. This revision simply addresses the application of the requirement based on the spacing between the two handrails. Where the spacing between the two adjacent handrails does not exceed 6 inches, then the full width of the stairway can be used to satisfy the required egress width. Where the space between the rails exceeds 6 inches, the width of the stairway must be considered as being reduced by that excess amount.

To better understand the application of this provision, it is important to recognize that stairway width is typically measured to the wall located on the side of the stairway. The code then allows for a handrail to project into that clear width up to 4½ inches. This projection into the required width is permitted "at or below the handrail height." It is this 4½-inch allowable projection that serves as the basis for this new provision. Assuming that the handrails are 1½ inches in diameter, the code now allows the inside edge of the handrail to be located 3 inches away from the wall. Allowing that 3-inch dimension on each side of an intermediate handrail is how the 6-inch dimension identified by the code has been determined.

When applying this provision, building officials might want to consider how much of a projection is allowed on a stairway without creating a reduction in egress width. While the 6-inch clear space between the inside edges of the handrail is based on handrails that are 1½ inches in diameter, it is important to recognize that the code allows handrails to have a circular cross section of "at least 1¼ inches and not greater than 2 inches" based on Section 1014.3.1. If a 1¼-inch handrail is installed, the space between the wall and the inside edge of a handrail is allowed to be 3¼ inches and therefore within the 4½-inch maximum allowable projection limit. Conversely, if a 2-inch handrail is provided, the clearance between the wall and the inside edge of the handrail is limited to 2½ inches in order to be within the 4½-inch limitation. Therefore, the gap between the inside edge of a pair of intermediate handrails can actually be as small as 5 inches or as large as 6½ inches depending on the size of the handrail used if the intent is to limit the allowable projection to 4½ inches into each side. Therefore, if applied in a code-literal fashion, a 6-inch spacing between a pair of 2-inch intermediate handrails will actually result in 5 inches of projection into each side of the stairway. Because of this discrepancy that can occur if handrails larger than 1½ inch diameter are installed, building officials may want to decide if they follow the 6-inch limitation between the pair of intermediate handrails or the 4½-inch limit from the earlier portion of the code section. It is expected that the 6-inch dimension as specified typically will be applied where an intermediate handrail consists of a pair of handrails with a gap between them.

Code users should also recognize that this new provision only applies to stairways and not to ramps. As stated in the code text, the clear width of a ramp is measured between the handrails, and projections are not allowed within that required width.

1016.2

Egress through Intervening Spaces

CHANGE TYPE: Modification

CHANGE SUMMARY: A means of egress is now permitted through an elevator lobby provided access to at least one exit is available without passing through the lobby.

2015 CODE: ~~1014.2~~ 1016.2 Egress through Intervening Spaces. Egress through intervening spaces shall comply with this section.

1. Exit access through an enclosed elevator lobby is permitted. Access to at least one of the required exits shall be provided without travel through the enclosed elevator lobbies required by Section 3006, 3007 or 3008. Where the path of exit access travel passes through an enclosed elevator lobby, the level of protection required for the enclosed elevator lobby is not required to be extended to the exit unless direct access to an exit is required by other sections of this code.

(*No changes for previously existing 4 items and exceptions.*)

~~1018.6~~ 1020.6 Corridor Continuity. Fire-resistance-rated corridors shall be continuous from the point of entry to an exit, and shall not be interrupted by intervening rooms. Where the path of egress travel within a fire-resistance-rated corridor to the exit includes travel along unenclosed exit access stairways or ramps, the fire-resistance-rating shall be continuous for the length of the stairway or ramp and for the length of the connecting corridor on the adjacent floor leading to the exit.

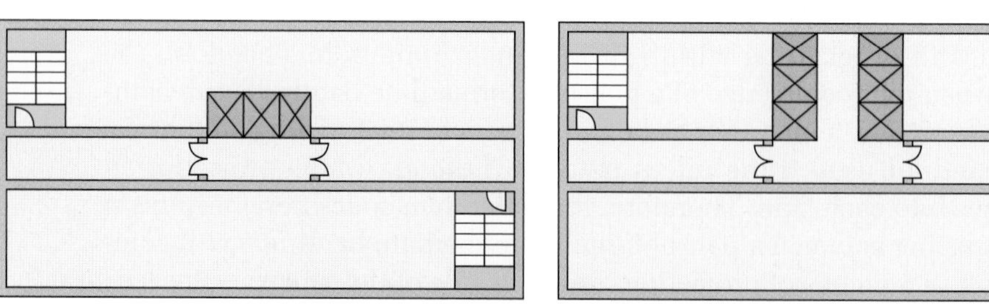

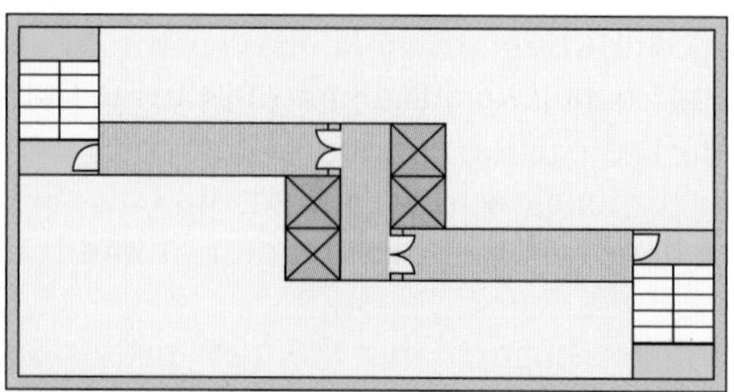

Exit access is permitted through an enclosed elevator lobby provided:

- Access to at least one exit shall be provided without travel through the lobby.
- Protection required for lobby is not required to extend to exit unless access to the exit is required by other sections (e.g., fire service access elevator lobby requires direct access to an exit stairway per Section 3007.6.1).

Egress through elevator lobby

Exceptions:

1. Foyers, lobbies or reception rooms constructed as required for corridors shall not be construed as intervening rooms.
2. Enclosed elevator lobbies as permitted by Item 1 of Section 1016.2 shall not be construed as intervening rooms.

~~713.14.1~~ 3006.4 Means of Egress. Elevator lobbies shall be provided with at least one means of egress complying with Chapter 10 and other provisions in this code. Egress through an elevator lobby shall be permitted in accordance with Item 1 of Section 1016.2.

~~3007.7~~ 3007.6 Fire Service Access Elevator Lobby. The fire service access elevator shall open into a fire service access elevator lobby in accordance with Sections 3007.6.1 through 3007.6.5. Egress is permitted through the elevator lobby in accordance with Item 1 of Section 1016.2.

Exception: Where a fire service access elevator has two entrances onto a floor, the second entrance shall be permitted to open into an elevator lobby in accordance with Section 3006.3.

~~3008.7~~ 3008.6 Occupant Evacuation Elevator Lobby. The occupant evacuation elevators shall open into an elevator lobby in accordance with Sections 3008.6.1 through 3008.6.6. Egress is permitted through the elevator lobby in accordance with Item 1 of Section 1016.2.

CHANGE SIGNIFICANCE: Fire-resistance-rated corridors have historically been required to be protected "continuous from the point of entry to an exit." Such corridors are not to be interrupted by intervening rooms; however, an exception has allowed interruption by "foyers, lobbies and reception areas." It has not previously been clear if an elevator lobby was intended to be included in the exception. With this revision and the companion changes made to the general elevator lobby provisions of Section 3006.4 (previously Section 713.14.1) and the lobby requirements of Sections 3007.6 and 3008.6 for fire service access or occupant evacuation elevators, the code now specifically allows travel through an elevator lobby and regulates how it is to be addressed.

Although egress is now permitted through an enclosed elevator lobby, every occupant must have access to at least one exit without passing through the lobby. If smoke were to spread through the elevator hoistway into the elevator lobby, at least one safe egress path would be available to the occupants.

1017.2.2

Travel Distance Increase for Groups F-1 and S-1

CHANGE TYPE: Modification

CHANGE SUMMARY: An increased exit access travel distance is now permitted for Groups F-1 and S-1 occupancies where specific requirements are met.

2015 CODE: ~~1016.2.2~~ 1017.2.2 Groups F-1 and S-1 Increase. The maximum exit access travel distance shall be 400 feet (122 m) in Group F-1 or S-1 occupancies where all of the following conditions are met:

1. The portion of the building classified as Group F-1 or S-1 is limited to one story in height.
2. The minimum height from the finished floor to the bottom of the ceiling or roof slab or deck is 24 feet (7315 mm).
3. The building is equipped throughout with an automatic fire sprinkler system in accordance with Section 903.3.1.1.

Excerpts from Table 1017.2 (showing portions related to this item).

TABLE ~~1016.2~~ 1017.2 Exit Access Travel Distance[a]

Occupancy	Without Sprinkler System (feet)	With Sprinkler System (feet)
A, E, F-1, M, R, S-1	200	250[b]
F-2, S-2, U	300	400[c]

For SI: 1 foot = 304.8 mm.

a. See the following sections for modifications to exit access travel distance requirements:
Section 412.7: For the distance limitations in aircraft manufacturing facilities.
Section 1017.2.2: For increased distance limitation in Groups F-1 and S-1.

b. Buildings equipped throughout with an automatic sprinkler system in accordance with Section 903.3.1.1 or 903.3.1.2. See Section 903 for occupancies where automatic sprinkler systems are permitted in accordance with Section 903.3.1.2.

c. Buildings equipped throughout with an automatic sprinkler system in accordance with Section 903.3.1.1.

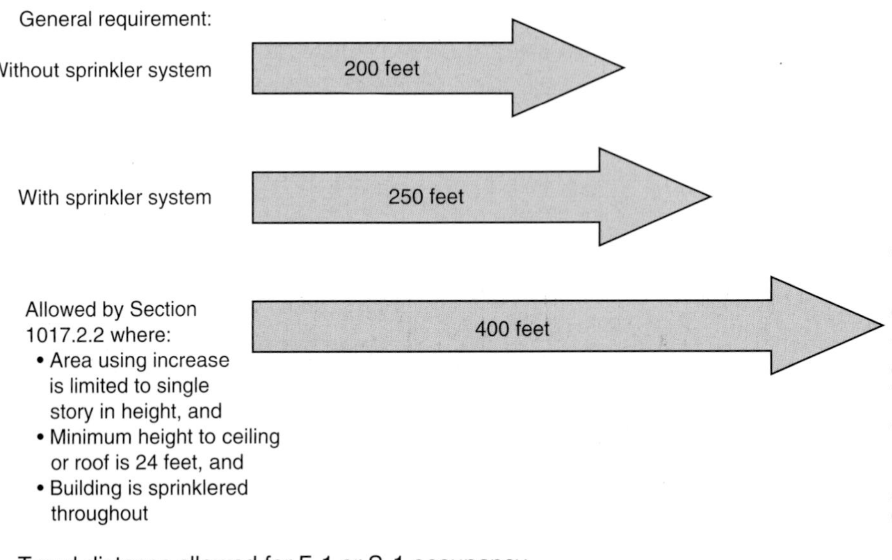
Exit access travel distance for F-1 or S-1 occupancy
General requirement:
Without sprinkler system
200 feet
With sprinkler system
250 feet
Allowed by Section 1017.2.2 where:
• Area using increase is limited to single story in height, and
• Minimum height to ceiling or roof is 24 feet, and
• Building is sprinklered throughout
400 feet
© International Code Council

Travel distance allowed for F-1 or S-1 occupancy

CHANGE SIGNIFICANCE: Travel distance limitations for Groups F-1 and S-1 occupancies have been extended to a maximum of 400 feet where compliance with three criteria set forth in Section 1017.2.2 is met. The Group F-1 or S-1 portion of the building is limited to one story in height and the space must have a minimum height of 24 feet measured to either the ceiling or to the roof deck. A third requirement mandates the installation of an automatic sprinkler system.

The 400-foot travel distance limitation was previously available in the 2006 and earlier editions of the IBC; however, it was tied to the requirement that the space have smoke and heat venting in accordance with Section 910. The 400-foot travel distance limit is unavailable in both the 2009 and 2012 codes because thermally activated vents were judged not to warrant such an increase. The reduction from 400 feet to 250 feet has dramatically affected the layout and design of large storage and factory buildings due to the need to either modify the proportions of the building or provide additional exits.

The increased travel distance is not based on the installation of smoke and heat vents as mandated by previous provisions, but is instead based on fire modeling and egress times. The study used to support this change was commissioned and published by the California State Fire Marshal's Office. The *Report to the California State Fire Marshal on Exit Access Travel Distance of 400 Feet by Task Group 400, December 20, 2010* and the subsequent *Fire Modeling Analysis Report* dated July 20, 2011, provide the technical basis for increasing the travel distance based on the required conditions.

The two criteria for the increased travel distance are (1) the increase is only applicable to portions of the building that are one story in height and (2) a 24-foot minimum ceiling height is provided. The first provision is intended to only limit the area utilizing the 400-foot travel distance to the single-story requirement. It is permissible for the building to have some areas such as an office area that are multistory, provided such areas are regulated by the general travel distance limits and the area utilizing the 400-foot distance limit is only one story in height. The ceiling or roof deck height requirement is based on the *Fire Modeling Analysis Report.* The substantial height provides a volume for the smoke to accumulate during the fire, therefore providing additional time for egress.

Code users should be aware of two other travel distance modifications in Table 1017.2. An additional reference has been inserted in footnote a that directs the user to Section 412.7 for aircraft manufacturing facilities. Section 412.7 now provides increased travel distance limits for aircraft manufacturing buildings of Type I or II construction based upon both the height and area of the space. These travel distance limits range from 400 to 1500 feet depending on the building's size. Footnote d has also been added to address the sprinkler requirements for Group H occupancies. Because Group H occupancies are only required to be sprinklered within the occupancy and not throughout the building, the footnote was added to distinguish the requirement from that of footnote c. This requirement

1017.2.2 continues

1017.2.2 continued should be considered in conjunction with the limitations of Section 1016.2, Item 2 and its exception that prohibit the Group H occupancy from passing through a more hazardous use. Given this restriction, the travel distance limitations were viewed as being acceptable whether the travel is all within the sprinklered Group H occupancy or if it passes through the Group H and continues through a nonsprinklered occupancy.

1018.3

Aisles in Groups B and M

CHANGE TYPE: Modification

CHANGE SUMMARY: The required width of aisles in Groups B and M occupancies is now consistent with the widths required for corridors and is no longer limited only to the capacity based on the occupant load served.

2015 CODE: **~~1017.3~~ 1018.3 Aisles in Groups B and M.** In Groups B and M occupancies, the minimum clear aisle width shall be determined by Section 1005.1 for the occupant load served, but shall ~~not~~ be not less than ~~36 inches (914 mm)~~ that required for corridors by Section 1020.2.

> **Exception:** Nonpublic aisles serving less than 50 people and not required to be accessible by Chapter 11 need not exceed 28 inches (711 mm) in width.

~~1017.5~~ 1018.5 Aisles in Other than Assembly Spaces and Groups B and M. In other than rooms or spaces used for assembly purposes and Group B and M occupancies, the minimum clear aisle ~~width~~ capacity shall be determined by Section 1005.1 for the occupant load served, but the width shall be not less than that required for corridors by Section 1020.2 ~~not be less than 36 inches (914 mm)~~.

> **Exception:** Nonpublic aisles serving less than 50 people and not required to be accessible by Chapter 11 need not exceed 28 inches (711 mm) in width.

1018.3 continues

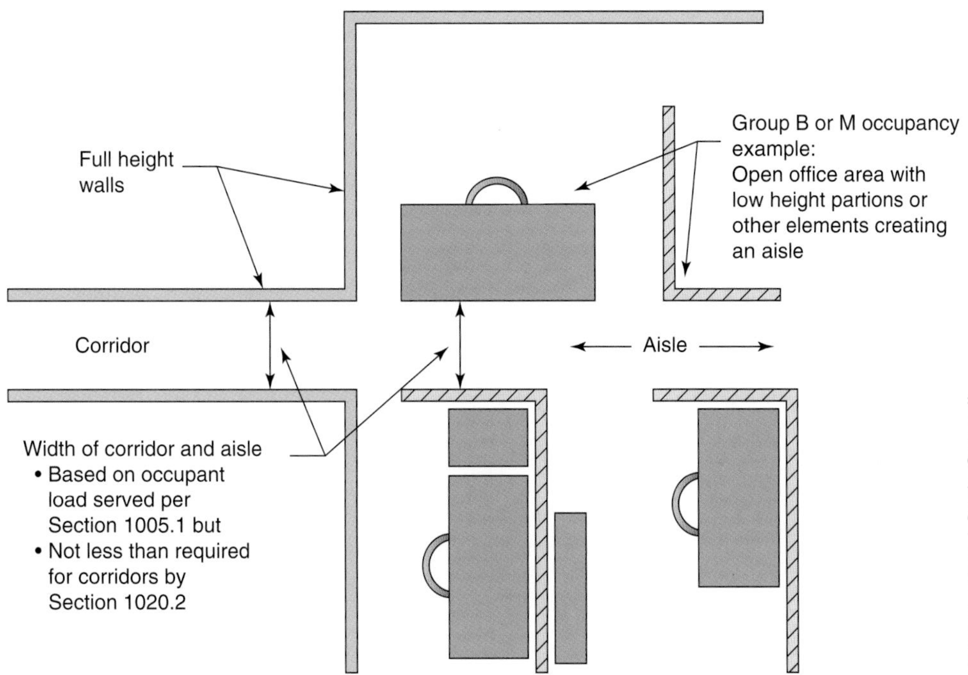

Minimum aisle widths in Group B and M occupancies

1018.3 continued

CHANGE SIGNIFICANCE: Historically, the minimum required width of aisles in Groups B and M occupancies have been based solely on the capacity required for the occupant load served, with a minimum required width of 36 inches. The new provision makes the minimum required width determination for aisles consistent with that required for corridors. A similar revision is also applicable to aisles in occupancies "other than assembly spaces and Groups B and M." Therefore, the minimum required aisle width will not only be based on the occupant load capacity (Section 1005.3), but also the minimum width for the component based on the references to Section 1005.1 (and therefore Section 1005.2) and Section 1020.2.

Most aisles that serve 50 or more occupants will be affected by the modification. Previously, the occupant load multiplied by the egress capacity factor (which is generally 0.2 inches per occupant) was solely applied in order to determine the minimum required aisle width. As an example, assuming an occupant load of 100 persons, the minimum aisle width based on the 2012 IBC is 36 inches, whereas the 2015 IBC requires the aisle to be a minimum of 44 inches in width.

Aisles are the main paths of egress through many types of spaces, such as cubicles in an open office area or between the merchandise pads in the display area of a store. Although not always confined by walls as for a corridor, it is reasonable that they should be sized consistently with corridors so that the occupants can exit the building safely. Having the aisle width coordinated with the corridor requirements ensures the occupants will have a consistent egress width as they move from aisles to corridors as might occur in an office building with both open office areas and other traditional support spaces.

The exception previously found within the code for nonpublic aisles in Groups B and M occupancies has been duplicated and placed in the section addressing other occupancies. Where the aisle serves a limited occupant load and is restricted it is reasonable to allow the width to be reduced. Including the exception within Section 1018.5 provides consistency between the various non-assembly occupancies.

1020.2

Corridor Width and Capacity

CHANGE TYPE: Clarification

CHANGE SUMMARY: A new exception helps to clarify the width requirements for corridors within Group I-2 occupancies for areas where bed or stretcher movement is not necessary.

2015 CODE: ~~1018.2~~ 1020.2 Width and Capacity. The ~~minimum width~~ required capacity of corridors shall be determined as specified in Section 1005.1, but the minimum width shall ~~not~~ be not less than that specified in Table ~~1018.2~~ 1020.2 ~~shall be as determined in Section 1005.1~~.

> **Exception:** In Group I-2 occupancies, corridors are not required to have a clear width of 96 inches (2438 mm) in areas where there will not be stretcher or bed movement for access to care or as part of the defend-in-place strategy.

1020.2 continues

TABLE ~~1018.2~~ 1020.2 Minimum Corridor Width

Occupancy	Width (minimum)
Any facilities not listed below	44 inches (1118 mm)
Access to and utilization of mechanical, plumbing or electrical systems or equipment	24 inches (610 mm)
With an ~~required occupancy capacity~~ occupant load of less than 50	36 inches (914 mm)
Within a dwelling unit	36 inches (914 mm)
In Group E with a corridor having an ~~required capacity~~ occupant load of 100 or more	72 inches (1829 mm)
In corridors and areas serving ~~gurney~~ stretcher traffic in ~~occupancies where patients receive outpatient medical care, which causes the patient to be incapable of self-preservation~~ ambulatory care facilities	72 inches (1829 mm)
Group I-2 in areas where required for bed movement	96 inches (2438 mm)

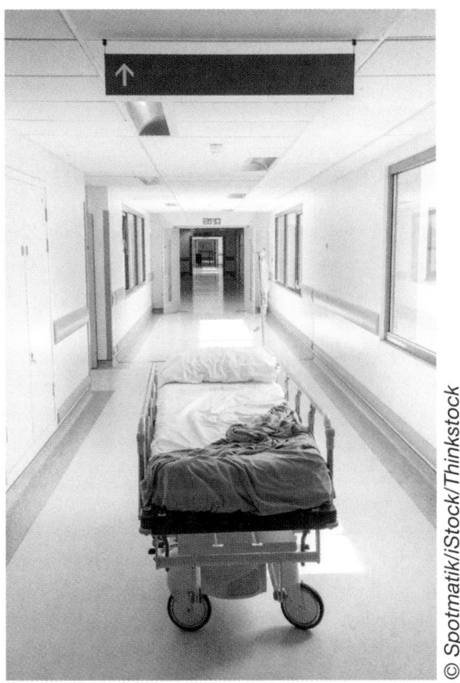

Bed movement in corridors requires larger widths

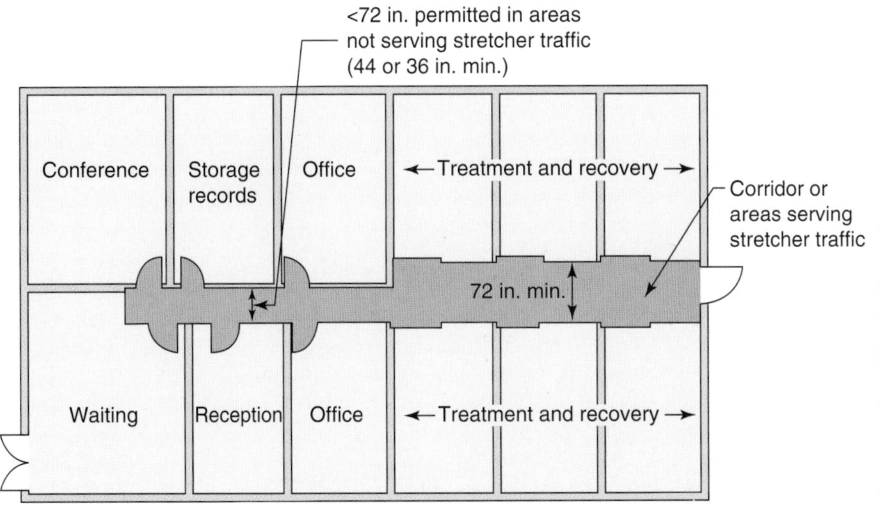

Reduced corridor width permitted in areas not serving stretcher traffic

1020.2 continued

CHANGE SIGNIFICANCE: Because hospitals typically include accessory spaces or mixed-occupancy areas that are not used for patient care, the building official must have the clear ability to apply judgment in determining the appropriate means-of-egress requirements. The revised provision serves as a reminder that consideration for Group I-2 corridor width should not only be given to general patient care areas, but also to areas that may provide access to the care areas or areas designated as patient relocation or refuge areas under the hospital's emergency plans. Therefore, where an area not normally used for patient bed movement is designated as a part of the refuge area required by Section 407.5.1, then the wider aisle width will be required within that space if there will be bed movement in the space.

The intent of providing a wider corridor is to allow for bed movement in opposite directions similar to the 44-inch general corridor provisions allowing people to pass in opposite directions. Therefore, where patients may be moved on beds, this wider corridor is needed. In other areas, the 44- or 36-inch minimum width corridors are acceptable if they adequately provide the required egress capacity.

1023.3.1

Stairway Extension

CHANGE TYPE: Modification

CHANGE SUMMARY: An interior exit stairway is now permitted to continue directly into an exit passageway without a required fire door assembly to separate the two elements.

2015 CODE: ~~**1022.3.1**~~ **1023.3.1 Extension.** Where interior exit stairways and ramps are extended to an exit discharge or a public way by an exit passageway, the interior exit stairway and ramp shall be separated from the exit passageway by a fire barrier constructed in accordance with Section 707 or a horizontal assembly constructed in accordance with Section 711, or both. The fire-resistance rating shall be not less than that

1023.3.1 continues

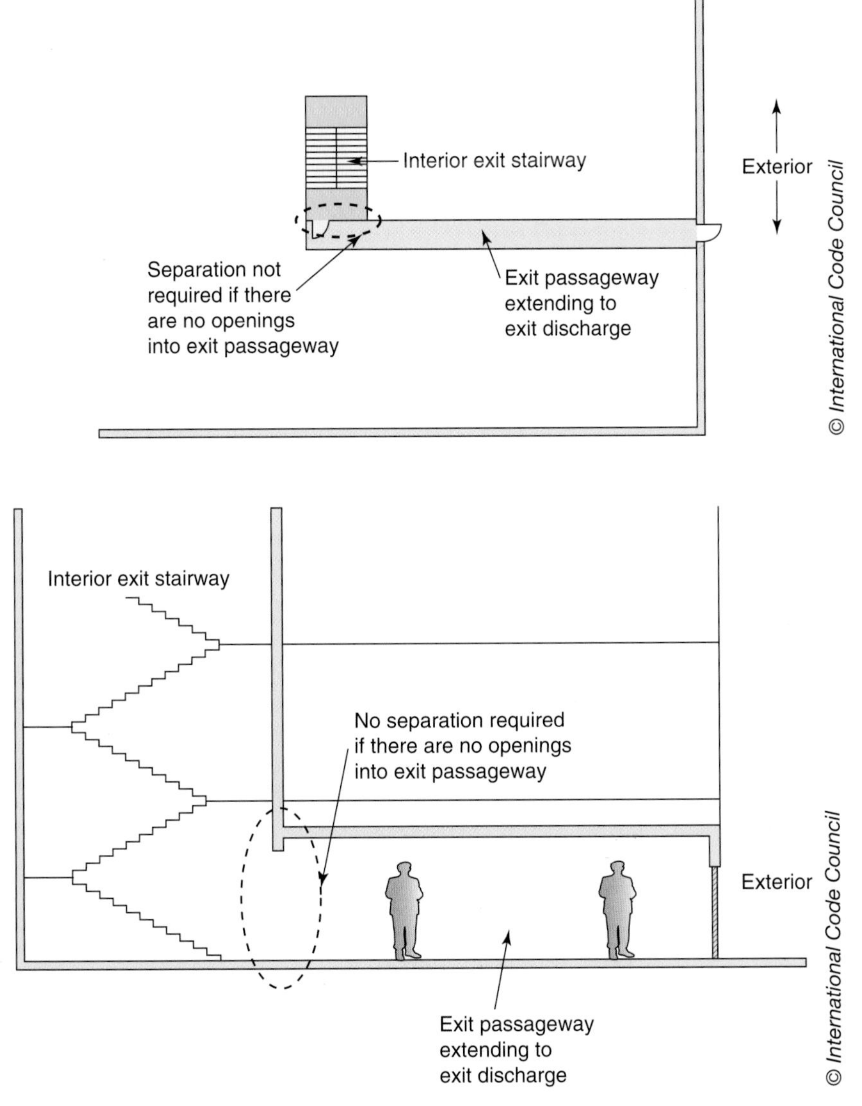

Separation between stairway and passageway can be eliminated

1023.3.1 continued

required for the interior exit stairway and ramp. A fire door assembly complying with Section 716.5 shall be installed in the fire barrier to provide a means of egress from the interior exit stairway and ramp to the exit passageway. Openings in the fire barrier other than the fire door assembly are prohibited. Penetrations of the fire barrier are prohibited.

Exceptions:

1. Penetrations of the fire barrier in accordance with Section ~~1022.5~~ 1023.5 shall be permitted.
2. Separation between an interior exit stairway or ramp and the exit passageway extension shall not be required where there are no openings into the exit passageway extension.

CHANGE SIGNIFICANCE: Exit passageways are often used to extend an interior exit stairway from a remote location within the building to the exterior of the building. Besides maintaining exit continuity, occupants are able to remain within the protected enclosure and maintain the required level of egress protection as is required by Section 1022.1. The base requirements of Section 1023.3.1 require that the interior exit stairway or ramp be separated from the exit passageway by a fire barrier and fire door assembly. The new exception will allow this separation between the horizontal and vertical exit elements to be eliminated where there are no openings into the exit passageway.

The general purpose of the separation in the base requirement is to prevent smoke from an open door within the passageway from traveling through the interior exit stairway to other levels of the building. Where there are no openings into the passageway, there is little concern of smoke or fire entering the passageway and then compromising the exit enclosure. In addition, the egress travel can proceed more quickly because there is no intermediate door through which the occupants must travel as they transition from the interior exit stairway to the horizontal travel within the exit passageway.

When applying this new exception, it is important to recognize that this specific requirement mandates that there are no openings into the exit passageway extension other than the connection from the interior exit stairway or ramp and the door providing egress from the passageway. This makes the exception more restrictive than the general exit passageway requirements found within Sections 1024 and 1024.5 that typically allow doors providing egress from normally occupied spaces to open into the passageway. Penetrations into the passageway are not restricted by the new exception but are limited as specified in Section 1024.6 to those elements serving the exit passageway itself. The same restriction is also applied to the ventilation of the exit passageway, as it must be regulated and limited as addressed in Section 1024.7.

CHANGE TYPE: Modification

1029.13.2.2.1

Stepped Aisle Construction Tolerances

CHANGE SUMMARY: The variation allowed between adjacent risers within a stepped aisle is now limited.

2015 CODE: ~~**1028.11.2**~~ **1029.13.2.2 Risers.** Where the gradient of ~~aisle stairs~~ stepped aisles is to be the same as the gradient of adjoining seating areas, the riser height shall ~~not~~ be not less than 4 inches (102 mm) nor more than 8 inches (203 mm) and shall be uniform within each flight.

Exceptions:

1. Riser height nonuniformity shall be limited to the extent necessitated by changes in the gradient of the adjoining seating area to maintain adequate sightlines. Where nonuniformities exceed $^3/_{16}$ inch (4.8 mm) between adjacent risers, the exact location of such nonuniformities shall be indicated with a distinctive marking stripe on each tread at the nosing or leading edge adjacent to the nonuniform risers. Such stripe shall be ~~a minimum of~~ not less than 1 inch (25 mm), and ~~a maximum of~~ not more than 2 inches (51 mm), wide. The edge marking stripe shall be distinctively different from the contrasting marking stripe.
2. Riser heights not exceeding 9 inches (229 mm) shall be permitted where they are necessitated by the slope of the adjacent seating areas to maintain sightlines.

1029.13.2.2.1 Construction Tolerances. The tolerance between adjacent risers on a stepped aisle that were designed to be equal height shall not exceed $^3/_{16}$ inch (4.8 mm). Where the stepped aisle is designed in accordance with Exception 1 of Section 1029.13.2.2, the stepped aisle shall be constructed so that each riser of unequal height, determined in the direction of descent, is not more than $^3/_8$ inch (10 mm) in height different from adjacent risers where stepped aisle treads are less than 22 inches

1029.13.2.2.1 continues

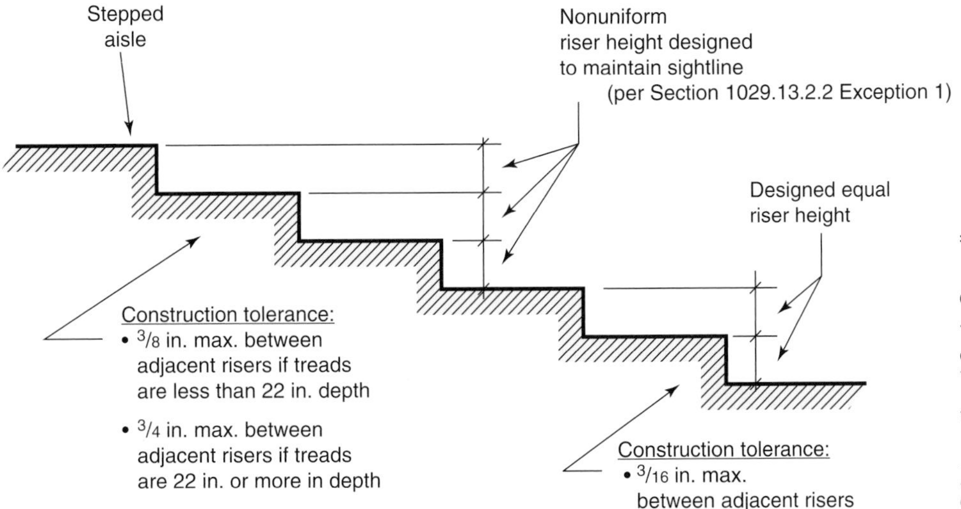

Limitations on riser height variations

1029.13.2.2.1 continued (560 mm) in depth and ¾ inch (19 mm) in height different from adjacent risers where stepped aisle treads are 22 inches (560 mm) or greater in depth.

CHANGE SIGNIFICANCE: Although construction tolerances for stairways are specified in Section 1011.5.4, they are not applicable to the stepped aisles regulated by Section 1029. Riser heights within a stepped aisle are allowed by Exception 1 of Section 1029.13.2.2 to vary "to the extent necessitated by changes in the gradient of the adjoining seating area to maintain adequate sightlines." However, the code has never limited the actual difference between adjacent risers caused by variations in the construction. New provisions now limit the variation between risers that are intended to be equal in height, as well as those that are intended to vary in order to maintain the sightlines.

The new provisions will effectively place limits on the variation that is allowed by Exception 1 of Section 1029.13.2.2. Previously, because there were no limitations, the exception would allow any amount of variation between adjacent risers provided the distinctive marking stripe was placed at the nosing where variations exceeded 3⁄16 inch. The exception supported the lack of construction quality or diligence by permitting striping to be used where risers that were designed to be equal in height varied by more than 3⁄16 inch when they were actually constructed.

Under the new limitations established in Section 1029.13.2.2.1, risers that "were designed to be equal height" will be limited to having a 3⁄16-inch construction tolerance similar to what is allowed for stairways by Section 1011.5.4. Where the stepped aisle risers are varied using Exception 1 in Section 1029.13.2.2 "to maintain adequate sightlines," the difference between adjacent risers is limited to either 3⁄8 inch or ¾ inch depending on the tread depth. Previously, the code would not have placed any restriction on the height change between adjacent risers other than the code text about being "limited to the extent necessitated. . . . to maintain adequate sightlines." So whereas riser heights previously could have varied by an inch or more if that coincided with the gradient of the adjacent seating area, they will now have a maximum limit and be held to the 3⁄8- or ¾-inch variation.

Having the construction tolerance and the maximum allowable designed variation limited will help to improve the safety of occupants using the stepped aisles for access to seating or for egress. It will also help ensure that the distinctive marking stripe required by Exception 1 in Section 1029.13.2.2 is more useful and limited to those situations where the riser heights are designed to vary due to the sightlines.

PART **5**

Accessibility

Chapter 11

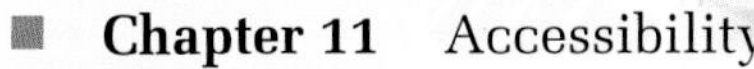

■ **Chapter 11** Accessibility

Chapter 11 is intended to address the accessibility and usability of buildings and their elements to persons having physical disabilities. The provisions within the chapter are generally considered as scoping requirements that state what and where accessibility is required or how many accessible features or elements must be provided. The technical requirements, addressing how accessibility is to be accomplished, are found in ICC A117.1, as referenced by Chapter 11. The concept of the code is to initially mandate that all buildings and building elements be accessible and then to reduce the required accessibility where logical and reasonable. ■

1103.2.8

Areas in Places of Religious Worship

1104.4

Multistory Buildings and Facilities

1107.3, 1107.4

Accessible Spaces and Routes

1107.5.1.1, 1107.6.4.1

Accessible Units in Assisted Living Facilities

1107.6.1.1

Group R—Accessible Units

1109.2

Accessible Water Closet Compartments

1109.2.3

Accessible Lavatories

1110

Recreational Facilities

1103.2.8

Areas in Places of Religious Worship

CHANGE TYPE: Modification

CHANGE SUMMARY: Small areas used for religious ceremonies are now exempt from the access requirements.

2015 CODE: 1103.2 General Exceptions. Sites, buildings, structures, facilities, elements and spaces shall be exempt from this chapter to the extent specified in this section.

1103.2.8 Areas in Places of Religious Worship. Raised or lowered areas, or portions of areas, in places of religious worship that are less than 300 square feet (30 m^2) in area and located 7 inches (178 mm) or more above or below the finished floor and used primarily for the performance of religious ceremonies are not required to comply with this chapter.

CHANGE SIGNIFICANCE: Unlike the federal Americans with Disabilities Act (ADA), the IBC requires religious facilities to be accessible. The IBC addresses these types of uses as it would any typical assembly space and therefore looks for access not only to the main seating areas, but also to presentation and performance areas, speaker platforms, and other raised areas. Areas such as altars, bimahs, baptisteries, pulpits, minibars and minarets are examples of the types of areas that may be addressed by this new provision and exempted from access.

Many religious architectural building features are based on traditions or rituals that are often conducted from either raised or recessed areas within the sanctuary or worship area. Because these elements were not exempted by the IBC access requirements, it often created difficulty in making modifications to historic or traditional elements in an attempt to

Raised area for religious worship that is not required to be accessible

make these features accessible. In addition, because the IBC did not provide any modified scoping and the A117.1 standard has never specifically developed separate technical requirements, these elements were required to be fully compliant. As addressed previously, the uncertainty as to what was properly required was compounded due to the fact that the ADA exempts the entire building used for religious worship or controlled by religious organizations. The new allowance will provide guidance and create consistency by specifically providing the details of where access is not required.

Where the elements are used "primarily for the performance of religious ceremonies," are of a limited size (less than 300 square feet in area), and are either raised or lowered 7 inches or more from the accessible level, a general exemption is now provided that removes the access requirement. Code users should note that the provision does not indicate that this is an aggregate area limitation, so it would be permissible to have several elements such as a high altar, baptistery and pulpit all addressed separately. The size and elevation differences selected for this exemption were based upon existing provisions with the code and the ADA that exempt employee work areas and raised areas used for specialized purposes. Having a size limitation and elevation threshold will allow an exemption for these reasonably sized areas of the religious practice but will not exempt the entire building as done by the ADA. They will also ensure that elements used by a large number of people, such as choir seating areas, that exceed the size limit will be accessible. It is possible that someone in that group may need the access, and it does not typically directly affect the traditional religious elements of the building.

1104.4

Multistory Buildings and Facilities

Government buildings require access to all levels

CHANGE TYPE: Modification

CHANGE SUMMARY: A distinction has been made between the requirements for access within a story and those with greater level changes, such as between stories or mezzanines.

2015 CODE: **1104.4 Multi~~level~~ Story Buildings and Facilities.** At least one accessible route shall connect each accessible ~~level~~ story~~, including mezzanines~~ and mezzanine in multilevel buildings and facilities.

Exceptions:

1. An accessible route is not required to stories and mezzanines that have an aggregate area of not more than 3000 square feet (278.7 m^2) and are located above and below accessible levels. This exception shall not apply to:
   - **1.1.** Multiple tenant facilities of Group M occupancies containing five or more tenant spaces used for the sales or rental of goods and where at least one such tenant space is located on a floor level above or below the accessible levels;
   - **1.2.** ~~Levels~~ Stories or mezzanines containing offices of health care providers (Group B or I); ~~or~~
   - **1.3.** Passenger transportation facilities and airports (Group A-3 or B); or
   - **1.4.** Government buildings.
2. ~~Levels~~ Stories or mezzanines that do not contain accessible elements or other spaces as determined by Sections 1107 or 1108 are not required to be served by an accessible route from an accessible level.
3. In air traffic control towers, an accessible route is not required to serve the cab and the floor immediately below the cab.
4. Where a two-story building or facility has one story or mezzanine with an occupant load of five or fewer persons that does not contain public use space, that story or mezzanine shall not be required to be connected by an accessible route to the story above or below.
5. ~~Vertical access to elevated employee work stations within a courtroom is not required at the time of initial construction, provided a ramp, lift or elevator can be installed without requiring reconfiguration or extension of the courtroom or extension of the electrical system.~~ *(Relocated to Section 1104.3.)*

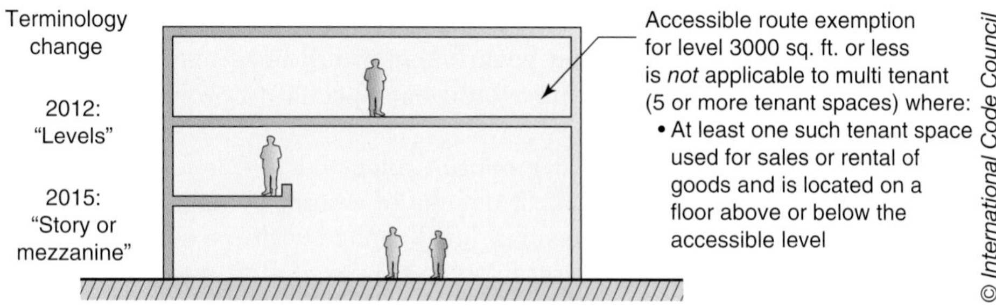
Terminology change
2012: "Levels"
2015: "Story or mezzanine"
Accessible route exemption for level 3000 sq. ft. or less is *not* applicable to multi tenant (5 or more tenant spaces) where:
• At least one such tenant space used for sales or rental of goods and is located on a floor above or below the accessible level
© International Code Council

Multilevel building

CHANGE SIGNIFICANCE: Several subtle changes have been made within Section 1104.4. Through format changes, an effort was made to coordinate the IBC provisions with the access provisions of the ADA. As a part of this process, Section 1104.3 was modified to address connecting all accessible spaces, including those within a story, while Section 1104.4 is referenced and applicable for changes in elevation between stories or mezzanines. Section 1104.4 addresses changes in elevation where the accessible route is typically done by means of an elevator.

Item 1.1 of the first exception has been modified to coordinate better with the ADA and clarify that the provision only applies when one or more of the tenants are located on a non-accessible level. Although not often misapplied, the previous provision could have been viewed to imply that where no more than four tenant spaces are located on the non-accessible level, the exception would apply. The intent is that in retail facilities where five or more tenant spaces exist within the building, the exception cannot be used if one or more tenant spaces are located on the non-accessible level.

Item 1.2 in Exception 1 as well as Exceptions 2 and 4 were clarified as to their application to both stories and mezzanines. Previously the code used the less clear wording of "levels" or only addressed the story. The application to mezzanines is now clearly addressed and included as a part of the base paragraph requirement or excluded by the applicable exceptions. The IBC and ADA refer to stories and mezzanines but with a difference in terminology. By revising the exceptions and the base paragraph to clearly include mezzanines, there is now a clear distinction between these larger elevation changes and those of only a few steps that may have been included in the previous "levels." As mentioned previously, the intent of Section 1104.4 is to deal with changes in elevation that are handled typically by means of an elevator and not just a small level change of a few feet.

Item 1.4 has been added to Exception 1 to exclude the use of the 3000-square-foot exemption from applying to governmental buildings. Governmental buildings are regulated by Title II of the ADA and have additional requirements that mandate access to services and programs that may not be possible if Exception 1 is applied. Instead of going into a lengthy exclusion related to the funding source or program aspects that may be a part of the ADA, it was far simpler to exclude the government buildings from using this 3000-square-foot exception, which may have

1104.4 continues

1104.4 continued

ultimately placed the public entity in conflict with the ADA. The net result of this added item is that governmental buildings may not use the 3000-square-foot exception and would be expected to provide access to every story or mezzanine.

The previous Exception 5 has been relocated to Section 1104.3 because it is not addressing a level change of a story or mezzanine but is instead only dealing with the small level change within a courtroom and the access to elevated work areas. When this exception was relocated to Section 1104.3, it was also modified to reference the specific requirements within Section 1108.4.1.4 so the requirements would be clear and coordinated between the various provisions.

1107.3, 1107.4

Accessible Spaces and Routes

CHANGE TYPE: Modification

CHANGE SUMMARY: The provisions for connecting all spaces within a building have been modified to clearly identify the distinction for those with a change of elevation between stories or mezzanines.

2015 CODE: 1107.3 Accessible Spaces. Rooms and spaces available to the general public or available for use by residents and serving Accessible units, Type A units or Type B units shall be accessible. Accessible spaces shall include toilet and bathing rooms, kitchen, living and dining areas and any exterior spaces, including patios, terraces and balconies.

Exceptions:

1. Stories and mezzanines exempted by Section 1107.4.
2. ~~1.~~ Recreational facilities in accordance with Section ~~1109.15~~ 1110.2.
~~2.~~ ~~In Group I-2 facilities, doors to sleeping units shall be exempted from the requirements for maneuvering clearance at the room side provided the door is a minimum of 44 inches (1118 mm) in width.~~ *(Relocated to Section 1107.5.3.1.)*
3. Exterior decks, patios or balconies that are part of Type B units and have impervious surfaces, and that are not more than 4 inches (102 mm) below the finished floor level of the adjacent interior space of the unit.

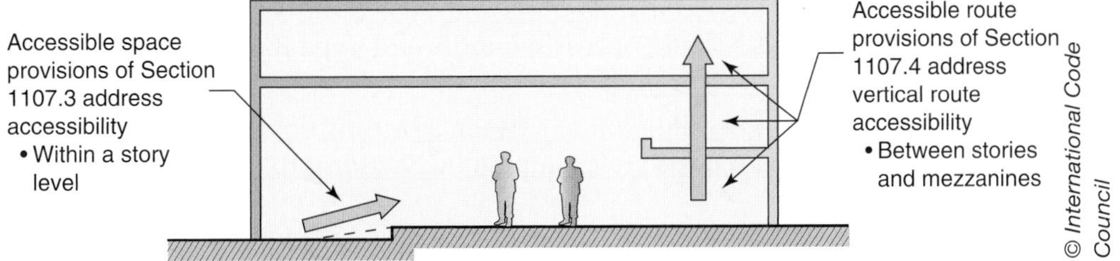
Accessible space provisions of Section 1107.3 address accessibility
• Within a story level
Accessible route provisions of Section 1107.4 address vertical route accessibility
• Between stories and mezzanines
© International Code Council

Access within spaces versus accessible route

1107.4 Accessible Route. At least one accessible route shall connect accessible building or facility entrances with the primary entrance of each Accessible unit, Type A unit and Type B unit within the building or facility and with those exterior and interior spaces and facilities that serve the units.

Exceptions:

1. If due to circumstances outside the control of the owner, either the slope of the finished ground level between accessible facilities and buildings exceeds one unit vertical in 12 units horizontal (1:12), or where physical barriers or legal restrictions prevent the installation of an accessible route, a vehicular route with parking that complies with Section 1106 at each public or common use facility or building is permitted in place of the accessible route.

1107.3, 1107.4 continues

1107.3, 1107.4 continued

2. ~~Exterior decks, patios or balconies that are part of Type B units and have impervious surfaces, and that are not more than 4 inches (102 mm) below the finished floor level of the adjacent interior space of the unit.~~ *(Relocated to Section 1107.3)*
2. In Group I-3 facilities, an accessible route is not required to connect stories or mezzanines where Accessible units, all common use areas serving Accessible units and all public use areas are on an accessible route.
3. In Group R-2 facilities with Type A units complying with Section 1107.6.2.2.1, an accessible route is not required to connect stories or mezzanines where Type A units, all common use areas serving Type A units and all public use areas are on an accessible route.
4. In other than Group R-2 dormitory housing provided by places of education, in Group R-2 facilities with Accessible units complying with Section 1107.6.2.3.1, an accessible route is not required to connect stories or mezzanines where Accessible units, all common use areas serving Accessible units and all public use areas are on an accessible route.
5. In Group R-1, an accessible route is not required to connect stories or mezzanines within individual units, provided the accessible level meets the provisions for Accessible units and sleeping accommodations for two persons minimum and a toilet facility are provided on that level.
6. In congregate residences in Groups R-3 and R-4, an accessible route is not required to connect stories or mezzanines where Accessible units or Type B units, all common use areas serving Accessible units and Type B units and all public use areas serving Accessible units and Type B units are on an accessible route.
7. An accessible route between stories is not required where Type B units are exempted by Section 1107.7.

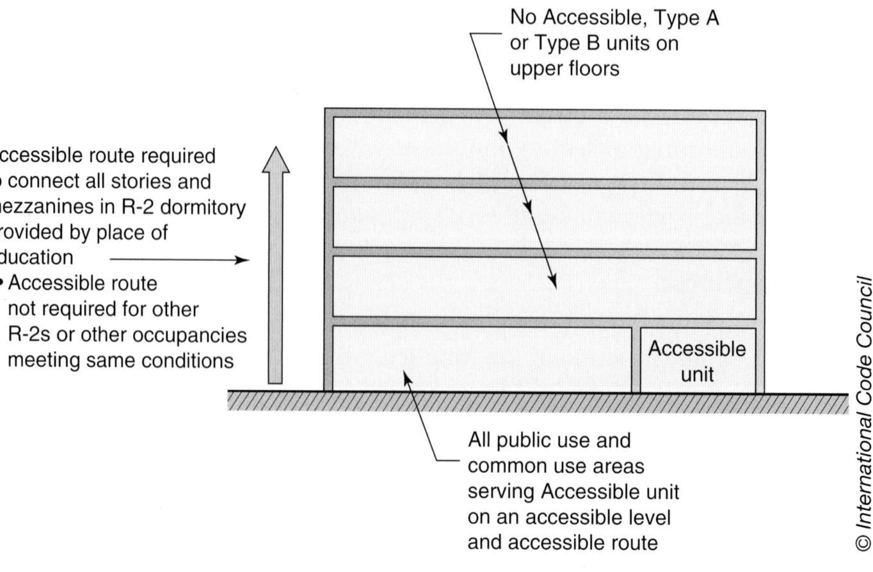

Accessible route requirement for R-2 dormitory

CHANGE SIGNIFICANCE: Although similar to the changes to Section 1104.4, this series of revisions specifically addresses the accessible routes between levels within residential and institutional occupancies, rather than the provisions of Sections 1104.3 and 1104.4 that are general access requirements applicable to all occupancies. Sections 1107.3 and 1107.4 have been modified to address the general access and vertical access within a floor and between stories. Section 1107.3 is intended to deal with connecting all accessible spaces within a building, and a reference to Section 1107.4 directs users to that section, which addresses changes in elevation between stories or to a mezzanine where the route is typically by means of an elevator.

Section 1107.3 has been modified by the inclusion of a new Exception 1 that directs users to Section 1107.4 and the various exceptions within that portion of the code that might eliminate the need for an elevator or other means of access to stories or mezzanines within the living facilities. This new exception is intended to coordinate with the requirements of Section 206.2.4 of the ADA and Exception 3 within that section.

The previous second exception that addressed doors to Group I-2 sleeping units has been deleted and relocated to Section 1107.5.3.1. By relocating the requirement as an exception specifically for the hospital rooms that are required to be accessible (Section 1107.5.3.1), it is now clear that accessible patient sleeping rooms are the rooms to which the exception applies; it also makes it clear that the intent is to allow these rooms to not meet the unit entry requirements in ICC A117.1 Section 1002.5. The 2012 code text would have applied to all Group I-2 occupancies, whereas the revised provision is only applicable to hospitals.

The new Exception 3 has been relocated to this section from Section 1107.4, Exception 2, where it had previously been located. This change coordinates with the formatting change because this is an elevation change within the story and is not a change involving a story or mezzanine where an elevator would typically be used for the transition.

The majority of the exceptions within Section 1107.4 provide exemptions for the accessible route to upper levels when the Accessible units, all common-use areas, and all public-use areas are on an accessible level. Exceptions 2, 3, 4 and 6 fall under this concept. Exception 2 addresses Group I-3 jails and the Accessible units, common-use areas and public-use areas within them. Exception 3 is similar but applicable to larger Group R-2 apartments, monasteries and convents (more than 20 dwelling units or sleeping units, per Section 1107.6.2.2.1). Exception 6 also is conceptually similar but it now applies to smaller group homes or congregate facilities such as fraternities and sororities that can fall within the Group R-3 or R-4 classification.

With the exclusion of dormitories, Exception 4 in Section 1107.4 picks up all of the other Group R-2 uses (see Section 1107.6.2.3) that have Accessible units per Section 1107.6.2.3.1 and were not included in the "apartment houses, monasteries and convent" provisions of Section 1107.6.2.2 and the previously discussed Exception 3. This would include uses such as fraternities, sororities, larger non-transient congregate living facilities, vacation time-share properties, and non-transient hotels and motels. When applying Exception 4 it is important to recognize that dormitories provided by places of education are excluded from using the exception. Based upon the Department of Justice Regulations specific to dormitories and other housing typically associated with universities, dorms in places of education such as ADA Title II buildings are required

1107.3, 1107.4 continues

1107.3, 1107.4 continued

to have an accessible route to all levels (ADA Section 206.2.3, Exception 4). The 2015 IBC does not differentiate between public schools (ADA Title II) or private schools (ADA Title III), but simply requires the accessible route to all levels of the dorm regardless of where the Accessible units might be located. This allows people who may need an accessible route to visit friends or other people living on the other floors even where the Accessible units are located on a different level.

Exception 5 is only applicable within an individual unit of a hotel where the unit is either multistory or contains a mezzanine. In this situation, if a sleeping area and toilet are located on the accessible level and constructed to comply with the Accessible unit requirements, then an accessible route is not required to connect to other stories or a mezzanine. This exception was added to the IBC to coordinate with ADA Section 206.2.3, Exception 5.

The last exception within Section 1107.4, Exception 7, is intended to coordinate with the requirements of the Fair Housing Act and to provide an exemption for the accessible route to other levels when the building is not provided with an elevator. The requirements of Section 1107.7 provide a number of exceptions that coordinate with the Fair Housing Act and exempt access to other levels.

1107.5.1.1, 1107.6.4.1

Accessible Units in Assisted Living Facilities

CHANGE TYPE: Modification

CHANGE SUMMARY: The minimum number of Accessible units required in assisted living facilities now varies depending on the capabilities of the residents.

2015 CODE: 1107.5.1 Group I-1. Accessible units and Type B units shall be provided in Group I-1 occupancies in accordance with Sections 1107.5.1.1 and 1107.5.1.2.

1107.5.1.1 Accessible Units. In Group I-1 Condition 1, at least 4 percent, but not less than one, of the dwelling units and sleeping units shall be Accessible units. In Group I-1 Condition 2, at least 10 percent, but not less than one, of the dwelling units and sleeping units shall be Accessible units.

1107.6.4 Group R-4. Accessible units and Type B units shall be provided in Group R-4 occupancies in accordance with Sections 1107.6.4.1 and 1107.6.4.2.

1107.6.4.1 Accessible Units. In Group R-4 Condition 1, at least one of the dwelling units or sleeping units shall be an Accessible unit. In Group R-4 Condition 2, at least two of the dwelling units or sleeping units

1107.5.1.1, 1107.6.4.1 continues

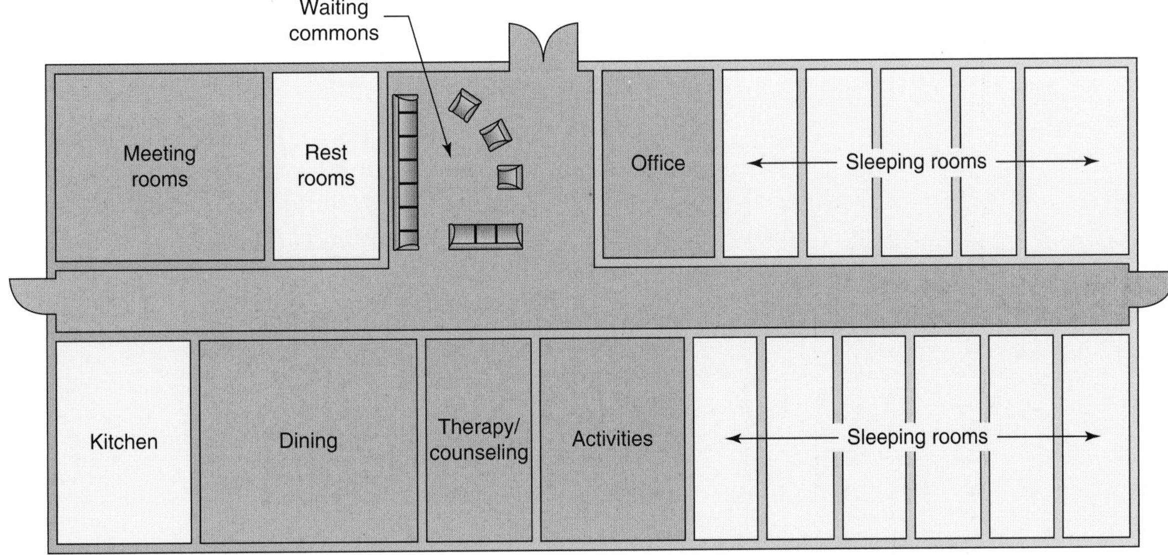

Accessible Units in Group I-1 and R-4 Occupancies			
I-1 Occupancies		R-4 Occupancies	
Condition 1	Condition 2	Condition 1	Condition 2
≥4% but not less than 1	≥10% but not less than 1	At least 1	At least 2 units*

* Bedrooms within Group R-4 facilities shall be counted as sleeping units for the purpose of determining the number of units.

Residents' capabilities affect the number of Accessible units required

1107.5.1.1, 1107.6.4.1 continued

shall be an Accessible unit. Bedrooms within Group R-4 facilities shall be counted as sleeping units for the purpose of determining the number of units.

CHANGE SIGNIFICANCE: Where assisted-living-type facilities are classified as either a Group I-1 or R-4 occupancy, the code now establishes the minimum required number of Accessible units based on the capabilities of the residents. For "Condition 1" facilities where the residents can evacuate without assistance, a smaller number or percentage of the units must be Accessible units. Where the occupancy is a "Condition 2" facility and the residents may need limited assistance in evacuation, a greater number of Accessible units is now mandated based on the assumption that there will be a greater need for them.

Within the Group I-1 occupancies, the revised requirement will keep the minimum required percentage of Accessible units at the previously required 4 percent level where the residents are capable of evacuating on their own. Where the facility is considered a "Condition 2" use and the residents do need assistance, a minimum of 10 percent of the units are now required to be accessible. The 10 percent requirement is based on the anticipated need in these types of facilities. Although the demand for the higher percentage is likely to exceed the 4 percent general requirement for "Condition 1" facilities, the use is considered "custodial care," not nursing care, and therefore it is reasonable to mandate less than the 50 percent required by Section 1107.5.2.1 for Group I-2 nursing homes.

Where the care recipients in a Group R-4 facility are in need of additional assistance and classified as a "Condition 2" occupancy, a minimum of at least two Accessible units must be provided. Previously, Group R-4 facilities only required a single Accessible unit, which will continue to be the expectation where the residents can evacuate without assistance and the facility is a "Condition 1" use.

The last portion of the text within Section 1107.6.4.1 is important to note because it now identifies the bedrooms within small Group R-4 facilities as separate sleeping units. This specific requirement will override the general provision from the definition of "sleeping unit," which typically would not view sleeping rooms within a dwelling unit as being separate sleeping units. Understanding this distinction is necessary in order to ensure that the proper number of Accessible units is provided.

1107.6.1.1

Group R—Accessible Units

CHANGE TYPE: Modification

CHANGE SUMMARY: The method by which multiple buildings on a site are reviewed when determining the required number of Accessible units has been revised to consider building size in addition to the total number of units on the site.

2015 CODE: 1107.6.1.1 Accessible Units. Accessible dwelling units and sleeping units shall be provided in accordance with Table 1107.6.1.1. Where buildings contain more than 50 dwelling units or sleeping units, the number of Accessible units shall be determined per building. Where buildings contain 50 or fewer dwelling units or sleeping units, all dwelling units and sleeping units on a site shall be considered to determine the total number of Accessible units. Accessible units shall be dispersed among the various classes of units. ~~Roll-in showers provided in Accessible units shall include a permanently mounted folding shower seat.~~

CHANGE SIGNIFICANCE: The IBC has always included all of the units on a site in the determination of the minimum number of Accessible units that must be provided for a Group R-1 occupancy. This method will continue to be the appropriate means for calculating the requirements, but only for those buildings that contain 50 or fewer units. For larger buildings, those with more than 50 units, the minimum number of required Accessible units must be determined on an individual building basis by applying Table 1107.6.1.1 to that individual building.

1107.6.1.1 continues

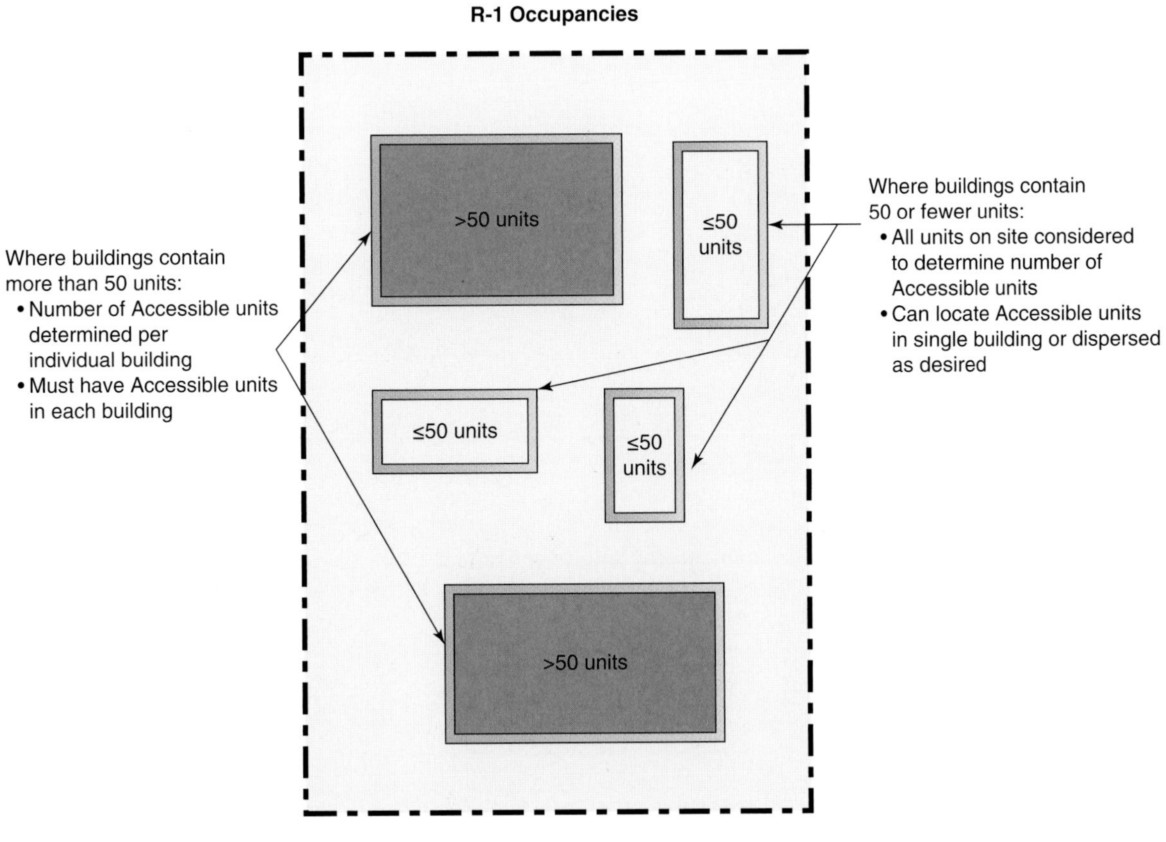

R-1 Occupancies

>50 units

≤50 units

Where buildings contain more than 50 units:
- Number of Accessible units determined per individual building
- Must have Accessible units in each building

Where buildings contain 50 or fewer units:
- All units on site considered to determine number of Accessible units
- Can locate Accessible units in single building or dispersed as desired

≤50 units

≤50 units

>50 units

Building size affects the required number and location of Accessible units

1107.6.1.1 continued

The provisions are now coordinated with Department of Justice (DOJ) regulations applicable to places of lodging that address multiple buildings on a site depending on the size of the building. The revised IBC requirement is now consistent with the DOJ regulations in Section 36.406(c). Where the buildings have 50 or fewer units within each building, the aggregate number of units on the site is determined prior to applying Table 1107.6.1.1 and calculating the number of required Accessible units. Where a building has more than 50 units, those buildings are treated separately for the purpose of determining the required number of Accessible units and the type of accessible bathing facilities.

The requirement for a folding shower seat within a roll-in shower has also been eliminated. The 2003 edition of A117.1 allowed shower seats in roll-in showers to be removable and a lodging facility could provide the seat when requested. Because the 2009 edition of the standard now requires a folding seat as a minimum requirement for every roll-in shower, it ensures that the seats are available to all users who may need them and eliminates the need for the IBC to address this requirement.

1109.2

Accessible Water Closet Compartments

CHANGE TYPE: Modification

CHANGE SUMMARY: In larger toilet rooms, a minimum of 5 percent of the water closet compartments must be wheelchair accessible and another minimum of 5 percent must be ambulatory accessible compartments.

2015 CODE: **1109.2 Toilet and Bathing Facilities.** Each toilet room and bathing room shall be accessible. Where a floor level is not required to be connected by an accessible route, the only toilet rooms or bathing rooms provided within the facility shall not be located on the inaccessible floor. Except as provided for in Sections 1109.2.2 and 1109.2.3, at least one of each type of fixture, element, control or dispenser in each accessible toilet room and bathing room shall be accessible.

Exceptions:

1. ~~In~~ Toilet rooms or bathing rooms accessed only through a private office, not for common or public use and intended for use by a single occupant, ~~any of the following alternatives are allowed:~~ shall be permitted to comply with the specific exceptions in ICC A117.1.
   - ~~1.1 Doors are permitted to swing into the clear floor space, provided the door swing can be reversed to meet the requirements in ICC A117.1;~~
   - ~~1.2 The height requirements for the water closet in ICC A117.1 are not applicable;~~
   - ~~1.3 Grab bars are not required to be installed in a toilet room, provided that reinforcement has been installed in the walls and located so as to permit the installation of such grab bars; and~~
   - ~~1.4 The requirement for height, knee and toe clearance shall not apply to a lavatory.~~
2. This section is not applicable to toilet and bathing rooms that serve dwelling units or sleeping units that are not required to be accessible by Section 1107.
3. Where multiple single-user toilet rooms or bathing rooms are clustered at a single location, at least 50 percent but not less than one room for each use at each cluster shall be accessible.
4. Where no more than one urinal is provided in a toilet room or bathing room, the urinal is not required to be accessible.
5. Toilet rooms or bathing rooms that are part of critical care or intensive care patient sleeping rooms serving Accessible units are not required to be accessible.
6. Toilet rooms or bathing rooms designed for bariatric patients are not required to comply with the toilet room and bathing room requirement in ICC A117.1. The sleeping units served by bariatric toilet or bathing rooms shall not count toward the required number of Accessible sleeping units.

~~6.~~ 7. Where toilet facilities are primarily for children's use, required accessible water closets, toilet compartments and lavatories shall be permitted to comply with children's provision of ICC A117.1.

1109.2 continues

1109.2 continued

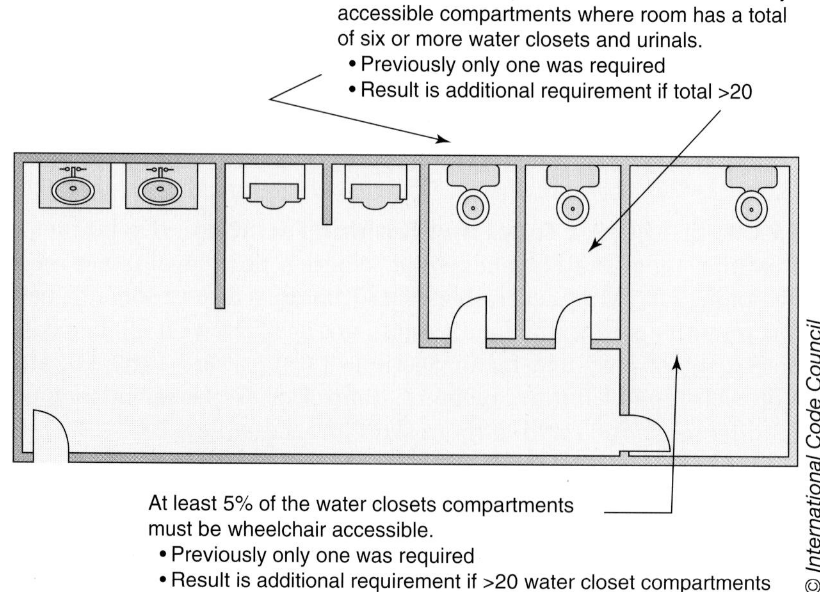

At least 5% of compartments must be ambulatory accessible compartments where room has a total of six or more water closets and urinals.
• Previously only one was required
• Result is additional requirement if total >20
At least 5% of the water closets compartments must be wheelchair accessible.
• Previously only one was required
• Result is additional requirement if >20 water closet compartments
© International Code Council

Additional accessible compartments are required in larger toilet rooms

1109.2.2 Water Closet Compartment. Where water closet compartments are provided in a toilet room or bathing room, at least ~~one~~ 5 percent of the total number of compartments shall be wheelchair-accessible ~~compartment shall be provided~~. Where the combined total water closet compartments and urinals provided in a toilet room or bathing room is six or more, at least ~~one~~ 5 percent of the total number of compartments shall be ambulatory-accessible ~~water closet compartment shall be~~ provided in addition to the wheelchair-accessible compartment.

CHANGE SIGNIFICANCE: For most toilet rooms, the revisions to Section 1109.2.2 will not result in a change in the requirements, as only a single wheelchair-accessible compartment will continue to be required. However, where large toilet rooms are provided with more than 20 compartments, the new 5 percent minimum requirement now mandates that additional compartments comply as wheelchair-accessible. Ambulatory-accessible compartments are also affected by a new 5 percent minimum requirement. Where the total number of toilet compartments and urinals exceeds 20, additional ambulatory-accessible compartments must be provided in order to meet the 5 percent minimum requirement. Previously, only a single ambulatory-accessible compartment was required for any toilet room containing six or more water closet compartments and/or urinals.

Proportionate accessibility is now provided in large toilet rooms that will accommodate the increasing number of people with ambulatory disabilities. In toilet rooms with more than 20 fixtures within them, it is likely that more than one user within the room will be using a wheelchair, crutches, walker, or a scooter. Therefore, it is appropriate to provide additional wheelchair compartments and ambulatory-accessible compartments that allow improved access and support options.

The revisions within Exception 1 of Section 1109.2 have simply deleted provisions that are no longer needed within the IBC because they are located within the technical requirements of the A117.1 standard. The

A117.1 standard addresses these requirements for privately accessed toilet and bathing rooms within exceptions found in Sections 603.2.2, 604.4, 604.5, 606.2 and 606.3 of that document. These requirements are also covered within the ADA as modifications that can be made to accommodate an employee under exceptions within Sections 603.2.3, 604.4, 604.5, 606.2 and 606.3 of that document.

The new Exception 6 within Section 1109.2 provides an exclusion for facilities within Accessible units serving bariatric patients. Although this exemption does not exist within the ADA, it was determined that these unique facilities should be excluded from compliance with the general accessibility provisions. Because of the physical size of bariatric patients, many are not able to use a water closet that is located with the centerline 16 to 18 inches from an adjacent wall. Not only is the space not adequate for some of the patients, but it may not allow a nurse to get next to the patient to offer assistance with rising or sitting down, or allow for lifts and other elements to be maneuvered into the space to assist the user. The same issues would occur with the 36-inch by 36-inch transfer shower. So although designing toilet and bathing rooms for bariatric patients will not make them accessible in accordance with the A117.1 standard, they will be designed to be accessible and usable for the population they are intended to serve.

1109.2.3

Accessible Lavatories

CHANGE TYPE: Modification

CHANGE SUMMARY: In order to prevent the placement of the only accessible lavatory within an accessible stall where it would not be available to all users, the required accessible lavatory must now be located in a common area of the toilet room or bathing facility.

2015 CODE: **1109.2.3 Lavatories.** Where lavatories are provided, at least 5 percent, but not less than one, shall be accessible. Where an accessible lavatory is located within the accessible water closet compartment at least one additional accessible lavatory shall be provided in the multi-compartment toilet room outside the water closet compartment. Where the total lavatories provided in a toilet room or bathing facility is six or more, at least one lavatory with enhanced reach ranges shall be provided.

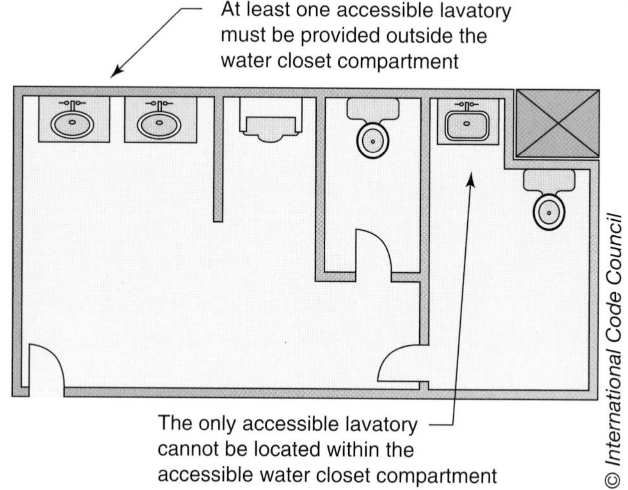
At least one accessible lavatory must be provided outside the water closet compartment
The only accessible lavatory cannot be located within the accessible water closet compartment
© International Code Council

© Matthew Valentine/iStock/Thinkstock

An accessible lavatory must be available to all users

CHANGE SIGNIFICANCE: To ensure that the accessible lavatory is available to all users of the toilet room at any time the room is open, it is now stated that an accessible lavatory must be located outside of a water closet compartment. If the only accessible lavatory is placed within one of the toilet compartments (including the wheelchair-accessible compartment), then others in the bathroom would not have access to that lavatory when the stall is in use.

The modification does not prevent the inclusion of an accessible lavatory within the accessible water closet compartment, but it would simply mandate that at least one accessible lavatory be located where it is available without going into a water closet compartment. The A117.1 standard is conceptually equivalent in regard to the turning space within the restroom. Section 603.2.1 of the standard prohibits the required turning space from being located within a toilet compartment. Including this requirement in the code coordinates the IBC with the requirements found in Section 213.3.4 of the ADA.

1110

Recreational Facilities

CHANGE TYPE: Modification

CHANGE SUMMARY: More detailed scoping requirements for recreational facilities have been included within the new Section 1110 to coordinate with the ADA and provide the scoping for technical requirements found within Chapter 11 of the A117.1 standard.

2015 CODE: **~~1109.15~~ 1110.1 ~~Recreational and Sports Facilities.~~** (Relocated to Section 1110.1.)

~~1109.15.1~~ 1110.2.2 ~~Facilities Serving in a Single Building~~. (Relocated to Section 1110.2.2.)

~~1109.15.2~~ 1110.2.3 ~~Facilities Serving Multiple Buildings~~. (Relocated to Section 1110.2.3.)

~~1109.15.3~~ 1110.3 ~~Other Occupancies~~. (Relocated to Section 1110.3.)

~~1109.15.4~~ 1110.4 ~~Recreational and Sports Facilities Exceptions~~. (Relocated to Section 1110.4.)

SECTION 1110
RECREATIONAL FACILITIES

1110.1 ~~1109.15~~ General ~~Recreational and Sports Facilities.~~ Recreational ~~and sports~~ facilities shall be provided with accessible features in accordance with Sections 1110.2 ~~1109.15.1~~ through 1110.4 ~~1109.15.4~~.

1110.2 Facilities Serving Group R-2, R-3 and R-4 Occupancies. Recreational facilities that serve Group R-2, R-3 and Group R-4 occupancies shall comply with Sections 1110.2.1 through 1110.2.3 as applicable.

1110.3 ~~1109.15.3~~ Other Occupancies. ~~All~~ Recreational facilities not falling within the purview of Section 1110.2 ~~1109.15.1 or 1109.15.2~~ shall be accessible.

1110.4 ~~1109.15.4~~ Recreational ~~and Sports~~ Facilities ~~Exceptions~~. Recreational ~~and sports~~ facilities shall be ~~required to be~~ accessible ~~shall be exempt from this chapter to~~ and shall be on an accessible route to the extent specified in this section.

1110.4.1 Area of Sport Activity. Each area of sport activity shall be on an accessible route and shall not be required to be accessible except as provided for in Sections 1110.4.2 through 1110.4.14.

1110.4.2 ~~1108.2.2.4~~ Team or Player Seating.

1110.4.3 ~~1109.15.4.1~~ Bowling Lanes.

1110.4.4 ~~1109.15.4.2~~ Court Sports.

1110.4.5 ~~1109.15.4.3~~ Raised Boxing or Wrestling Rings.

1110.4.6 ~~1109.15.4.4~~ Raised Refereeing, Judging and Scoring Areas.

1110.4.7 Animal Containment Areas.

1110.4.8 Amusement Rides.

1110.4.9 Recreational Boating Facilities.

1110.4.10 Exercise Machines and Equipment.

1110 continues

Scoping for recreational facilities coordinates with the ADA

1110 continued

1110.4.11 Fishing Piers and Platforms.

1110.4.12 Miniature Golf Facilities.

1110.4.13 Swimming Pools, Wading Pools, Hot Tubs and Spas.

1110.4.14 Shooting Facilities with Firing Positions.

(*Because this code change affected substantial portions of Chapter 11, the entire code change text is too extensive to be included here. Refer to Code Changes E-208, E-209, E-210, E-211, E-212, E-213, E-214, E-216, and E-217 in the* 2015 IBC Code Changes Resource Collection *for the complete text and history of the code change.*)

CHANGE SIGNIFICANCE: In order to coordinate with the ADA, the scoping for a variety of recreational items has been relocated, revised or added to a new Section 1110 for recreational facilities. The overall intent is to provide access to recreational facilities so that persons with mobility impairments can participate to the best of their ability. The requirements are not intended to change any essential aspects of that recreational activity. Users should be somewhat familiar with these requirements because they are currently required by the ADA and will provide more detailed scoping and guidance as to the amount and level of access that is required by the technical requirements found within Chapter 11 of the A117.1 standard. Because the A117.1 standard was previously modified to coordinate with the ADA recreational requirements, additional information and discussion can be found by reviewing the requirements for A117.1 Chapter 11 in *Significant Changes to the A117.1 Accessibility Standard, 2009 Edition*.

The revisions for Section 1110 are a compilation of several code changes that addressed the various general requirements, as well as the individual changes, that affected each of the various types of recreational facilities. For a more complete analysis, refer to code changes E208 (general section format and reorganization), E209 (area of sport activity), E210 (animal containment areas), E211 (amusement rides), E212 (recreational boating), E213 (exercise machines and equipment), E214 (miniature golf facilities), E215 (swimming pools, wading pools, hot tubs and spas), and E217 (shooting facilities with firing positions) on the ICC website and in the *2015 IBC Code Changes Resource Collection*. For additional information and guidance, code users should also review the materials available from the Department of Justice and from the U.S. Access Board (www.access-board.gov) because the new recreation provisions are generally intended to coordinate with the ADA requirements.

Code changes related to the recreational provisions can also be found within the *International Existing Building Code* (IEBC) and within Appendix E of the IBC.

PART 6

Building Envelope, Structural Systems, and Construction Materials

Chapters 12 through 26

- **Chapter 12** Interior Environment No changes addressed
- **Chapter 13** Energy Efficiency No changes addressed
- **Chapter 14** Exterior Walls
- **Chapter 15** Roof Assemblies and Rooftop Structures No changes addressed
- **Chapter 16** Structural Design
- **Chapter 17** Structural Tests and Special Inspections
- **Chapter 18** Soils and Foundations
- **Chapter 19** Concrete
- **Chapter 20** Aluminum No changes addressed
- **Chapter 21** Masonry
- **Chapter 22** Steel
- **Chapter 23** Wood
- **Chapter 24** Glass and Glazing
- **Chapter 25** Gypsum Board and Plaster
- **Chapter 26** Plastic

The interior environment provisions of Chapter 12 include requirements for lighting, ventilation, and sound transmission. Chapter 13 provides a reference to the *International Energy Conservation Code* for provisions governing energy efficiency. Regulations governing the building envelope are located in Chapters 14 and 15, addressing exterior wall coverings and roof coverings, respectively. Structural systems are regulated through the structural design provisions of Chapter 16, whereas structural testing and special inspections are addressed in Chapter 17. The provisions of Chapter 18 apply to soils and foundation systems. The requirements for materials of construction, both structural and non-structural, are located in Chapters 19 through 26. Structural materials regulated by the code include concrete, lightweight metals, masonry, steel, and wood. Glass and glazing, gypsum board, plaster, and plastics are included as regulated non-structural materials. ■

1405.3
Vapor Retarders

1602.1
Definitions and Notations

1603
Construction Documents

1603.1.7
Flood Design Data

1603.1.8
Special Loads

1604.3
Serviceability

1604.5
Risk Category

1607.5
Partition Loads

1607.9

Impact Loads for Façade Access Equipment

1607.10.2

Alternative Uniform Live Load Reduction

1607.12

Roof Loads

1607.12.5

Photovoltaic Panel Systems

1609.1.1

Determination of Wind Loads

1613.3.1

Mapped Acceleration Parameters

1613.5

Amendments to ASCE 7

1613.6

Ballasted Photovoltaic Panel Systems

1704.5

Submittals to the Building Official

1705.2

Steel Construction

1705.2.3

Open Web Steel Joists and Joist Girders

TABLE 1705.3

Required Special Inspections of Concrete Construction

1705.11

Special Inspection for Wind Resistance

1705.12

Special Inspection for Seismic Resistance

1708.3.2

Static Load Testing

1709.5

Exterior Window and Door Assemblies

1711

Material and Test Standards

1803.5

Investigated Conditions

1804.1

Excavation Near Foundations

1808.3

Design Surcharge Loads

1810.2.5

Group Effects

1810.3

Design and Detailing

1901.3

Anchoring to Concrete

1901.4

Composite Structural Steel and Concrete Structures

1904

Durability Requirements

1905.1.3

Modifications to ACI 318, Section 18.5

1905.1.8

Modifications to ACI 318, Section 17.2.3

2101.2

Masonry Design Methods

2103

Masonry Construction Materials

2104

Masonry Construction

2105

Quality Assurance

2111, 2113

Masonry Fireplaces and Chimneys

2210

Cold-Formed Steel

2211

Cold-Formed Steel Light-Frame Construction

continues

2303.1.4

Structural Glued Cross-Laminated Timber

2303.1.13

Engineered Wood Rim Board

2304.6

Exterior Wall Sheathing

2304.10.6

Load Path

2304.12

Protection Against Decay and Termites

2308

Conventional Light-Frame Construction

2308.2.5

Allowable Roof Span

2308.7

Roof and Ceiling Framing

2309

Wood Frame Construction Manual

2406.4.7

Safety Glazing Adjacent to Bottom Stair Landing

CHAPTER 25

Gypsum Panel Products

2612

Plastic Composites

1405.3

Vapor Retarders

CHANGE TYPE: Modification

CHANGE SUMMARY: The required types and locations appropriate for each class of vapor retarder have been revised to also indicate where certain vapor retarders are not allowed to be installed.

2015 CODE: **1405.3 Vapor Retarders.** Vapor retarders as described in Section 1405.3.3 shall be provided in accordance with Sections 1405.3.1 and 1405.3.2, or an approved design using accepted engineering practice for hygrothermal analysis.

~~**1405.3**~~ **1405.3.1 Class I and II Vapor Retarders.** Class I and II vapor retarders shall not be provided on the interior side of frame walls in Zones 1 and 2. Class I vapor retarders shall not be provided on the interior side of frame walls in Zones 3 and 4. Class I or II vapor retarders shall be provided on the interior side of frame walls in Zones 5, 6, 7, 8 and Marine 4. The appropriate zone shall be selected in accordance with Chapter 3 of the *International Energy Conservation Code*.

Exceptions:

1. Basement walls.
2. Below-grade portion of any wall.
3. Construction where moisture or its freezing will not damage the materials.
4. Conditions where Class III vapor retarders are required in Section 1405.3.2.

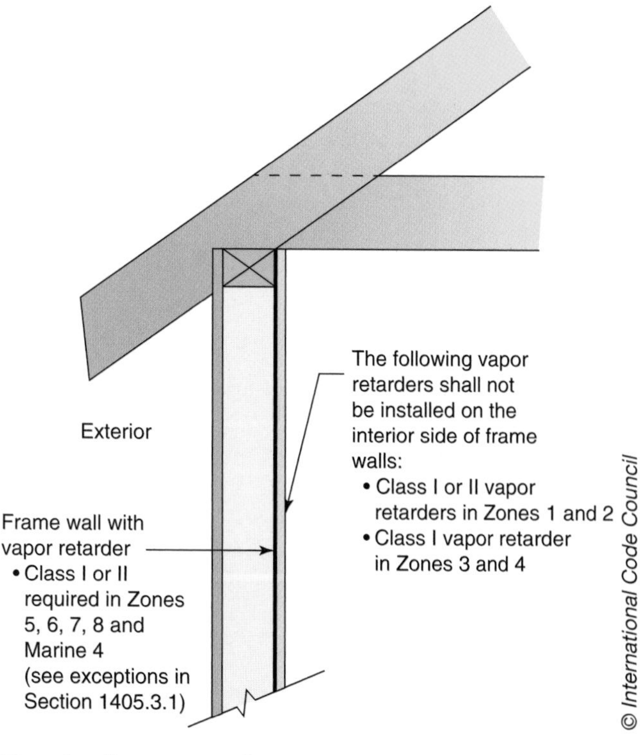
Exterior
Frame wall with vapor retarder
• Class I or II required in Zones 5, 6, 7, 8 and Marine 4 (see exceptions in Section 1405.3.1)
The following vapor retarders shall not be installed on the interior side of frame walls:
• Class I or II vapor retarders in Zones 1 and 2
• Class I vapor retarder in Zones 3 and 4
© International Code Council

Class I or II vapor retarders

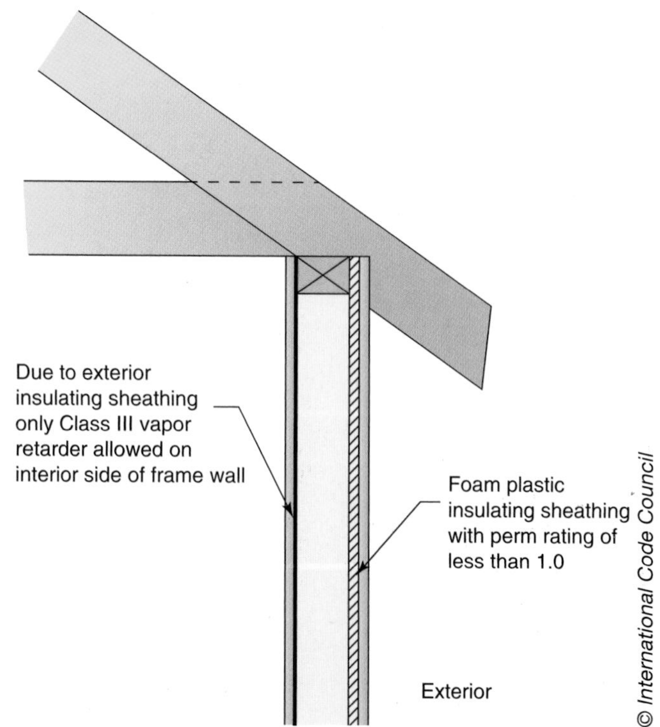
Due to exterior insulating sheathing only Class III vapor retarder allowed on interior side of frame wall
Foam plastic insulating sheathing with perm rating of less than 1.0
Exterior
© International Code Council

Class III vapor retarder

~~1405.3.1~~ 1405.3.2 Class III Vapor Retarders. Class III vapor retarders shall be permitted where any one of the conditions in Table ~~1405.3.1~~ 1405.3.2 is met. Only Class III vapor retarders shall be used on the interior side of frame walls where foam plastic insulating sheathing with perm rating of less than 1 is applied in accordance with Table 1405.3.2 on the exterior side of the frame wall.

TABLE ~~1405.3.1~~ 1405.3.2 Class III Vapor Retarders

(No changes to Table or footnote.)

~~1405.3.2~~ 1405.3.3 Material Vapor Retarder Class. The vapor retarder class shall be based on the manufacturer's certified testing or a tested assembly. The following shall be deemed to meet the class specified:

Class I: Sheet polyethylene, nonperforated aluminum foil with a perm rating of less than or equal to 0.1.

Class II: Kraft-faced fiberglass batts or paint with a perm rating greater than 0.1 and less than or equal to 1.0.

Class III: Latex or enamel paint with a perm rating of greater than 1.0 and less than or equal to 10.0.

CHANGE SIGNIFICANCE: Moisture being trapped within the walls of a structure or being allowed to travel through an assembly where it can condense within the wall continues to create problems. The revised provisions not only provide specific details regarding the type of vapor retarder required within each climate zone, but also indicate which classes of systems are not to be installed. Situations that contribute to these moisture problems can be eliminated by addressing systems that effectively create a double vapor barrier where insulating sheathing is used on the exterior side of the wall in cold climates, or where low-permeability vapor retarders are installed on the interior side in warmer climates.

Section 1405.3.1 now specifically prohibits the use of certain low-permeability vapor retarders within the warmer climate zones. By prohibiting the Class I and II retarders on the interior side of walls in Climate Zones 1 and 2, and the Class I retarders in Climate Zones 3 and 4, the code helps to address the problem where warmer moist air from the exterior enters into the wall and then is trapped and condenses against the cooler interior side of the wall. It should be noted that the provision does not prohibit these particular vapor retarders in the specified climate zones, but that it is simply mandating that they are not placed on the interior side of the wall. If the better vapor retarders are installed on the interior side of the wall in the warmer climates, a reversed vapor retarder is essentially created, which can lead to the moisture being trapped and condensed within the wall. As expected, the intent is to keep the moisture from entering the wall so it cannot come into contact with the cool side of the wall, and to also allow the wall to dry as needed.

The new provisions in Section 1405.3.2 related to the use of Class III vapor retarders are intended to avoid situations where the wall would have effective vapor retarders on both sides of the wall and could therefore trap the moisture within the wall. Where foam plastic insulated sheathing is installed on the exterior side of a wall, it creates an effective vapor retarder on that surface and creates a "warm wall." If a Class I or II

1405.3 continues

1405.3 continued

vapor retarder is then added to the interior side of the wall, it creates the double vapor barrier system and could lead to problems. Therefore, only Class III vapor retarders with a perm rating greater than 1 and no more than 10 are permitted on the interior side of the wall so that the moisture can escape back into the interior of the building.

Section 1405.3.3 now provides better clarification as to the types of vapor retarders by specifically including a perm rating with each class in addition to a general description of some of the materials. Having a perm rating specified will make determination of the appropriate classification easier.

1602.1

Definitions and Notations

CHANGE TYPE: Modification

CHANGE SUMMARY: The definitions of "flexible diaphragm" and "rigid diaphragm" have been deleted from the code and replaced by a reference to the procedures for classifying diaphragms in the 2010 edition of the national load standard, *Minimum Design Loads for Buildings and Other Structures* (ASCE/SEI 7-10).

2015 CODE:

SECTION 202
DEFINITIONS

DIAPHRAGM. A horizontal or sloped system acting to transmit lateral forces to ~~the~~ vertical~~-resisting~~ elements of the lateral force-resisting system. When the term "diaphragm" is used, it shall include horizontal bracing systems.

~~**DIAPHRAGM, FLEXIBLE.** A diaphragm is flexible for the purpose of distribution of story shear and torsional moment where so indicated in Section 12.3.1 of ASCE 7.~~

~~**DIAPHRAGM, RIGID.** A diaphragm is rigid for the purpose of distribution of story shear and torsional moment when the lateral deformation of the diaphragm is less than or equal to two times the average story drift.~~

SECTION 1602
DEFINITIONS AND NOTATIONS

1602.1 Definitions. The following terms are defined in Chapter 2:

DIAPHRAGM.
Diaphragm, Blocked.
Diaphragm Boundary.
Diaphragm Chord.
~~**Diaphragm, Flexible.**~~
~~**Diaphragm, Rigid.**~~
Diaphragm, Unblocked.

(Portions of text not shown remain unchanged.)

1602.1 continues

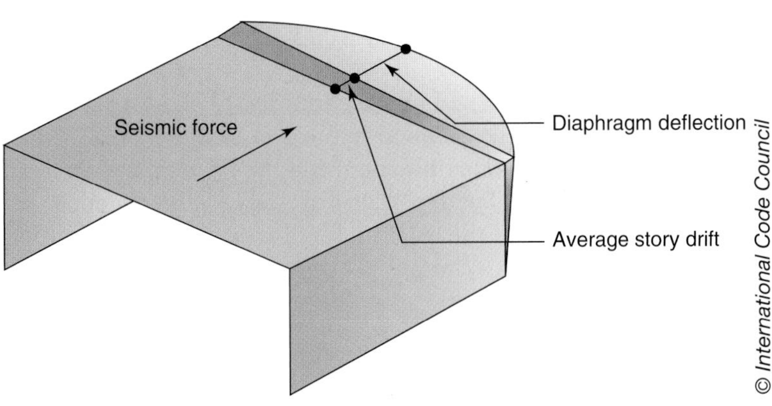

Flexible diaphragm

1602.1 continued

1604.4 Analysis. Load effects on structural members and their connections shall be determined by methods of structural analysis that take into account equilibrium, general stability, geometric compatibility and both short- and long-term material properties.

Members that tend to accumulate residual deformations under repeated service loads shall have included in their analysis the added eccentricities expected to occur during their service life.

Any system or method of construction to be used shall be based on a rational analysis in accordance with well-established principles of mechanics. Such analysis shall result in a system that provides a complete load path capable of transferring loads from their point of origin to the load-resisting elements.

The total lateral force shall be distributed to the various vertical elements of the lateral force-resisting system in proportion to their rigidities, considering the rigidity of the horizontal bracing system or diaphragm. Rigid elements assumed not to be a part of the lateral force-resisting system are permitted to be incorporated into buildings provided their effect on the action of the system is considered and provided for in the design. A diaphragm is rigid for the purpose of distribution of story shear and torsional moment when the lateral deformation of the diaphragm is less than or equal to two times the average story drift. Where required by ASCE 7, ~~Except where diaphragms are flexible, or are permitted to be analyzed as flexible,~~ provisions shall be made for the increased forces induced on resisting elements of the structural system resulting from torsion due to eccentricity between the center of application of the lateral forces and the center of rigidity of the lateral force-resisting system.

Every structure shall be designed to resist the overturning effects caused by the lateral forces specified in this chapter. See Section 1609 for wind loads, Section 1610 for lateral soil loads and Section 1613 for earthquake loads.

1613.3.5.1 Alternative Seismic Design Category Determination. Where S_1 is less than 0.75, the seismic design category is permitted to be determined from Table 1613.3.5(1) alone when all of the following apply:

1. In each of the two orthogonal directions, the approximate fundamental period of the structure, T_a, in each of the two orthogonal directions determined in accordance with Section 12.8.2.1 of ASCE 7, is less than 0.8 T_s determined in accordance with Section 11.4.5 of ASCE 7.
2. In each of the two orthogonal directions, the fundamental period of the structure used to calculate the story drift is less than T_s.
3. Equation 12.8-2 of ASCE 7 is used to determine the seismic response coefficient, C_S.
4. The diaphragms are rigid or are permitted to be idealized as rigid in accordance with ~~as defined in~~ Section 12.3.1 of ASCE 7 or, for diaphragms that are ~~flexible,~~ permitted to be idealized as flexible in accordance with Section 12.3.1 of ASCE 7, the distances between vertical elements of the seismic force-resisting system do not exceed 40 feet (12 192 mm).

CHANGE SIGNIFICANCE: Potential conflicts between the definitions related to flexible diaphragms and rigid diaphragms in the IBC and referenced standard, ASCE 7-10, have been resolved. It was deemed inappropriate to include enforceable code requirements or references to standards as part of a definition. In actual practice, the code user must use the requirements of ASCE-7 to categorize the diaphragm for distribution of seismic forces, so there is no need or advantage to include the diaphragm definitions in the IBC. Thus, the definitions for flexible and rigid diaphragms were deleted from the IBC. Chapter 16 of the IBC contains sufficient references to the seismic and wind provisions in ASCE 7 to determine when diaphragms can be idealized as flexible, rigid or semi-rigid. In addition to deleting the definitions, the criteria pertaining to rigid diaphragms have been added and the reference to flexible diaphragms has been deleted from Section 1604.4, Analysis. A relatively minor modification in language is made to Section 1613.3.5.1 as it relates to alternative Seismic Design Category (SDC) determination.

1603

Construction Documents

CHANGE TYPE: Modification

CHANGE SUMMARY: Two additional items related to snow load drifting are now required to be identified on the construction documents.

2015 CODE: **1603.1.3 Roof Snow Load Data.** The ground snow load, P_g, shall be indicated. In areas where the ground snow load, P_g, exceeds 10 pounds per square foot (psf) (0.479 kN/m^2), the following additional information shall also be provided, regardless of whether snow loads govern the design of the roof:

1. Flat-roof snow load, P_f.
2. Snow exposure factor, C_e.
3. Snow load importance factor, I.
4. Thermal factor, C_t.
5. Drift surcharge loads, p_d, where the sum of p_d and P_f exceeds 20 psf (0.96 kN/m^2).
6. Width of snow drift(s), w.

CHANGE SIGNIFICANCE: Section 1603 requires specific structural-related information to be shown on the construction documents. Section 107 contains administrative provisions for submittal of documents that include construction documents as defined in Section 202. Including design assumptions and loading information on construction drawings is valuable to the owners and engineers who are tasked with evaluating or reevaluating existing structures. The two additional items pertaining to snow drifts supplement the snow load information already required and clarify how the registered design professional interpreted the design requirements related to snow drifting. In addition to other criteria, Section 1603 now requires snow drift surcharge and drift width to be indicated on the construction documents where applicable.

Photo Courtesy of Peter Kulczyk

Snow drift on roof

1603.1.7

Flood Design Data

CHANGE TYPE: Clarification

CHANGE SUMMARY: The term "subject to high-velocity wave action" in regard to flood hazard areas has been replaced with "coastal high hazard areas" in several chapters and sections of the code.

2015 CODE: **1603.1.7 Flood Design Data.** For buildings located in whole or in part in flood hazard areas as established in Section 1612.3, the documentation pertaining to design, if required in Section 1612.5, shall be included and the following information, referenced to the datum on the community's Flood Insurance Rate Map (FIRM), shall be shown, regardless of whether flood loads govern the design of the building:

1. Flood design class assigned according to ASCE 24.

~~1~~2. In flood hazard areas ~~not subject to high-velocity wave action~~ other than coastal high hazard areas or coastal A zones, the elevation of the proposed lowest floor, including the basement.

~~2~~3. In flood hazard areas ~~not subject to high-velocity wave action~~ other than coastal high hazard areas or coastal A zones, the elevation to which any nonresidential building will be dry flood proofed.

~~3~~4. In ~~flood hazard areas subject to high-velocity wave action~~ coastal high hazard areas and coastal A zones, the proposed elevation of the bottom of the lowest horizontal structural member of the lowest floor, including the basement.

1603.1.7 continues

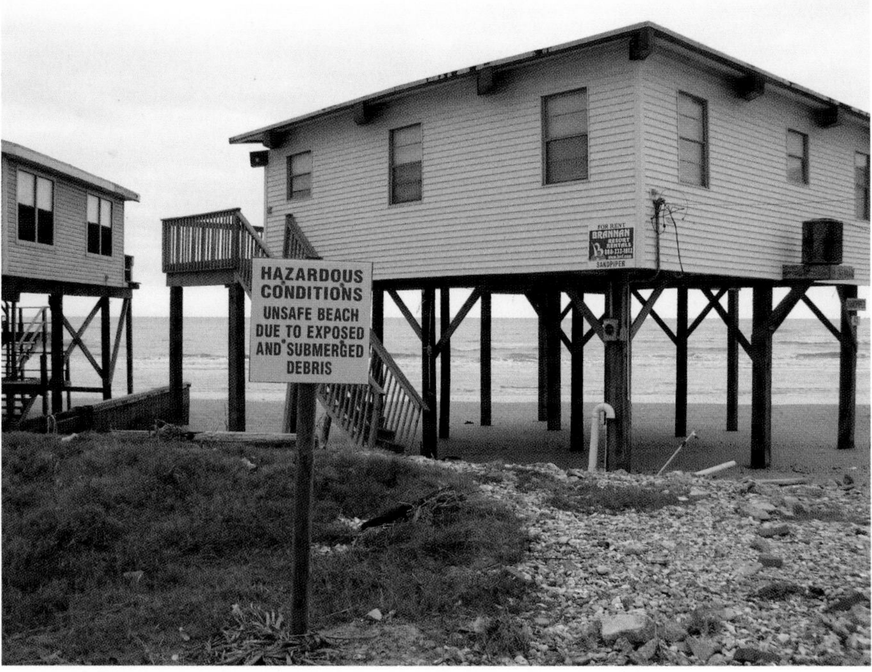

Photo Courtesy of NOAA

Coastal high hazard area

1603.1.7 continued

(Because this code change revised several chapters and sections of the code, the entire code change text is too extensive to be included here. Refer to Code Change S103-12 in the 2015 IBC Code Changes Resource Collection *for the complete text and history of the code change.)*

CHANGE SIGNIFICANCE: In various provisions of the code involving flood hazard areas, the term "subject to high-velocity wave action" has been replaced with "coastal high hazard areas" to be consistent with the terminology used in ASCE 24, *Flood Resistant Design and Construction,* which is referenced in Section 1612.4. The term "flood hazard area subject to high-velocity wave action" describes flood hazard areas designated as Zone V on Flood Insurance Rate Maps (FIRM), but that term is not used by the National Flood Insurance Program (NFIP), ASCE 24 or the IRC. The NFIP regulations define "coastal high hazard area" in the Code of Federal Regulations, 40 CFR 59.1.

1603.1.8

Special Loads

CHANGE TYPE: Addition

CHANGE SUMMARY: The dead load of any rooftop-mounted photovoltaic (PV) solar panels must now be identified on the construction documents.

2015 CODE: 1603.1.8 Special Loads. Special loads that are applicable to the design of the building, structure or portions thereof shall be indicated along with the specified section of this code that addresses the special loading condition.

1603.1.8.1 Photovoltaic Panel Systems. The dead load of rooftop-mounted photovoltaic panel systems, including rack support systems, shall be indicated on the construction documents.

CHANGE SIGNIFICANCE: With the increasing use of photovoltaic (solar) panels on rooftops, it has become clear that there is a need for specific code requirements related to these panels. Such panels are considered fixed equipment and as such would fall under the definition of dead load. Rooftop PV panels are considered special loads and their dead load should be indicated on the construction documents. The code language was revised to clarify that the provisions apply to roof loads for the design of the roof structure, not to the design of the photovoltaic panels or modules themselves. A new definition of "photovoltaic panel system" has been added to Section 202, and new Section 1607.12.5 provides loading requirements for roof structures supporting PV panel systems.

Photo Courtesy of Peter Kulczyk

Rooftop-mounted solar panels

1604.3

Serviceability

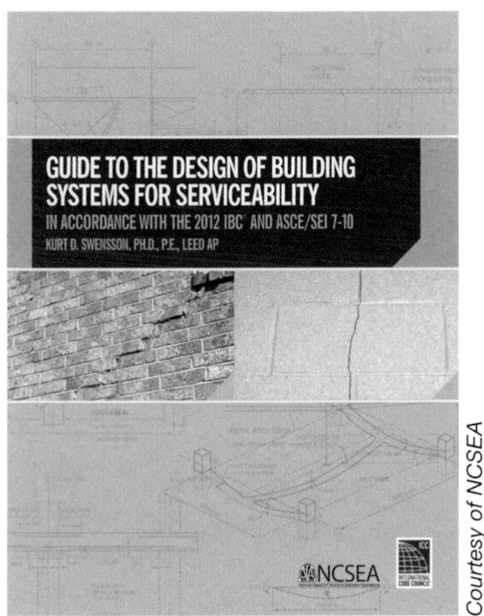

Courtesy of NCSEA

Guide to the Design of Building Systems for Serviceability

CHANGE TYPE: Modification

CHANGE SUMMARY: Modifications have been made to the deflection limits established in Table 1604.3 for interior partitions, wood members, and wind loads to both clarify and update the provisions.

2015 CODE: 1604.3 Serviceability. Structural systems and members thereof shall be designed to have adequate stiffness to limit deflections and lateral drift. See Section 12.12.1 of ASCE 7 for drift limits applicable to earthquake loading.

1604.3.1 Deflections. The deflections of structural members shall not exceed the more restrictive of the limitations of Sections 1604.3.2 through 1604.3.5 or that permitted by Table 1604.3.

1604.3.2 Reinforced Concrete. The deflection of reinforced concrete structural members shall not exceed that permitted by ACI 318.

1604.3.3 Steel. The deflection of steel structural members shall not exceed that permitted by AISC 360, AISI S100, ASCE 8, SJI CJ~~-1.0~~, SJI JG~~-1.1~~, SJI K~~-1.1~~ or SJI LH/DLH~~-1.1~~, as applicable.

TABLE 1604.3 Deflection Limits[a, b, c, h, i]

Construction	L	S or W^f	$D + L^{d,g}$
Roof members:[e]			
Supporting plaster or stucco ceiling	l/360	l/360	l/240
Supporting nonplaster ceiling	l/240	l/240	l/180
Not supporting ceiling	l/180	l/180	l/120
Floor members	l/360	—	l/240
Exterior walls ~~and interior partitions~~:			
With plaster or stucco finishes	—	l/360	—
With other brittle finishes	—	l/240	—
With flexible finishes	—	l/120	—
Interior Partitions:[b]			
With plaster or stucco finishes	l/360	—	—
With other brittle finishes	l/240	—	—
With flexible finishes	l/120	—	—
Farm buildings	—	—	l/180
Greenhouses	—	—	l/120

For SI: 1 foot = 304.8 mm.

a. For structural roofing and siding made of formed metal sheets, the total load deflection shall not exceed l/60. For secondary roof structural members supporting formed metal roofing, the live load deflection shall not exceed l/150. For secondary wall members supporting formed metal siding, the design wind load deflection shall not exceed l/90. For roofs, this exception only applies when the metal sheets have no roof covering.

b. ~~Interior partitions not exceeding 6ft in height and~~ Flexible, folding and portable partitions are not governed by the provisions of this section. The deflection criterion for interior partitions is based on the horizontal load defined in Section 1607.14.

c. See Section 2403 for glass supports.

d. For wood structural members having a moisture content of less than 16 percent at time of installation and used under dry conditions, the deflection resulting from $L + 0.5D$ is permitted to be substituted for the deflection resulting from $L + D$.
d. The deflection limit for the $D+L$ load combination only applies to the deflection due to the creep component of long-term dead load deflection plus the short-term live load deflection. For wood structural members that are dry at time of installation and used under dry conditions in accordance with the AWC NDS, the creep component of the long-term deflection shall be permitted to be estimated as the immediate dead load deflection resulting from $0.5D$. For wood structural members at all other moisture conditions, the creep component of the long-term deflection is permitted to be estimated as the immediate dead load deflection resulting from D. The value of $0.5D$ shall not be used in combination with AWC NDS provisions for long-term loading.
e. The above deflections do not ensure against ponding. Roofs that do not have sufficient slope or camber to assure adequate drainage shall be investigated for ponding. See Section 1611 for rain and ponding requirements and Section 1503.4 for roof drainage requirements.
f. The wind load is permitted to be taken as 0.42 times the "component and cladding" loads for the purpose of determining deflection limits herein. Where members support glass in accordance with Section 2403 using the deflection limit therein, the wind load shall be no less than 0.6 times the "component and cladding" loads for the purpose of determining deflection.
g. For steel structural members, the dead load shall be taken as zero.
h. For aluminum structural members or aluminum panels used in skylights and sloped glazing framing, roofs or walls of sunroom additions or patio covers, not supporting edge of glass or aluminum sandwich panels, the total load deflection shall not exceed $1/60$. For continuous aluminum structural members supporting edge of glass, the total load deflection shall not exceed $1/175$ for each glass lite or $1/60$ for the entire length of the member, whichever is more stringent. For aluminum sandwich panels used in roofs or walls of sunroom additions or patio covers, the total load deflection shall not exceed $1/120$.
i. For cantilever members, 1 shall be taken as twice the length of the cantilever.

CHANGE SIGNIFICANCE: Various changes have been made to Table 1604.3 dealing with allowable deflection limits. A new group of requirements pertaining to deflection of interior partitions has been added to the table, and the reference to interior partitions was deleted from the group of requirements for exterior walls, allowing the deflection criteria to be correctly based on live load. The 5-psf live load for interior walls and partitions is specified in Section 1607.14. This clarifies that deflection limits do apply to interior walls and provides the applicable criteria of 1/360, 1/240 or 1/120 depending on the type of finish material. Brittle finishes require higher stiffness, which is accomplished by stricter deflection limits. The reference to interior partitions previously found in footnote b was deleted because the 6-foot height threshold is already included in Section 1607.14.

Footnote d relating to deflection of wood members originated with the 1964 UBC, which used a K factor of 0.5 applied to the dead load for "seasoned" wood, indicating it had a moisture content of less than 16 percent at the time of installation and was used in dry conditions in service. The footnote was subsequently carried over into the IBC but the underlying intent was not entirely clear. The purpose of footnote d is to limit the total deflection based on the combination of live load deflection and a creep component of the dead load deflection. As previously written, the footnote could be misinterpreted to mean the creep component of dead load deflection is calculated using AWC NDS provisions and the reduced dead load of $0.5D$. The deflection limit for the $D+L$ load combination only applies to the deflection due to the creep component of long-term dead load deflection plus the short-term live load deflection. For wood

1604.3 continues

1604.3 continued

structural members that are dry at time of installation and used under dry conditions in accordance with the AWC NDS, the creep component of the long-term deflection is estimated as the immediate dead load deflection resulting from 0.5*D*. For wood structural members at all other moisture conditions, the creep component of the long-term deflection is estimated as the immediate dead load deflection resulting from *D*. The value of 0.5*D* is not used in combination with AWC NDS provisions for long-term loading. The revised footnote d ensures calculation of *D*+*L* deflection for comparison with the limit in Table 1604.3 is consistent with the provisions in NDS Section 3.5.2 for long-term loading and consistent with similar provisions in ACI 318 as described in the ACI 318 Commentary. Note that for steel members, the dead load is taken as zero because steel does not exhibit a long-term (creep) component of deflection as do concrete and wood.

Footnote f was modified in the 2012 IBC to be 0.42 ($0.7 \times 0.6 = 0.42$) because serviceability (deflection) calculations are done at an allowable stress design level and wind pressures in ASCE 7-10 are ultimate loads with a wind load factor of 0.6 for allowable stress design. The newly added sentence to footnote f ensures that for design of members supporting glass in accordance with Section 2403 using the deflection limits specified therein, the wind load is 0.6 times the "component and cladding" loads for determining deflection.

1604.5

Risk Category

CHANGE TYPE: Clarification

CHANGE SUMMARY: In the application of assigning the appropriate risk category for a structure, Section 1604.5 has been revised to clarify that where standards refer to ASCE 7 Table 1.5-1, IBC Table 1604.5 should be used instead. In addition, descriptions for Risk Category III structures have been revised to include occupancy classifications to help clarify the intent.

2015 CODE: 1604.5 Risk Category. Each building and structure shall be assigned a risk category in accordance with Table 1604.5. Where a referenced standard specifies an occupancy category, the risk category shall not be taken as lower than the occupancy category specified therein. Where a referenced standard specifies that the assignment of a risk category be in accordance with ASCE 7, Table 1.5-1, Table 1604.5 shall be used in lieu of ASCE 7, Table 1.5-1.

1604.5.1 Multiple Occupancies. Where a building or structure is occupied by two or more occupancies not included in the same risk category, it shall be assigned the classification of the highest risk category corresponding to the various occupancies. Where buildings or structures have two or more portions that are structurally separated, each portion shall be separately classified. Where a separated portion of a building or structure provides required access to, required egress from or shares life safety components with another portion having a higher risk category, both portions shall be assigned to the higher risk category.

TABLE 1604.5 Risk Category of Buildings and Other Structures

Risk Category	Nature of Occupancy
I	Buildings and other structures that represent a low hazard to human life in the event of failure, including but not limited to: • Agricultural facilities. • Certain temporary facilities. • Minor storage facilities.
II	Buildings and other structures except those listed in Risk Categories I, III and IV
III	Buildings and other structures that represent a substantial hazard to human life in the event of failure, including but not limited to: • Buildings and other structures whose primary occupancy is public assembly with an occupant load greater than 300. • Buildings and other structures containing ~~elementary school, secondary school or day care facilities~~ Group E occupancies with an occupant load greater than 250. • Buildings and other structures containing ~~adult education facilities, such as colleges and universities,~~ educational occupancies for students above the 12th grade with an occupant load greater than 500. • Group I-2 occupancies with an occupant load of 50 or more resident care recipients but not having surgery or emergency treatment facilities. • Group I-3 occupancies. • Any other occupancy with an occupant load greater than 5,000[a]. • Power-generating stations, water treatment facilities for potable water, waste water treatment facilities and other public utility facilities not included in Risk Category IV. • Buildings and other structures not included in Risk Category IV containing quantities of toxic or explosive materials that: Exceed maximum allowable quantities per control area as given in Table 307.1(1) or 307.1(2) or per outdoor control area in accordance with the *International Fire Code*; and Are sufficient to pose a threat to the public if released[b].

1604.5 continues

1604.5 continued

TABLE 1604.5 Risk Category of Buildings and Other Structures

Risk Category	Nature of Occupancy
IV	Buildings and other structures designated as essential facilities, including but not limited to: • Group I-2 occupancies having surgery or emergency treatment facilities. • Fire, rescue, ambulance and police stations and emergency vehicle garages. • Designated earthquake, hurricane or other emergency shelters. • Designated emergency preparedness, communications and operations centers and other facilities required for emergency response. • Power-generating stations and other public utility facilities required as emergency backup facilities for Risk Category IV structures. • Buildings and other structures containing quantities of highly toxic materials that: Exceed maximum allowable quantities per control area as given in Table 307.1(2) or per outdoor control area in accordance with the *International Fire Code*; and Are sufficient to pose a threat to the public if released[b]. • Aviation control towers, air traffic control centers and emergency aircraft hangars. • Buildings and other structures having critical national defense functions. • Water storage facilities and pump structures required to maintain water pressure for fire suppression.

a. For purposes of occupant load calculation, occupancies required by Table 1004.1.2 to use gross floor area calculations shall be permitted to use net floor areas to determine the total occupant load.

b. Where approved by the building official, the classification of buildings and other structures as Risk Category III or IV based on their quantities of toxic, highly toxic or explosive materials is permitted to be reduced to Risk Category II, provided it can be demonstrated by a hazard assessment in accordance with Section 1.5.3 of ASCE 7 that a release of the toxic, highly toxic or explosive materials is not sufficient to pose a threat to the public.

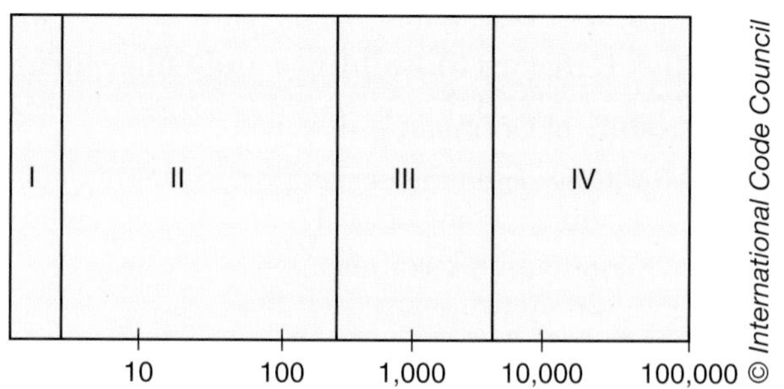

Approximate relationship between number of lives placed at risk by a failure and occupancy category per ASCE 7

CHANGE SIGNIFICANCE: IBC Table 1604.5 and ASCE 7 Table 1.5-1 both contain descriptions used to determine the risk category of structures, but these descriptions are not always identical. According to Section 102.4.1, where there is a conflict between the provisions of the IBC and the provisions in a referenced standard, the requirements of the code govern. In some cases, other referenced standards refer to ASCE 7 Table 1.5-1 to determine risk category, and code users may be unaware that IBC Table 1604.5 should be used instead. Section 1604.5 is revised to clarify that where a referenced standard specifies that the assignment of risk category is to be in accordance with ASCE 7 Table 1.5-1, IBC Table 1604.5 should be used instead.

Two of the descriptions in Table 1604.5 for Risk Category III have been modified to refer specifically to the occupancy classifications from IBC Chapter 3 to clarify the intent and facilitate proper application. The term "elementary school, secondary school or day care facilities" has been replaced with a simple reference to Group E occupancies, and the term "adult education facilities, such as colleges and universities" has been replaced with "educational occupancies for students above the 12th grade." This change establishes a specific threshold for student classification and clarifies that trade schools are included.

1607.5

Partition Loads

CHANGE TYPE: Modification

CHANGE SUMMARY: In office buildings and in other buildings where the location of partitions is subject to change, partition loads are to be considered unless the floor is designed for an 80-psf or greater live load.

2015 CODE: 1607.5 Partition Loads. In office buildings and in other buildings where partition locations are subject to change, provisions for partition weight shall be made, whether or not partitions are shown on the construction documents, unless the specified live load ~~exceeds~~ is 80 psf (3.83 kN/m^2) or greater. The partition load shall not be less than a uniformly distributed live load of 15 psf (0.72 kN/m^2).

CHANGE SIGNIFICANCE: The IBC has previously required floors in office buildings and other buildings where partition locations are subject to change to be designed to include the partition live load unless the floor was designed for a live load that exceeds 80 psf. Item 22 of Table 1607.1 establishes a minimum live load of 80 psf for corridors above the first story. Even though office floors have a minimum required live load of 50 psf, it is a common practice to design the entire upper floors for a live load of 80 psf so there is no concern about future changes in the locations of the corridors. In the 2012 IBC, a floor would have to be designed for a live load *greater than 80 psf* to take advantage of the exception in regard to ignoring partition loads. Otherwise, one would have to include a 15-psf partition live load in addition to the 80-psf corridor live load. Due to the new modification, the additional partition load need not be included if the floor is designed for 80 psf or greater. With this change, the IBC is no longer consistent with Section 4.3.2 of ASCE 7-10, which has an exception stating that the partition live load is not required where the live load *exceeds* 80 psf. In addition, Section 12.7.2 of ASCE 7 requires the effective seismic weight, W, to include 10 psf where the partition load of Section 4.2.2 applies. With this change to Section 1607.5, the IBC no longer requires the partition load if the floor is designed for 80 psf, which could be interpreted to mean the effective seismic weight need not include the additional 10 psf because the partition gravity load is not required.

Partition walls

1607.9

Impact Loads for Façade Access Equipment

CHANGE TYPE: Addition

CHANGE SUMMARY: Provisions addressing impact loads for elements supporting façade access equipment and lifeline anchorages have been established.

2015 CODE: **1607.9.3 Elements Supporting Hoists for Façade Access Equipment.** In addition to any other applicable live loads, structural elements that support hoists for façade access equipment shall be designed for a live load consisting of the larger of the rated load of the hoist times 2.5 and the stall load of the hoist.

1607.9.4 Lifeline Anchorages for Façade Access Equipment. In addition to any other applicable live loads, lifeline anchorages and structural elements that support lifeline anchorages shall be designed for a live load of at least 3100 pounds (13.8 kN) for each attached lifeline, in every direction that a fall arrest load may be applied.

CHANGE SIGNIFICANCE: Although the federal government has safety requirements for façade access equipment and regulates it through the Occupational Safety and Health Administration (OSHA), the code has historically been silent in this area. Because OSHA requirements are not written in code language or engineering language, new code provisions addressing façade access equipment have been established while maintaining consistency with OSHA requirements for loading on suspended platforms and lifeline anchorages. A design live load of 2.5 times the

1607.9 continues

Suspended façade access platform

1607.9 continued

rated load, when combined with a live load factor of 1.6, results in a total factored load of 4.0 times the rated load, which is consistent with OSHA's requirements for the design of scaffolds used for building maintenance. A design live load of 3100 pounds combined with a live load factor of 1.6 results in a total factored load of 4960 pounds, which is essentially the same as OSHA's requirements for the design of lifeline anchorages. Although these overall factors may seem excessive, they are intended to address accidental hang-up-and-fall scenarios as well as the starting and stopping forces that the platforms typically experience on a day-to-day basis. The provisions also address fall arrest loads that can occur in typical lanyards for body harnesses. OSHA allows stopping forces as high as 2540 pounds to be generated by a person free-falling 6 feet. Actual workers may weigh more than the weight assumed by OSHA, and they may fall more than 6 feet. Because lifeline anchorages are required in case there is a problem with the primary suspension system, the effective factor of safety of two (from a design load of 2540 pounds to an ultimate load of 5000 pounds) is deemed necessary to provide an acceptable level of safety. To address these safety issues, these loading requirements have been added to Section 1607.9, which covers impact load requirements.

1607.10.2

Alternative Uniform Live Load Reduction

CHANGE TYPE: Modification

CHANGE SUMMARY: The alternate live load reduction method has been corrected to be consistent with the original intent as it appeared in the *Uniform Building Code* (UBC).

2015 CODE: **1607.10.2 Alternative Uniform Live Load Reduction.** As an alternative to Section 1607.10.1 and subject to the limitations of Table 1607.1, uniformly distributed live loads are permitted to be reduced in accordance with the following provisions. Such reductions shall apply to slab systems, beams, girders, columns, piers, walls and foundations.

3. For live loads not exceeding 100 psf (4.79 kN/m²), the design live load for any structural member supporting 150 square feet (13.94 m²) or more is permitted to be reduced in accordance with Equation 16-24.

$$R = 0.08(A - 150) \qquad \textbf{(Equation 16-24)}$$

$$\text{For SI: } R = 0.861(A - 13.94)$$

Such reduction shall not exceed the smallest of:

1. 40 percent for <s>horizontal</s> members <u>supporting one floor</u>;
2. 60 percent for <s>vertical</s> members <u>supporting two or more floors</u>; or
3. R as determined by the following equation.

$$R = 23.1(1 + D/L_o) \qquad \textbf{(Equation 16-25)}$$

where:

A = Area of floor supported by the member, square feet (m²).
D = Dead load per square foot (m²) of area supported.
L_o = Unreduced live load per square foot (m²) of area supported.
R = Reduction in percent.

1607.10.2 continues

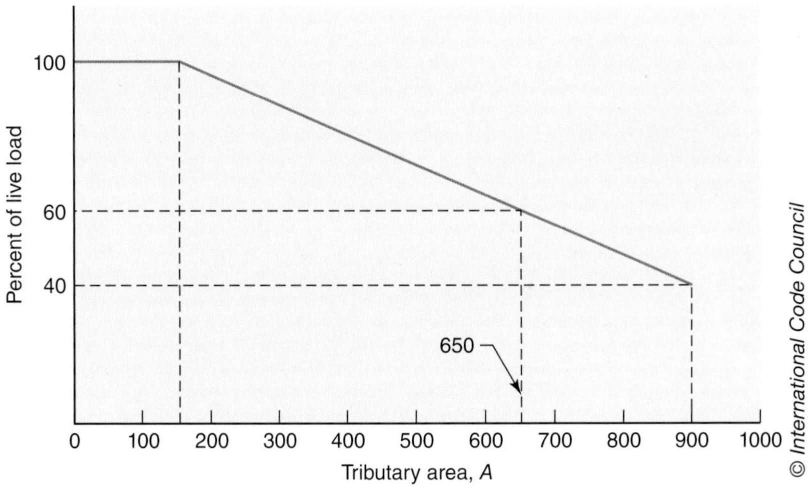

Alternate live load reduction

1607.10.2 continued

CHANGE SIGNIFICANCE: The alternate live load reduction method in Section 1607.9.2 originated with the UBC and was the primary live load reduction formula used in the western United States for decades. When this live load reduction method was incorporated into the 2000 IBC, it was considered as an alternate to the general live load reduction procedure in Section 1607.9.1. During incorporation into the IBC, the maximum reductions were changed from "40 percent for members receiving load from one level only" and "60 percent for other members" as they appeared in the 1997 UBC to "40 percent for horizontal members" and "60 percent for vertical members." This differentiation is inconsistent with the original UBC wording and intent because some horizontal members receive live load from more than one floor and because many vertical elements do not receive live load from more than one floor. Further, it is not consistent with the differentiation presented in Section 1607.9.1, which, like the UBC, differentiates reductions based on whether a member supports one floor or more than one floor. Basing the allowable live load reductions on number of floors supported as opposed to whether a member is horizontal or vertical is appropriate because the premise behind differentiating between supporting one floor or more than one floor is basically probability based, reasonably assuming that the probability of two or more floors having a relatively large live load is smaller than that of a single floor having a relatively large live load—hence the larger live load reduction for elements that support more than one floor. The same argument cannot be made for differentiating live load reductions based on whether or not the element under consideration is oriented horizontally or vertically, which is what was considered based on the previous code text. The change to Section 1607.9.2 restores the original intent of the UBC provision and makes the alternate live load reduction provision more consistent with the general provisions of Section 1607.9.1.

1607.12

Roof Loads

CHANGE TYPE: Addition

CHANGE SUMMARY: The term "vegetative roof" has been defined in Section 202 and a reference to ASTM E 2397 has been added to Section 1607.

2015 CODE:

SECTION 202
DEFINITIONS

VEGETATIVE ROOF. An assembly of interacting components designed to waterproof and normally insulate a building's top surface that includes, by design, vegetation and related landscape elements.

1607.12.3 Occupiable Roofs. Areas of roofs that are occupiable, such as vegetative roofs, roof gardens or for assembly or other similar purposes, and marquees are permitted to have their uniformly distributed live loads reduced in accordance with Section 1607.10.

1607.12.3.1 Vegetative and Landscaped Roofs. ~~The uniform design live load in unoccupied landscaped areas on roofs shall be 20 psf (0.958 kN/m²).~~ The weight of all landscaping materials shall be considered as dead load and shall be computed on the basis of saturation of the soil as determined in accordance with ASTM E 2397. The uniform design live load in unoccupied landscaped areas on roofs shall be 20 psf (0.958 kN/m²). The uniform design live load for occupied landscaped areas on roofs shall be determined in accordance with Table 1607.1.

1607.12 continues

Green roof

1607.12 continued

CHANGE SIGNIFICANCE: The definition of "vegetative roof" has been added to Section 202 to be consistent with the definition in the ICC *International Green Construction Code* and ASTM D1079, *Standard Terminology Relating to Roofing and Waterproofing.* A reference to ASTM E 2397, *Standard Practice for Determination of Dead Loads and Live Loads Associated with Vegetative (Green) Roof Systems,* has been added for the purpose of determining soil loads on vegetative roofs. The provisions pertaining to landscaped roofs have been reorganized, and a reference to Table 1607.1 has been added for live loads on occupiable landscaped areas of roofs.

1607.12.5

Photovoltaic Panel Systems

CHANGE TYPE: Addition

CHANGE SUMMARY: Design requirements for roof structures supporting photovoltaic (PV) solar panels and modules have been added to Section 1607.

2015 CODE: **1607.12.5 Photovoltaic Panel Systems.** Roof structures that provide support for photovoltaic panel systems shall be designed in accordance with Sections 1607.12.5.1 through 1607.12.5.4, as applicable.

1607.12.5.1 Roof Live Load. Roof surfaces to be covered by solar photovoltaic panels or modules shall be designed for the roof live load, L_r, assuming that the photovoltaic panels or module are not present. The roof photovoltaic live load in areas covered by solar photovoltaic panels or modules shall be in addition to the panel loading unless the area covered by each solar photovoltaic panel or module is inaccessible. Areas where the clear space between the panels and the rooftop is 24 inches (610 mm) or less shall be considered inaccessible. Roof surfaces not covered by photovoltaic panels shall be designed for the roof live load.

1607.12.5.2 Photovoltaic Panels or Modules. The structure of a roof that supports solar photovoltaic panels or modules shall be designed to accommodate the full solar photovoltaic panels or modules and ballast dead load, including concentrated loads from support frames in combination with the loads from Section 1607.12.5.1 and other applicable loads. Where applicable, snow drift loads created by the photovoltaic panels or modules shall be included.

1607.12.5 continues

Photo Courtesy of Peter Kulczyk

Roof-mounted photovoltaic panel

1607.12.5 continued

1607.12.5.3 Photovoltaic Panels or Modules Installed as an Independent Structure. Solar photovoltaic panels or modules that are independent structures and do not have accessible/occupied space underneath are not required to accommodate a roof photovoltaic live load, provided the area under the structure is restricted to keep the public away. All other loads and combinations per Section 1605 shall be accommodated.

Solar photovoltaic panels or modules that are designed to be the roof, and span to structural supports, and have accessible/occupied space underneath shall have the panels or modules and all supporting structure designed to support a roof photovoltaic live load, as defined in Section 1607.12.5.1 in combination with other applicable loads. Solar photovoltaic panels or modules in this application are not permitted to be classified as "not accessible" per Section 1607.12.5.1.

1607.12.5.4 Ballasted Photovoltaic Panel Systems. Roof structures that provide support for ballasted photovoltaic panel systems shall be designed, or analyzed, in accordance with Section 1604.4; checked in accordance with Section 1604.3.6 for deflections; and checked in accordance with Section 1611 for ponding.

CHANGE SIGNIFICANCE: With the increasing use of photovoltaic (PV) solar panels and modules mounted on rooftops, there is a need for specific design loading requirements related to these panels. Roof surfaces that are covered by solar PV panels must be designed for the roof live load, L_r, in addition to the panel loading unless the area covered by each solar PV panel or module is considered inaccessible. Areas where the clear space between the panels and the rooftop is 24 inches or less are considered inaccessible. The structure of a roof that supports solar panels must be designed for the full panel and ballast dead load, including concentrated loads from support frames in combination with the roof live load as well as any other applicable loads, including any loads due to snow drifting. Panels that are independent structures and do not have accessible or occupied space underneath are not required to accommodate the PV live load, provided the area under the structure is restricted to prevent public access. Solar PV panels designed to be the roof itself must be designed to support the roof live load in combination with any other applicable loads where the panels span to structural supports and have accessible or occupied space underneath. Roof structures that provide support for ballasted photovoltaic panel systems must be analyzed and designed in accordance with the general requirements of Section 1604.4 and meet the deflection limits of Section 1604.3.6. Where applicable, the roof system must be checked for progressive deflection due to ponding in accordance with Section 1611.

The definition of a photovoltaic panel system has been added to Section 202, defined as a system of discrete photovoltaic panels that converts solar radiation into electricity, and includes rack support systems.

1609.1.1

Determination of Wind Loads

CHANGE TYPE: Modification

CHANGE SUMMARY: A reference to the new wind tunnel testing standard ASCE 49 has been added to Section 1609.1.1, Exception 6.

2015 CODE: 1609.1.1 Determination of Wind Loads. Wind loads on every building or structure shall be determined in accordance with Chapters 26 to 30 of ASCE 7 or provisions of the alternate all-heights method in Section 1609.6. The type of opening protection required, the ultimate design wind speed, V_{ult}, and the exposure category for a site is permitted to be determined in accordance with Section 1609 or ASCE 7. Wind shall be assumed to come from any horizontal direction and wind pressures shall be assumed to act normal to the surface considered.

Exceptions:

(No changes to Exceptions 1 through 5.)

6. Wind tunnel tests in accordance with ~~Chapter 31 of ASCE 7~~ ASCE 49 and with Sections 31.4 and 31.5 of ASCE 7.

The wind speeds in Figures 1609.3(1), 1609.3(2) and 1609.3(3) are ultimate design wind speeds, V_{ult}, and shall be converted in accordance with Section 1609.3.1 to nominal design wind speeds, V_{asd}, when the provisions of the standards referenced in Exceptions ~~1 through 5~~ 4 and 5 are used.

1609.1.1 continues

Photo Courtesy of Rowan Williams Davies & Irwin Inc. (RWDI)

Wind tunnel

1609.1.1 continued

CHAPTER 35

ASCE/SEI. ASCE 49—07 Wind Tunnel Testing for Buildings and Other Structures.

CHANGE SIGNIFICANCE: Section 1609.1.1 requires wind loads on buildings and other structures to be determined in accordance with Chapters 26 through 30 of ASCE 7, with some exceptions. Exception 6 refers to Chapter 31 of ASCE 7, which covers the wind tunnel procedure. The exception now includes a reference to the new standard, *ASCE 49 Wind Tunnel Testing for Buildings and Other Structures*, which provides minimum requirements for wind tunnel tests to determine wind loads and structural response of buildings and other structures. ASCE 49 includes wind loads for main wind-force-resisting systems and individual structural components and cladding. Because of the limited scope of ASCE 49, Section 31.4 (Load Effects) and ASCE 7 Section 31.5 (Wind-Borne Debris) are still required for determining wind loads; these provisions have been retained in the exception. The 2012 edition of ASCE 49 is a new referenced standard added to Chapter 35. Because the 2013 edition of ICC 600, the 2012 edition of the AWC Wood Frame Construction Manual (WFCM), and Supplement 3-12 of AISI S230 are all based on ultimate wind speeds, design wind speeds should be converted in accordance with Section 1609.3.1 to nominal design wind speeds when the provisions of the standards referenced in Exception 4 (NAAMM FP 1001) and Exception 5 (TIA-222) are used.

1613.3.1

Mapped Acceleration Parameters

CHANGE TYPE: Addition

CHANGE SUMMARY: The U.S. Geological Survey (USGS) recently developed seismic hazard and Risk-Targeted Maximum Considered Earthquake (MCER) ground motion maps for Guam and American Samoa, which have now been included in the IBC.

2015 CODE: 1613.3.1 Mapped Acceleration Parameters. The parameters S_S and S_1 shall be determined from the 0.2- and 1-second spectral response accelerations shown on Figures 1613.3.1(1) through 1613.3.1(~~6~~8). Where S_1 is less than or equal to 0.04 and S_S is less than or equal to 0.15, the structure is permitted to be assigned Seismic Design Category A. ~~The parameters S_s and S_1 shall be, respectively, 1.5 and 0.6 for Guam and 1.0 and 0.4 for American Samoa.~~

1613.3.1 continues

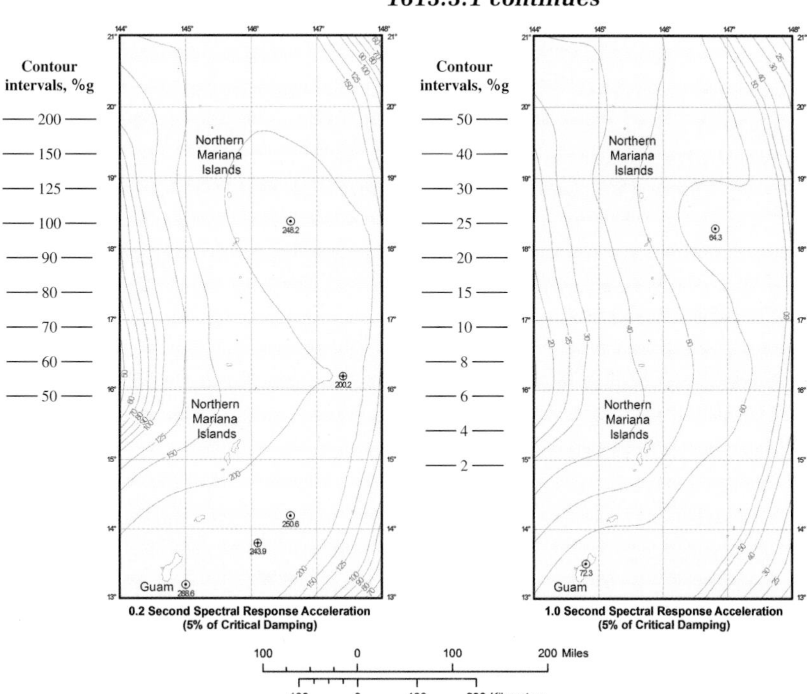

Explanation

Contours of spectral response acceleration expressed as a percent of gravity.

—10—

—10—

Point values of spectral response acceleration expressed as a percent of gravity.

⊕ 200.2 Local minimum

⊙ 250.6 Local maximum

⊕ 243.9 Saddle point

DISCUSSION

Maps prepared by United States Geological Survey (USGS) in collaboration with the Federal Emergency Management Agency (FEMA)-funded Building Seismic Safety Council (BSSC). The basis is explained in commentary prepared by BSSC and in the references.

Ground motion values contoured on these maps incorporate:

- a target risk of structural collapse equal to 1% in 50 years based upon a generic structural fragility
- a factor of 1.1 and 1.3 for 0.2 and 1.0 sec, respectively, to adjust from a geometric mean to the maximum response regardless of direction
- deterministic upper limits imposed near large, active faults, which are taken as 1.8 times the estimated median response to the characteristic earthquake for the fault (1.8 is used to represent the 84th percentile response), but not less than 150% and 60% g for 0.2 and 1.0 sec, respectively.

As such, the values are different from those on the uniform-hazard 2012 USGS National Seismic Hazard Maps for Guam and the Northern Mariana Islands posted at http://earthquake.usgs.gov/hazmaps.

Larger, more detailed versions of these maps are not provided because it is recommended that the corresponding USGS web tool (http://earthquake.usgs.gov/designmaps) be used to determine the mapped value for a specified location.

REFERENCES

Building Seismic Safety Council, 2009, NEHRP Recommended Seismic Provisions for New Buildings and Other Structures: FEMA P-750/2009 Edition, Federal Emergency Management Agency, Washington, DC.

Huang, Yin-Nan, Whittaker, A.S., and Luco, Nicolas, 2008, Maximum spectral demands in the near-fault region, Earthquake Spectra, Volume 24, Issue 1, pp. 319-341.

Luco, Nicolas, Ellingwood, B.R., Hamburger, R.O., Hooper, J.D., Kimball, J.K., and Kircher, C.A., 2007, Risk-Targeted versus Current Seismic Design Maps for the Conterminous United States, Structural Engineers Association of California 2007 Convention Proceedings, pp. 163-175.

Mueller, C.S., Haller, K.M., Luco, Nicolas, Petersen, M.D., and Frankel, A.D., 2012 Seismic Hazard Assessment for Guam and the Northern Mariana Islands: U.S. Geological Survey Open-File Report 2012-1015.

Risk-Targeted Maximum Considered Earthquake (MCER) response accelerations for Guam

1613.3.1 continued

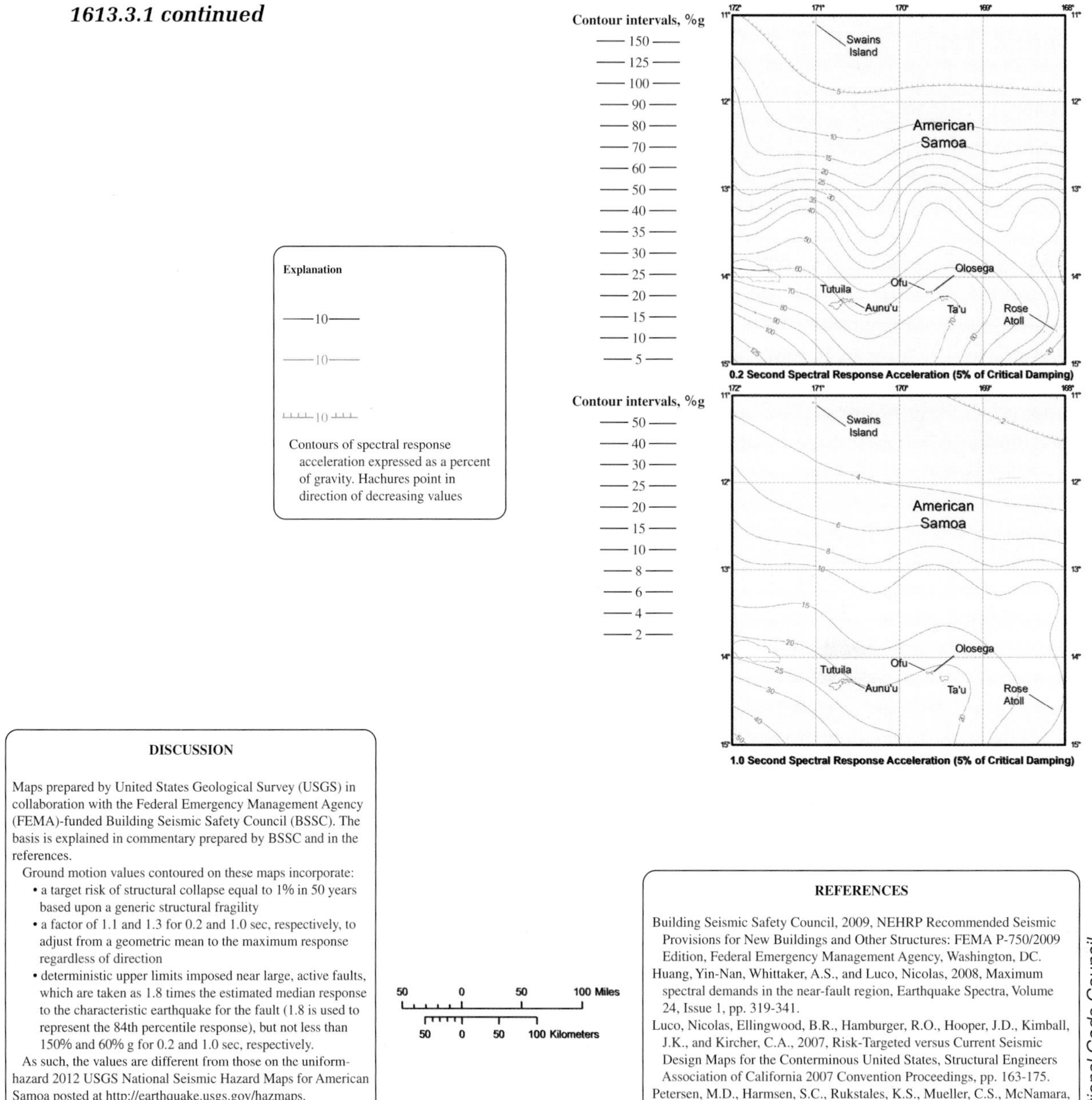

Risk-Targeted Maximum Considered Earthquake (MCER) response accelerations for American Samoa

CHANGE SIGNIFICANCE: Under the National Earthquake Hazards Reduction Program (NEHRP), the U.S. Geological Survey (USGS) is charged with the responsibility to develop and maintain seismic hazard maps that are the basis of the Risk-Targeted Maximum Considered Earthquake (MCER) ground motion maps used in the IBC. As part of that responsibility, the USGS recently developed seismic hazard and MCER ground motion maps for Guam and American Samoa using the same methodology as for the conterminous United States, Hawaii, Alaska, Puerto Rico and the U.S. Virgin Islands. The MCER ground motion maps for Guam and American Samoa are now included as Figures 1613.3.1(7) and 1613.3.1(8) of the IBC.

1613.5

Amendments to ASCE 7

CHANGE TYPE: Addition

CHANGE SUMMARY: An amendment to the diaphragm anchorage requirements of Section 12.11.2 of ASCE 7 clarifies that the 2.5-to-1 aspect ratio applies to wood, wood structural panel or untopped steel-deck-sheathed subdiaphragms.

2015 CODE: **1613.5 Amendments to ASCE 7.** The provisions of Section 1613.5 shall be permitted as an amendment to the relevant provisions of ASCE 7.

1613.5.1 Transfer of Anchorage Forces into Diaphragm. Modify ASCE 7 Section 12.11.2.2.1 as follows:

12.11.2.2.1 Transfer of Anchorage Forces into Diaphragm. Diaphragms shall be provided with continuous ties or struts between diaphragm chords to distribute these anchorages forces into the diaphragms. Diaphragm connections shall be positive, mechanical or welded. Added chords are permitted to be used to form subdiaphragms to transmit the anchorage forces to the main continuous cross-ties. The maximum length-to-width ratio of a wood, wood structural panel or untopped steel deck sheathed structural subdiaphragm that serves as part of the continuous tie system shall be 2.5 to 1. Connections and anchorages capable of resisting the prescribed forces shall be provided between the diaphragm and the attached components. Connections shall extend into the diaphragm a sufficient distance to develop the force transferred into the diaphragm.

CHANGE SIGNIFICANCE: When limitations on subdiaphragms were first proposed for the 1997 UBC in the form of a limitation on allowable shear, the reasoning primarily focused on tilt-up buildings with nailed wood structural panel diaphragms after poor performance was observed in the aftermath of the 1971 San Fernando (Sylmar) Earthquake. When

1613.5 continues

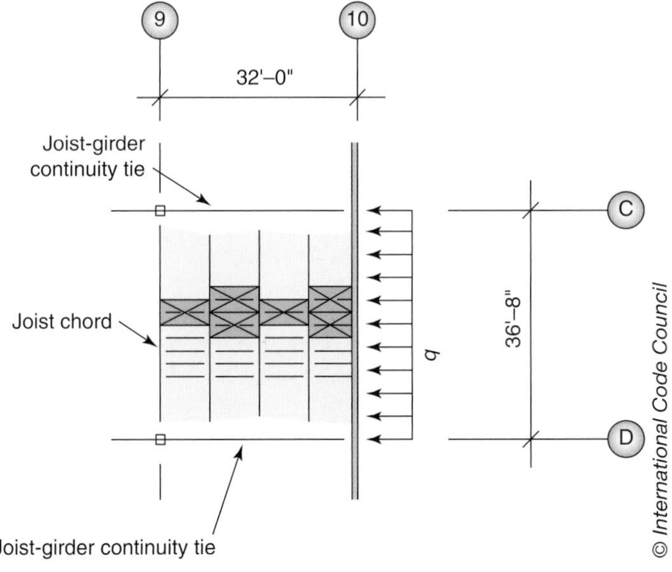

Typical subdiaphragm

1613.5 continued

approved for the 1997 UBC, the application of the aspect ratio limitation specifically applied only to wood structural subdiaphragms. In the subsequent process of being incorporated into the IBC and ASCE 7, the code text limiting the application only to wood subdiaphragms was apparently discontinued, making the requirement applicable to all subdiaphragms. This new amendment to the diaphragm anchorage requirements of ASCE 7 Section 12.11.2 reintroduces the limitation to wood subdiaphragms because they are the original systems of concern and observed poor performance. The provision includes untopped steel deck diaphragms due to their similarities in construction and perceived structural behavior. This aspect ratio limit is not necessary for good performance of other diaphragm types. With this aspect ratio limit removed for concrete, composite deck, and other diaphragm types, other diaphragm limitations within the referenced material standards govern their design. The language in the code section that differs from ASCE 7 Section 12.11.2 is shown here in italics: "The maximum length-to-width ratio of *a wood, wood structural panel, or untopped steel deck sheathed* structural subdiaphragm *that serves as part of the continuous tie system* shall be 2.5 to 1."

1613.6

Ballasted Photovoltaic Panel Systems

CHANGE TYPE: Addition

CHANGE SUMMARY: Seismic requirements for ballasted roof-mounted photovoltaic (PV) solar panels have been added to Section 1613.6.

2015 CODE: **1613.6 Ballasted Photovoltaic Panel Systems.** Ballasted, roof-mounted photovoltaic panel systems need not be rigidly attached to the roof or supporting structure. Ballasted non-penetrating systems shall be design and installed only on roofs with slopes not more than one unit vertical in 12 units horizontal. Ballasted non-penetrating systems shall be designed to resist sliding and uplift resulting from lateral and vertical forces as required by Section 1605, using a coefficient of friction determined by acceptable engineering principles. In structures assigned to Seismic Design Category C, D, E or F, ballasted non-penetrating the systems shall be designed to accommodate seismic displacement determined by nonlinear response-history analysis or shake-table testing, using input motions consistent with ASCE 7 lateral and vertical seismic forces for non-structural components on roofs.

CHANGE SIGNIFICANCE: Ballasted roof-mounted photovoltaic (PV) panel systems are not required to be rigidly attached to the roof or supporting structure. Non-penetrating ballasted panel systems are only allowed to be installed on roofs with slopes less than or equal to 1:12. The systems must be designed to resist sliding and uplift resulting from lateral and vertical forces as required by the applicable load combinations using a coefficient of friction determined by acceptable engineering principles of mechanics. When installed on structures in Seismic Design Category C, D, E or F,

1613.6 continues

Photo Courtesy of Peter Kulczyk

Ballasted photovoltaic panels

1613.6 continued ballasted non-penetrating systems must be designed to accommodate seismic displacement determined by nonlinear response-history analysis or shake-table testing, using input motions consistent with the non-structural component requirements of ASCE 7. A definition of "photovoltaic panel system" has been added to Section 202, and new Section 1603.1.8 adds requirements for including dead load of rooftop-mounted PV solar panels on the construction documents. A new Section 1607.12.5 adds loading requirements for roof structures supporting PV panel systems.

1704.5

Submittals to the Building Official

CHANGE TYPE: Addition

CHANGE SUMMARY: Requirements for submittal of reports and certificates related to construction that is subject to special inspections and tests are now clearly specified.

2015 CODE: **1704.5 Submittals to the Building Official.** In addition to the submittal of reports of special inspections and tests in accordance with Section 1704.2.4, reports and certificates shall be submitted by the owner or the owner's authorized agent to the building official for each of the following:

1. Certificates of compliance for the fabrication of structural, load-bearing or lateral load-resisting members or assemblies on the premises of a registered and approved fabricator in accordance with Section 1704.2.5.1;
2. Certificates of compliance for the seismic qualification of nonstructural components, supports and attachments in accordance with Section 1705.13.2;
3. Certificates of compliance for designated seismic systems in accordance with Section 1705.13.3;
4. Reports of preconstruction tests for shotcrete in accordance with Section 1908.5;
5. Certificates of compliance for open web steel joists and joist girders in accordance with Section 2207.5;
6. Reports of material properties verifying compliance with the requirements of AWS D1.4 for weldability as specified in Section 26.5.4 of ACI 318 for reinforcing bars in concrete complying with a standard other than ASTM A 706 that are to be welded; and

1704.5 continues

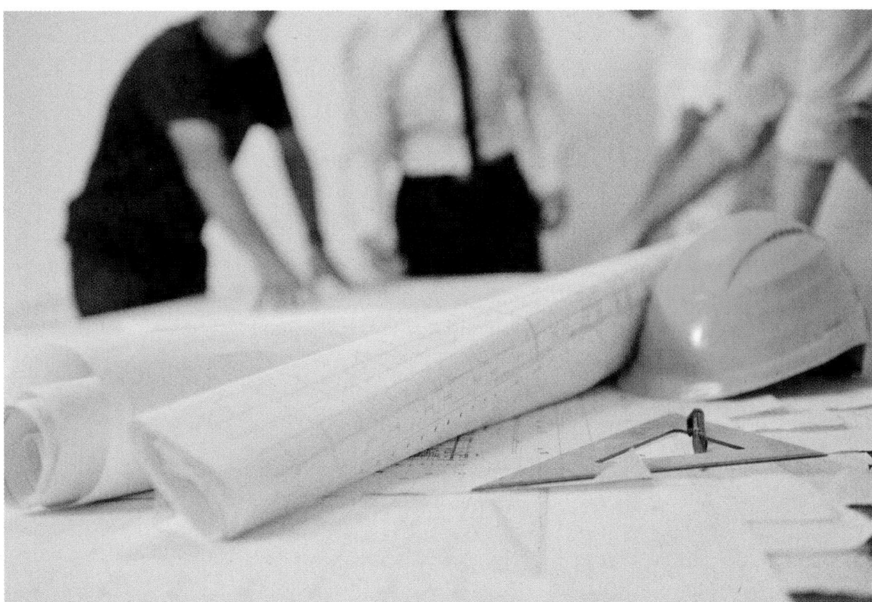

Construction documents

1704.5 continued

7. Reports of mill tests in accordance with Section 20.2.2.5 of ACI 318 for reinforcing bars complying with ASTM A 615 and used to resist earthquake-induced flexural or axial forces in the special moment frames, special structural walls, or coupling beams connecting special structural walls, of seismic force-resisting systems in structures assigned to Seismic Design Category B, C, D, E or F.

CHANGE SIGNIFICANCE: New Section 1704.5 consolidates and clarifies the requirements for submittal of reports and certificates for items subject to special inspections and tests to verify that materials or products meet certain special code requirements, or are alternatives to the general requirements of the code. The items addressed in Section 1704.5 are typically references to provisions found elsewhere in the code or in a referenced standard. Items 1 and 5 require certificates of compliance for fabricated structural elements and open web steel bar joists and girders. Items 2 and 3 require certificates of compliance for seismic qualification of nonstructural components and elements of designated seismic systems as required by Section 1705.13. Item 4 requires preconstruction test reports for shotcrete as required by Chapter 19. Item 6 requires material property reports (carbon equivalence) to verify the weldability of reinforcing bars complying with a standard other than ASTM A706. Item 7 requires mill test reports for reinforcing bars compliant with ASTM A 615 used in certain elements of the seismic force-resisting system in structures in a Seismic Design Category (SDC) other than SDC A.

1705.2

Steel Construction

CHANGE TYPE: Modification

CHANGE SUMMARY: The special inspection requirements for structural steel elements and cold-formed steel decks have been modified to coordinate the provisions with the new terminology used for structural steel elements within IBC Chapter 22, AISC 360 and the new SDI standard.

2015 CODE: **1705.2 Steel Construction.** The special inspections ~~for~~ and nondestructive testing of steel ~~elements of~~ construction in buildings, ~~and~~ structures, and portions thereof shall be ~~as required~~ in accordance with this section.

> **Exception:** Special inspections of the steel fabrication process shall not be required where the fabricator does not perform any welding, thermal cutting or heating operation of any kind as part of the fabrication process. In such cases, the fabricator shall be required to submit a detailed procedure for material control that demonstrates the fabricator's ability to maintain suitable records and procedures such that, at any time during the fabrication process, the material specification, and grade for the main stress-carrying elements are capable of being determined. Mill test reports shall be identifiable to the main stress-carrying elements when required by the approved construction documents.

1705.2.1 Structural Steel. Special inspections and nondestructive testing ~~for structural steel~~ of structural steel elements in buildings, structures and portions thereof shall be in accordance with the quality assurance inspection requirements of AISC 360.

1705.2 continues

Photo Courtesy of Mortenson Construction

Structural steel construction

1705.2 continued

Exception: Special inspection of railing systems composed of structural steel elements shall be limited to welding inspection of welds at the base of cantilevered rail posts.

1705.2.2 Cold-Formed Steel ~~Construction Other Than Structural Steel~~ Deck. Special inspections ~~for steel construction other than structural steel shall be in accordance with Table 1705.2.2 and this section.~~ and qualification of welding special inspectors for cold-formed steel floor and roof deck shall be in accordance with the quality assurance inspection requirements of SDI QA/QC.

~~**1705.2.2.1 Welding.** Welding inspection and welding inspector qualification shall be in accordance with this section.~~

~~**1705.2.2.1.1 Cold-Formed Steel.** Welding inspection and welding inspector qualification for coldformed steel floor and roof decks shall be in accordance with AWS D1.3.~~

1705.2.3 Open Web Steel Joists and Joist Girders. Special inspections of open web steel joists and joist girders in buildings, structures and portions thereof shall be in accordance with Table 1705.2.3.

~~1705.2.2.2~~ 1705.2.4 Cold-Formed Steel Trusses Spanning 60 Feet or Greater. Where a cold-formed steel truss clear span is 60 feet (18 288 mm) or greater, the special inspector shall verify that the temporary installation restraint/bracing and the permanent individual truss member restraint/bracing are installed in accordance with the approved truss submittal package.

CHAPTER 35

Steel Deck Institute. SDI QA/QC-2011, Standard for Quality Control and Quality Assurance for Installation of Steel Deck.

CHANGE SIGNIFICANCE: Modifications to the special inspection and nondestructive testing requirements for structural steel elements and cold-formed steel decks coordinate the provisions with the new terminology for structural steel elements within IBC Chapter 22 and clarify that structural steel elements in buildings, structures and portions thereof are to be inspected and tested in accordance with the quality assurance requirements in AISC 360. A new SDI *Standard for Quality Control and Quality Assurance for Installation of Steel Deck* is referenced for the inspection of steel floor and roof decks. The new language of Section 1705.2 includes a number of editorial modifications: (1) adds a reference to "nondestructive testing" to clarify that the quality assurance provisions of AISC 360 and AISC 341 cover not only special inspections but also the testing of welds—the use of "nondestructive" is the appropriate industry terminology; (2) modifies the term "steel elements" to "steel construction" in order to be consistent with the terminology used in IBC Chapter 22; (3) clarifies that special inspection and testing may be required in buildings, structures, or portions thereof; (4) changes the title in

Section 1705.2.2 to specifically recognize that the provision covers cold-formed steel decks rather than using the generic term "structural steel;" and (5) relocates Section 1705.2.2.2 on cold-formed steel trusses to a new subsection, Section 1705.2.3, because the Section 1705.2.2 provisions are limited to cold-formed steel decks.

A reference is made to the new SDI QA/QC-2011 standard, which contains provisions for quality assurance inspection of steel floor and roof deck, and is coordinated with the requirements of AISC 360. The SDI standard clarifies the scope of required inspections and the responsibilities of both the installer's quality control personnel and the quality assurance inspector. It contains tables of inspection tasks that specifically list inspection requirements for material verification, deck installation, welding and mechanical fastening. The tables amplify and clarify the basic special inspection requirements for steel decks that were contained in the 2012 IBC, and bring all special inspection requirements for steel decks into a single location. The standard references AWS D1.3 for weld quality and the required material verification.

See the discussion of Section 1705.2.3 for new requirements related to special inspection of open web steel joists and joist girders.

1705.2.3

Open Web Steel Joists and Joist Girders

CHANGE TYPE: Addition

CHANGE SUMMARY: Special inspections are now required during the installation of open web steel joists and joist girders, and a new table specifies the type of inspection and applicable referenced standard.

2015 CODE: ~~**1705.2.2**~~ **1705.2.3 Open-Web Steel Joists and Joist Girders.** Special inspections of open web steel joists and joist girders in buildings, structures and portions thereof shall be in accordance with Table 1705.2.3.

TABLE ~~1705.2.2~~ 1705.2.3 Required Special Inspections of Open Web Steel Joists and Joist Girders

Type	Continuous	Periodic	Referenced Standard[a]
1. Installation of open web steel joists and joist girders			
a. End connections – welding or bolted		X	SJI specifications listed in Section 2207.1
b. Bridging – horizontal or diagonal			
1. Standard bridging		X	SJI specifications listed in Section 2207.1
2. Bridging that differs from the SJI specifications listed in Section 2207.1		X	

For SI: 1 inch = 25.4 mm.

a. Where applicable, see also Section 1705.12, Special inspection for seismic resistance.

Photo Courtesy of Steel Joist Institute

Open web steel joists

CHANGE SIGNIFICANCE: The structural design and installation of open web steel joists and joist girder systems is complex enough to warrant special inspection by personnel with sufficient expertise and approval by the building official as having the necessary competence to inspect the installation of steel joist systems. Examples of steel framing items that warrant special inspection of the installation are bearing seat attachments, field splices and bridging attachments. Where steel joist systems are used in roof and floor diaphragms, chords and collectors also become important critical elements of the lateral-force-resisting system.

The standard specifications for open web steel joists (SJI-K-2010 and SJI-LH/DLH-2010), joist girders (SJI-JG-2010) and composite steel joists (SJI-CJ-2010) by the Steel Joist Institute (SJI) contain provisions for inspections, but they are limited to quality control inspections by the manufacturer before shipment to verify compliance and workmanship with the specifications. (Refer to Section 5.12 of SJI-K-2010, Section 104.13 of SJI-LH/DLH-2010, Section 1004.10 of SJI-JG-2010 and Section 104.13 of SJI-CJ-2010.) The SJI standards—SJI CJ, SJI K, SJI LH/DLH and SJI JG—are referenced in Section 2207.1. Although the sections of the SJI standards just noted are also referenced in Section 4 of the *Code of Standard Practice for Steel Joists and Joist Girders* (SJI-COSP-2010) and the *Code of Standard Practice for Composite Steel Joists* (SJI-CJCOSP-2010), these SJI codes of standard practice are not referenced in the IBC.

Table 1705.3

Required Special Inspections of Concrete Construction

CHANGE TYPE: Modification

CHANGE SUMMARY: The requirement for special inspection of cast-in-place anchors in concrete where allowable loads have been increased or strength design is used has been deleted from Table 1705.3, specific requirements for the design and installation of adhesive anchors are now included in ACI 318, and continuous special inspection has been added for these types of anchors installed horizontally or in upwardly inclined orientations with sustained loads.

2015 CODE: 1705.3 Concrete Construction. Special inspections and ~~verifications for~~ tests of concrete construction shall be ~~as required by~~ performed in accordance with this section and Table 1705.3.

TABLE 1705.3 Required Special Inspections and Tests of Concrete Construction

Type	Continuous Special Inspection	Periodic Special Inspection	Referenced Standard[a]	IBC Reference
1. ~~Inspection of~~ Inspect reinforcement, including prestressing tendons, and verify placement.	—	X	ACI 318: ~~3.5, 7.1-7.7~~ Ch. 20, 25.2, 25.3, 26.5.1-26.5.3	~~1910.4~~ 1908.4
2. ~~Inspection of~~ Reinforcing ~~steel~~ bar welding ~~in accordance with Table 1705.2.2, Item 2b~~:			AWS D1.4; ACI 318: ~~3.5.2~~ 26.5.4	—
a. Verify weldability of reinforcing bars other than ASTM A706.	—	X		
b. Inspect single-pass fillet welds, maximum 5/16".	—	X		
c. All other welds.	X	—		
3. Inspect anchors cast in concrete ~~where allowable loads have been increased or where strength design is used~~.		X	ACI 318: ~~8.1.3, 21.2.8 D.9.2.~~ 17.8.2	~~1908.5, 1909.1~~
4. Inspect anchors post-installed in hardened concrete members[b]:				
a. Adhesive anchors installed in horizontally or upwardly inclined orientations to resist sustained tension loads.	X		ACI 318: ~~3.8.6, 8.1.3, 21.2.8 D.9.2.4~~ 17.8.2.4; ACI 318: ~~D.9.2~~ 17.8.2	~~1909.1~~
b. Mechanical anchors and adhesive anchors not defined in 4.a.		X		—

(Portions of table not shown remain unchanged.)

CHANGE SIGNIFICANCE: The reference to special inspection of anchors cast in concrete where allowable loads have been increased or strength design is used has been deleted because the allowable stress design procedure for anchorage to concrete has been deleted entirely from the IBC. Historically there have been two structural design procedures, working stress design and strength design. Over time, the term "working stress design" was replaced by "allowable stress design" to better describe the methodology. Over the years, the strength design procedure, called strength design for concrete and masonry and load and resistance factor design (LRFD) for steel and wood, has gradually replaced allowable stress design. The 2005 edition of the ACI 318 standard no longer included the so-called alternate design method (allowable stress design). Table 1705.3 requires periodic special inspection for anchors cast in concrete and references ACI 318 Section 17.8.2, which in turn references ACI 318 Section 1.3 and the legally adopted building code. ACI 318 Section 1.3 covers the purpose and scope of the ACI code.

The difficulty of properly installing adhesive anchors greatly increases when gravity works to drain the placed epoxy out of the predrilled hole. For proper and consistent installation, trained personnel are essential. Under sustained tension loads, epoxy will creep and debond, as evidenced by failure of the epoxy anchors that supported the ceiling panels in the I-90 connector tunnel in Boston. A proper installation is critical in this case and therefore the code requires continuous special inspection.

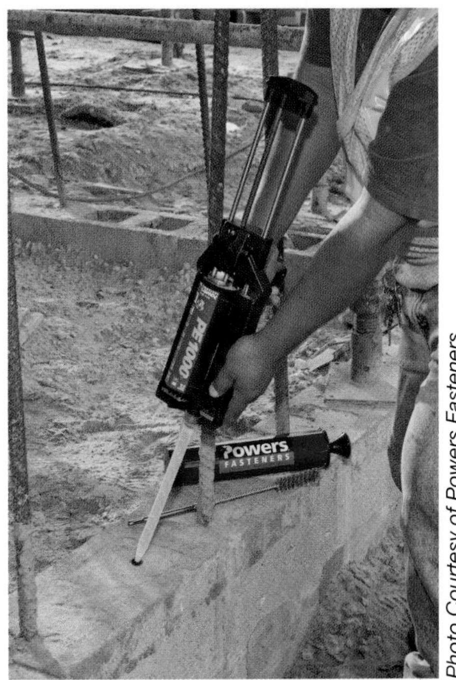

Photo Courtesy of Powers Fasteners

Adhesive anchor installation

1705.11

Special Inspection for Wind Resistance

CHANGE TYPE: Clarification

CHANGE SUMMARY: To better identify the intent, revisions have been made to the special inspection requirements for wind resistance. Specific requirements for the inspection of wind-resisting components have also been clearly identified.

2015 CODE: ~~1705.10~~ 1705.11 Special Inspections for Wind Resistance. Special inspections ~~itemized~~ for wind resistance specified in Sections ~~1705.10.1~~ 1705.11.1 through ~~1705.10.3~~ 1705.11.3, unless exempted by the exceptions to Section 1704.2, are required for buildings and structures constructed in the following areas:

1. In wind Exposure Category B, where V_{asd} as determined in accordance with Section 1609.3.1 is 120 miles per hour (52.8 m/sec) or greater.
2. In wind Exposure Category C or D, where V_{asd} as determined in accordance with Section 1609.3.1 is 110 mph (49 m/sec) or greater.

~~1705.10.1~~ 1705.11.1 Structural Wood. Continuous special inspection is required during field gluing operations of elements of the main wind-force-resisting system. Periodic special inspection is required for nailing, bolting, anchoring and other fastening of ~~components within~~ elements of the main wind-force-resisting system, including wood shear walls, wood diaphragms, drag struts, braces and hold-downs.

Photo Courtesy of APA—The Engineered Wood Association

High wind damage assessment

> **Exception:** Special inspections ~~is~~ are not required for wood shear walls, shear panels and diaphragms, including nailing, bolting, anchoring and other fastening to other ~~components~~ elements of the main wind-force-resisting system, where the fastener spacing of the sheathing is more than 4 inches (102 mm) on center.

~~1705.10.2~~ 1705.11.2 Cold-Formed Steel Light-Frame Construction. Periodic special inspection is required ~~during~~ for welding operations of elements of the main wind-force-resisting system. Periodic special inspection is required for screw attachment, bolting, anchoring and other fastening of ~~components within~~ elements of the main wind-force-resisting system, including shear walls, braces, diaphragms, collectors (drag struts) and hold-downs.

> **Exception:** Special inspections ~~is~~ are not required for cold-formed steel light-frame shear walls, ~~braces~~, and diaphragms, ~~collectors (drag struts) and hold downs,~~ including screwing, bolting, anchoring and other fastening to components of the wind-force resisting system where either of the following apply:
>
> 1. The sheathing is gypsum board or fiberboard.
> 2. The sheathing is wood structural panel or steel sheets on only one side of the shear wall, shear panel or diaphragm assembly and the fastener spacing of the sheathing is more than 4 inches (102 mm) on center (o.c.).

~~1705.10.3~~ 1705.11.3 Wind-Resisting Components. Periodic special inspection is required for fastening of the following systems and components:

1. Roof ~~cladding~~ covering, roof deck, and roof framing connections.
2. ~~Wall cladding~~ Exterior wall covering, and wall connections to roof and floor diaphragms and framing.

CHANGE SIGNIFICANCE: Although several of the revisions to the special inspection requirements for wind resistance are considered editorial, two of them are specifically to clarify the intent. The term "braces" was deleted from the exception to Section 1705.11.2 because the exception applies to shear walls, not braced walls. The terms "collectors (drag struts)" and "hold-downs" were deleted from the exception and replaced with a more general reference to "components of the wind-force resisting system." The modifications to the section on wind-resisting components are to clearly specify what is meant by "roof cladding" and "wall cladding," which are not terms defined in the code. The section now applies to fastening of the roof covering, roof deck, roof framing connections, exterior wall covering, wall connections to roof and floor diaphragms, and connections to framing.

1705.12

Special Inspection for Seismic Resistance

CHANGE TYPE: Addition

CHANGE SUMMARY: Periodic special inspection of cold-formed steel special bolted moment frames (CFS-SBMFs) is now mandated. In addition, several modifications to the special inspection requirements for seismic resistance have been made in order to clarify the intent.

2015 CODE: **~~1705.11~~ 1705.12 Special Inspections for Seismic Resistance.** Special inspections ~~itemized~~ for seismic resistance shall be required as specified in Sections 1705.12.1 through 1705.12.9, unless exempted by the exceptions of Section 1704.2. ~~are required for the following:~~

~~1. The seismic force-resisting systems in structures assigned to Seismic Design Category C, D, E or F in accordance with Sections 1705.11.1 through 1705.11.3, as applicable.~~

~~2. Designated seismic systems in structures assigned to Seismic Design Category C, D, E or F in accordance with Section 1705.11.4.~~

~~3. Architectural, mechanical and electrical components in accordance with Sections 1705.11.5 and 1705.11.6.~~

~~4. Storage racks in structures assigned to Seismic Design Category D, E or F in accordance with Section 1705.11.7.~~

~~5. Seismic isolation systems in accordance with Section 1705.11.8.~~

Photo Courtesy of FCP Inc. Buildings

Cold-formed steel special bolted moment frames (CFS-SBMFs)

Exception: The special inspections ~~itemized~~ specified in Sections 1705.12.1 through 1705.12.9 are not required for structures designed and constructed in accordance with one of the following:

1. The structure consists of light-frame construction; the design spectral response acceleration at short periods, S_{DS}, as determined in Section 1613.3.4, does not exceed 0.5; and the building height of the structure does not exceed 35 feet (10 668 mm).
2. The seismic force-resisting system of the structure consists of reinforced masonry or reinforced concrete; the design spectral response acceleration at short periods, S_{DS}, as determined in Section 1613.3.4, does not exceed 0.5; and the building height of the structure does not exceed 25 feet (7620 mm).
3. The structure is a detached one- or two-family dwelling not exceeding two stories above grade plane and does not have any of the following horizontal or vertical irregularities in accordance with Section 12.3 of ASCE 7:
   - 3.1. Torsional or extreme torsional irregularity.
   - 3.2. Nonparallel systems irregularity.
   - 3.3. Stiffness-soft story or stiffness-extreme soft story irregularity.
   - 3.4. Discontinuity in lateral strength-weak story irregularity.

1705.12.1 Structural Steel. Special inspections for seismic resistance shall be in accordance with Section 1705.12.1.1 or 1705.12.1.2, as applicable.

~~1705.11.1~~ 1705.12.1.1 Seismic Force-Resisting Systems. Special inspections ~~for structural steel~~ of structural steel in the seismic-force resisting systems of buildings and structures assigned to Seismic Design Category B, C, D, E or F shall be performed in accordance with the quality assurance requirements of AISC 341.

Exception: Special inspections ~~of structural steel~~ are not required in the seismic force-resisting systems of buildings and structures assigned to Seismic Design Category B or C that are not specifically detailed for seismic resistance, with a response modification coefficient, R, of 3 or less, excluding cantilever column systems.

1705.12.1.2 Structural Steel Elements. Special inspections of structural steel elements in the seismic force-resisting systems of buildings and structures assigned to Seismic Design Category B, C, D, E or F other than those covered in Section 1705.12.1.1, including struts, collectors, chords and foundation elements, shall be performed in accordance with the quality assurance requirements of AISC 341.

Exception: Special inspections of structural steel elements are not required in the seismic force-resisting systems of buildings and structures assigned to Seismic Design Category B or C with a response modification coefficient, R, of 3 or less.

1705.12 continues

1705.12 continued

~~1705.11.2~~ 1705.12.2 Structural Wood. For the seismic force-resisting systems of structures assigned to Seismic Design Category C, D, E or F:

1. Continuous special inspection shall be required during field gluing operations of elements of the seismic force-resisting system.
2. Periodic special inspection shall be required for nailing, bolting, anchoring and other fastening of ~~components within~~ elements of the seismic force-resisting system, including wood shear walls, wood diaphragms, drag struts, braces, shear panels and hold-downs.

Exception: Special inspections are not required for wood shear walls, shear panels and diaphragms, including nailing, bolting, anchoring and other fastening to other ~~components~~ elements of the seismic force-resisting system, where the fastener spacing of the sheathing is more than 4 inches (102 mm) on center (o.c.).

~~1705.11.3~~ 1705.12.3 Cold-Formed Steel Light-Frame Construction. For the seismic force-resisting systems of structures assigned to Seismic Design Category C, D, E or F, periodic special inspection shall be required:

1. ~~Periodic special inspection is required~~ For welding operations of elements of the seismic force-resisting system; and
2. ~~Periodic special inspection is required~~ For screw attachment, bolting, anchoring and other fastening of ~~components within~~ elements of the seismic force-resisting system, including shear walls, braces, diaphragms, collectors (drag struts) and hold-downs.

Exception: Special inspections ~~is~~ are not required for cold-formed steel light-frame shear walls and diaphragms, ~~collectors (drag struts) and hold downs~~ including screw installation, bolting, anchoring and other fastening to components of the seismic force-resisting system where either of the following applies:

1. The sheathing is gypsum board or fiberboard.
2. The sheathing is wood structural panel or steel sheets on only one side of the shear wall, shear panel or diaphragm assembly and the fastener spacing of the sheathing is more than 4 inches (102 mm) o.c.

~~1705.11.4~~ 1705.12.4 Designated Seismic Systems. For structures assigned to Seismic Design Category C, D, E or F, the special inspector shall examine designated seismic systems requiring seismic qualification in accordance with Section 13.2.2 of ASCE 7 and verify that the label, anchorage and mounting conforms to the certificate of compliance.

1705.12.9 Cold-Formed Steel Special Bolted Moment Frames. Periodic special inspection shall be provided for the installation of cold-formed steel special bolted moment frames in the seismic force-resisting systems of structures assigned to Seismic Design Category D, E or F.

(*Because the code changes revised substantial portions of Chapter 17, the entire code change text is too extensive to be included here. Refer to the* 2015 IBC Code Changes Resource Collection *for the complete text and history of the code changes.*)

CHANGE SIGNIFICANCE: Although many of the changes to the special inspection requirements for seismic resistance are editorial in nature in order to update the provisions and clarify the intent, a new section has been added that requires periodic special inspection to be provided for the installation of cold-formed steel special bolted moment frames (CFS-SBMFs) in structures assigned to Seismic Design Category D, E or F. The CFS-SBMF is a relatively new type of cold-formed steel moment frame seismic-force-resisting system designed to withstand anticipated seismic forces by the dissipation of energy through controlled inelastic deformation. The system is now listed in Table 12.2-1 of ASCE 7-10 (see Item C.12. $R = 3.5$, $\Omega_o = 3.0$, $C_d = 3.5$). Their structural design and installation is complex enough to warrant special inspection by personnel with sufficient expertise and approval by the building official as having the necessary competence to inspect the installation of these systems. Items that warrant special inspection include critical elements of the system such as the installation of the beam-to-column connections, as well as the anchorage to the foundation. The *Standard for Seismic Design of Cold-Formed Steel Structural Systems—Special Bolted Moment Frames, 2007 Edition* with Supplement No. 1 (AISI S110-07-S1-09) contains provisions for inspections and quality control by the fabricator.

1708.3.2

Static Load Testing

CHANGE TYPE: Modification

CHANGE SUMMARY: Static load test requirements have been revised to clarify the intent, the arbitrary factor of two has been removed, and the method for testing components that carry dynamic loads has been specified. Differences influenced by load duration effects when testing wood elements are now also addressed.

2015 CODE: **~~1709.3.2~~ 1708.3.2 Load Test Procedure Not Specified.** In the absence of applicable load test procedures contained within a standard referenced by this code or acceptance criteria for a specific material or method of construction, such existing structure shall be subjected to a test procedure developed by a registered design professional that simulates applicable loading and deformation conditions. For components that are not a part of the seismic load-resisting system, at a minimum the test load shall be equal to ~~two times the unfactored design loads~~ the specified factored design loads. For materials such as wood that have strengths that are dependent on load duration, the test load shall be adjusted to account for the difference in load duration of the test compared to the expected duration of the design loads being considered. For statically loaded components, the test load shall be left in place for a period of 24 hours. For components that carry dynamic loads (e.g., machine supports or fall arrest anchors), the load shall be left in place for a period

Photo Courtesy of Tinius Olsen

Load-testing machine

consistent with the component's actual function. The structure shall be considered to have successfully met the test requirements where the following criteria are satisfied:

1. Under the design load, the deflection shall not exceed the limitations specified in Section 1604.3.
2. Within 24 hours after removal of the test load, the structure shall have recovered not less than 75 percent of the maximum deflection.
3. During and immediately after the test, the structure shall not show evidence of failure.

CHANGE SIGNIFICANCE: The required static test load from the arbitrary value of two times the unfactored design load has been revised to a minimum of the specified factored design loads. By adding the phrase "at a minimum," and by referencing factored loads, the intent of the provision is made clear that the test load should be at least equal to the specified factored design load. Other referenced design standards such as AISC 360 and ACI 318 have been developed with the intent to ensure that structural elements are able to resist anticipated factored loads. When every element in a structure can resist anticipated factored loads, the structure's reliability is consistent with the intent of structural design standards. The load-testing provisions in the AISC and ACI standards make this clear by requiring proof loading to essentially full factored loads. The new language of the section is now consistent with other nationally recognized structural standards. The modified language also specifies how to test components that carry dynamic loads. When an element is designed to carry short-duration or dynamic loads, there is no need to sustain a proof test load for 24 hours. Finally, because wood elements respond differently depending on the duration of the loading, the test load should be adjusted to account for differences influenced by load duration effects where applicable.

1709.5

Exterior Window and Door Assemblies

CHANGE TYPE: Modification

CHANGE SUMMARY: The required design pressure ratings for exterior window and door assemblies are to be done on an allowable stress design basis.

2015 CODE: 1709.5 Exterior Window and Door Assemblies. The design pressure rating of exterior windows and doors in buildings shall be determined in accordance with Section 1709.5.1 or 1709.5.2. For the purposes of this section, the required design pressure shall be determined using the allowable stress design load combinations of Section 1605.3.

> **Exception:** Structural wind load design pressures for window units smaller than the size tested in accordance with Section 1709.5.1 or 1709.5.2 shall be permitted to be higher than the design value of the tested unit provided such higher pressures are determined by accepted engineering analysis. All components of the small unit shall be the same as the tested unit. Where such calculated design pressures are used, they shall be validated by an additional test of the window unit having the highest allowable design pressure.

CHANGE SIGNIFICANCE: With the new ultimate wind speeds and strength level wind loads established in ASCE 7-10, it is important to specify that the required design pressures for window and door assemblies are to be based on allowable stress design (ASD). It has been clarified that the ASD load combinations are to be used in the application of this section because the AAMA/WDMA/CSA, ASTM and ANSI/DASMA standards referenced in Section 1710.5 are based upon allowable stress design.

© Vadym Andrushchenko/iStock/Thinkstock

Exterior window assemblies

1711

Material and Test Standards

CHANGE TYPE: Deletion

CHANGE SUMMARY: The requirements for testing joist hangers in Section 1711.1 have been deleted entirely, and the requirements for testing concrete and clay roof tiles in Section 1711.2 have been relocated to Section 1504 addressing performance requirements for roof coverings and assemblies.

2015 CODE: ~~**1711.1 Joist Hangers.** Testing of joist hangers shall be in accordance with Sections 1711.1.1 through 1711.1.3, as applicable.~~

~~**1711.1.1 General.** The vertical load-bearing capacity, torsional moment capacity and deflection characteristics of joist hangers shall be determined in accordance with ASTM D 1761 using lumber having a specific gravity of 0.49 or greater, but not greater than 0.55, as determined in accordance with AF&PA NDS for the joist and headers.~~

> ~~**Exception:** The joist length shall not be required to exceed 24 inches (610 mm).~~

~~**1711.1.2 Vertical Load Capacity for Joist Hangers.** The vertical load capacity for the joist hanger shall be determined by testing a minimum of three joist hanger assemblies as specified in ASTM D 1761. If the ultimate vertical load for any one of the tests varies more than 20 percent~~

1711 continues

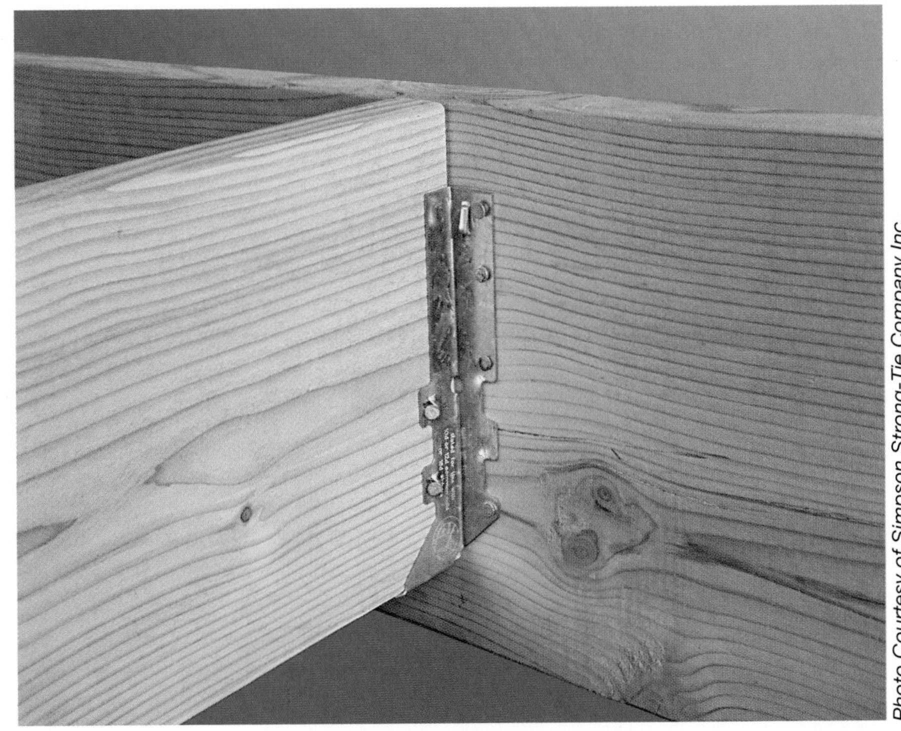

Photo Courtesy of Simpson Strong-Tie Company Inc.

Joist hanger

1711 continued ~~from the average ultimate vertical load, at least three additional tests shall be conducted. The allowable vertical load of the joist hanger shall be the lowest value determined from the following:~~

1. ~~The lowest ultimate vertical load for a single hanger from any test divided by three (where three tests are conducted and each ultimate vertical load does not vary more than 20 percent from the average ultimate vertical load).~~
2. ~~The average ultimate vertical load for a single hanger from all tests divided by three (where six or more tests are conducted).~~
3. ~~The average from all tests of the vertical loads that produce a vertical movement of the joist with respect to the header of $^{1}/_{8}$ inch (3.2 mm).~~
4. ~~The sum of the allowable design loads for nails or other fasteners utilized to secure the joist hanger to the wood members and allowable bearing loads that contribute to the capacity of the hanger.~~
5. ~~The allowable design load for the wood members forming the connection.~~

~~**1711.1.2.1 Design Value Modifications for Joist Hangers.** Allowable design values for joist hangers that are determined by Item 4 or 5 in Section 1711.1.2 shall be permitted to be modified by the appropriate duration of loading factors as specified in AF&PA NDS but shall not exceed the direct loads as determined by Item 1, 2 or 3 in Section 1711.1.2. Allowable design values determined by Item 1, 2 or 3 in Section 1711.1.2 shall not be modified by duration of loading factors.~~

~~**1711.1.3 Torsional Moment Capacity for Joist Hangers.** The torsional moment capacity for the joist hanger shall be determined by testing at least three joist hanger assemblies as specified in ASTM D 1761. The allowable torsional moment of the joist hanger shall be the average torsional moment at which the lateral movement of the top or bottom of the joist with respect to the original position of the joist is $^{1}/_{8}$ inch (3.2 mm).~~

~~**1711.2 Concrete and Clay Roof Tiles**~~ **1504.2.1 Testing.** Testing of concrete and clay roof tiles shall be in accordance with ~~Sections 1711.2.1 and 1711.2.2, as applicable.~~ Sections 1504.2.1.1 and 1504.2.1.2.

~~**1711.2.1**~~ **1504.2.1.1 Overturning Resistance.** Concrete and clay roof tiles shall be tested to determine their resistance to overturning due to wind in accordance with SBCCI SSTD 11 and Chapter 15.

~~**1711.2.2**~~ **1504.2.1.2 Wind Tunnel Testing.** Where concrete and clay roof tiles do not satisfy the limitations in Chapter 16 for rigid tile, a wind tunnel test shall be used to determine the wind characteristics of the concrete or clay tile roof covering in accordance with SBCCI SSTD 11 and Chapter 15.

CHAPTER 35

ASTM. ASTM D 7147-05, Specification for Testing and Establishing Allowable Loads of Joist Hangers.

CHANGE SIGNIFICANCE: Section 1711.1 has been deleted in its entirety because ASTM D1761-06 no longer contains provisions for the testing of joist hangers as the provisions were moved to ASTM D7147. The ASTM D7147 standard includes sampling and evaluation criteria as well as further refinements regarding the quality of test materials, adjustments for variation in test materials and limits on design values with materials other than those tested. In addition, because ASTM D7147 is specific to joist hangers used in wood construction and contains provisions that go beyond testing, the reference was appropriately relocated to Chapter 23. The remaining Section 1711 provisions in the 2012 IBC are now found in IBC Section 1504, which is more appropriate because it is applicable to performance requirements for roof coverings and assemblies.

1803.5

Investigated Conditions

CHANGE TYPE: Modification

CHANGE SUMMARY: The requirements addressing the evaluation of rock materials for foundation support have been updated to be more consistent with current geotechnical engineering practice. In addition, basic requirements for providing adequate underpinning and excavations have been added.

2015 CODE: **1803.5.6 Rock Strata.** Where subsurface explorations at the project site indicate variations ~~or doubtful characteristics~~ in the structure of ~~the~~ rock upon which foundations are to be constructed, a sufficient number of borings shall be drilled to sufficient depths to assess the competency of the rock ~~made to a depth of not less than 10 feet (3048 mm) below the level of the foundations to provide assurance of the soundness of the foundation bed~~ and its load-bearing capacity.

1803.5.7 Excavation Near Foundations. Where excavation will reduce ~~remove lateral~~ support from any foundation,~~an investigation shall be conducted to assess the potential consequences and address mitigation measures.~~ a registered design professional shall prepare an assessment of the structure as determined from examination of the structure, the review of available design documents and, if necessary, the excavation of test pits. The registered design professional shall determine the requirements for underpinning and protection and prepare site-specific plans, details and sequence of work for submission. Such support may be provided by underpinning, sheeting and bracing, or by other means acceptable to the building official.

Photo Courtesy of Crux Subsurface, Inc.

Geotechnical investigation

CHANGE SIGNIFICANCE: The past wording of Section 1803.5.6 suggested that it would be possible to provide "assurance of the soundness of rock" during the geotechnical evaluation phase, which may not necessarily be the case. Unfortunately, experience has shown that even at sites where rigorous evaluation of rock conditions is undertaken, it is often determined during construction that rock conditions between the locations sampled can vary significantly. Often the actual rock conditions at foundation locations are exposed or better defined (through excavation, proof-drilling, etc.) during construction, and interpretations of the conditions exposed during the construction process are necessary to complete the design of the foundation system. The modifications to Section 1803.5.6 express the characteristics necessary to assess the rock strata and estimate a load-bearing capacity based on observations and testing. Modifications to Section 1803.5.7 provide specific guidelines to identify responsibilities and basic requirements for providing safe and adequate underpinning and excavations. New Section 1804.2 was also added to provide specific requirements when underpinning is chosen to provide support for adjacent structures.

1804.1

Excavation Near Foundations

Foundation excavation

CHANGE TYPE: Addition

CHANGE SUMMARY: Basic requirements for providing safe and adequate underpinning at excavations have been added because the code was not specific on how to address excavations adjacent to structures.

2015 CODE: 1804.1 Excavation Near Foundations. Excavation for any purpose shall not ~~remove~~ reduce lateral support from any foundation or adjacent foundation without first underpinning or protecting the foundation against ~~settlement or lateral translation~~ detrimental lateral or vertical movement, or both.

1804.2 Underpinning. Where underpinning is chosen to provide the protection or support of adjacent structures, the underpinning system shall be designed and installed in accordance with provisions of this chapter and Chapter 33.

1804.2.1 Underpinning Sequencing. Underpinning shall be installed in a sequential manner that protects the neighboring structure and the working construction site. The sequence of installation shall be identified in the approved construction documents.

CHANGE SIGNIFICANCE: Specific requirements related to the excavation of foundations adjacent to structures had not previously been addressed in the IBC. Although Section 3307, Protection of Adjacent Property, requires adjoining public and private property, including footings, foundations, party walls and so forth, to be adequately protected from damage during construction, remodeling and demolition work, there were no specific details provided. Because the IBC contained very little detail, due diligence was required during excavations near neighboring structures to meet the intent of the code. Failures to perform proper pre-construction investigations and monitoring procedures have led to failures in construction during underpinning and excavation operations. Improperly constructed excavations have resulted in doors and windows that don't open, cracking of bearing walls and support members, failures of some critical structural members and even collapses resulting in fatalities.

Because the term "detrimental" is used to discuss settlement in other provisions of Chapter 18, as well as other chapters of the IBC, the term has been added here as well. The term "remove support" was changed to "reduce support," because the removal of support could lead to a failure. As indicated in Section 1803.5.7, underpinning is only one way of providing support; thus new Section 1804.2 provides requirements when underpinning is chosen to provide support.

1808.3

Design Surcharge Loads

CHANGE TYPE: Addition

CHANGE SUMMARY: Requirements pertaining to surcharge loads that could affect an adjacent structure have been added.

2015 CODE: **1808.3.2 Surcharge.** No fill or other surcharge loads shall be placed adjacent to any building or structure unless such building or structure is capable of withstanding the additional loads caused by the fill or the surcharge. Existing footings or foundations which will be affected by any excavation shall be underpinned or otherwise protected against settlement and shall be protected against detrimental lateral or vertical movement, or both.

> **Exception:** Minor grading for landscaping purposes shall be permitted where done with walk-behind equipment, where the grade is not increased more than one foot (305 mm) from original design grade, or where approved by the building official.

CHANGE SIGNIFICANCE: Although Chapter 33 covers safety requirements during construction, Chapter 18 has no specific provisions related to the effects of permanent loads that could surcharge a neighboring structure. Fill or other surcharge loads are not permitted to be placed adjacent to a building or structure unless it is capable of withstanding the additional loads caused by the surcharge load. Existing footings or foundations that could be affected by an excavation must be protected from detrimental lateral or vertical movement and settlement. When approved by the building official, minor grading for landscaping with limited grading heights using walk-behind equipment that does not induce high forces against an adjacent foundation or wall is permitted.

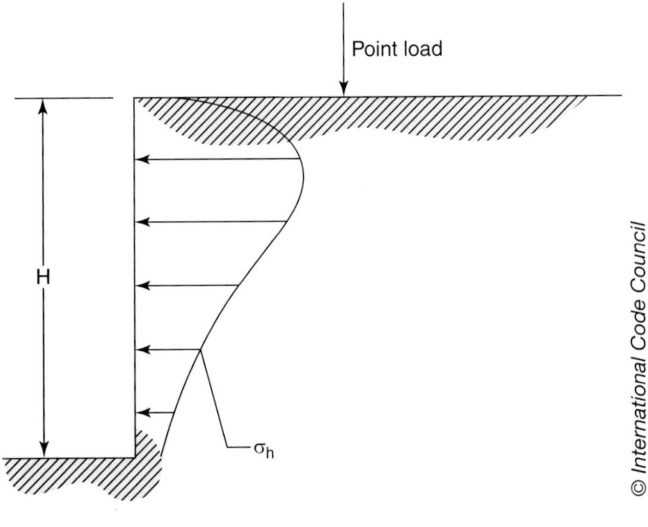

Surcharge loading

1810.2.5

Group Effects

Driving concrete pile foundations

CHANGE TYPE: Clarification

CHANGE SUMMARY: The requirements related to the evaluation of group effects on uplift of grouped deep foundation elements have been clarified.

2015 CODE: **1810.2.5 Group Effects.** The analysis shall include group effects on lateral behavior where the center-to-center spacing of deep foundation elements in the direction of lateral force is less than eight times the least horizontal dimension of an element. The analysis shall include group effects on axial behavior where the center-to-center spacing of deep foundation elements is less than three times the least horizontal dimension of an element. Group effects shall be evaluated using a generally accepted method of analysis; the analysis for uplift of grouped elements with center-to-center spacing less than three times the least horizontal dimension of an element shall be evaluated in accordance with Section 1810.3.3.1.6.

CHANGE SIGNIFICANCE: Group effects on uplift of deep foundations are to be evaluated where the spacing of elements is less than three times their least horizontal dimension. Although past provisions may have seemed clear without a change in language, Section 1810.3.3.1.6 is not specific as to the spacing that necessitates evaluation of group effects for uplift. Therefore, modifications were made to Section 1810.2.5 and cross referencing to Section 1810.3.3.1.6 (uplift capacity of grouped deep foundation elements) was added to help clarify the intent.

1810.3

Design and Detailing

CHANGE TYPE: Addition

CHANGE SUMMARY: Provisions addressing structural steel sheet piles have been added and the code provisions and standards related to steel deep foundation systems have been updated to clarify their intent.

2015 CODE: 1810.3.2.3 ~~Structural~~ Steel. Structural steel H-piles and structural steel sheet piling shall conform to the material requirements in ASTM A 6. Steel pipe piles shall conform to the material requirements in ASTM A 252 ~~and~~. Fully welded steel piles shall be fabricated from plates ~~shall~~ that conform to the material requirements in ASTM A 36, ~~ASTM A 252~~, ASTM A 283, ASTM A 572, ASTM A 588 or ASTM A 690 ~~ASTM A913 or ASTM A992~~.

1810.3.5.3 Steel. Steel deep foundation elements shall satisfy the requirements of this section.

1810.3.5.3.1 Structural Steel H-piles. Sections of structural steel H-piles shall comply with the requirements for HP shapes in ASTM A 6, or the following:

1. The flange projections shall not exceed 14 times the minimum thickness of metal in either the flange or the web and the flange widths shall not be less than 80 percent of the depth of the section.
2. The nominal depth in the direction of the web shall not be less than 8 inches (203 mm).
3. Flanges and web shall have a minimum nominal thickness of 3⁄8 inch (9.5 mm).

1810.3 continues

Sheet piling

TABLE 1810.3.2.6 Allowable Stresses for Materials Used in Deep Foundation Elements

Material Type and Condition	Maximum Allowable Stress[a]
3. ~~Structural steel~~ Steel in compression	
Cores within concrete-filled pipes or tubes	0.5 Fy ≤ 32,000 psi
Pipes, tubes or H-piles, where justified in accordance with Section 1810.3.2.8	0.5 Fy ≤ 32,000 psi
Pipes or tubes for micropiles	0.4 Fy ≤ 32,000 psi
Other pipes, tubes or H-piles	0.35 Fy ≤ 16,000 psi
Helical piles	0.6 Fy ≤ 0.5 Fu
5. ~~Structural steel~~ Steel in tension	
Pipes, tubes or H-piles, where justified in accordance with Section 1810.3.2.8	0.5 Fy ≤ 32,000 psi
Other pipes, tubes or H-piles	0.35 Fy ≤ 16,000 psi
Helical piles	0.6 Fy ≤ 0.5 Fu

a. f'_c is the specified compressive strength of the concrete or grout; *fpc* is the compressive stress on the gross concrete section due to effective prestress forces only; *fy* is the specified yield strength of reinforcement; *Fy* is the specified minimum yield stress of ~~structural~~ steel; *Fu* is the specified minimum tensile stress of ~~structural~~ steel.

(Portions of table not shown remain unchanged.)

1810.3 continued

1810.3.5.3.2 Fully Welded Steel Piles Fabricated from Plates. Sections of fully welded steel piles fabricated from plates shall comply with the following:

1. The flange projections shall not exceed 14 times the minimum thickness of metal in either the flange or the web and the flange widths shall not be less than 80 percent of the depth of the section.
2. The nominal depth in the direction of the web shall not be less than 8 inches (203 mm).
3. Flanges and web shall have a minimum nominal thickness of ⅜ inch (9.5 mm).

1810.3.5.3.3 Structural Steel Sheet Piling. Individual sections of structural steel sheet piling shall conform to the profile indicated by the manufacturer, and shall conform to the general requirements specified by ASTM A 6.

CHAPTER 35

ASTM. ASTM A 6/A 6M-11 Standard Specifications for General Requirements for Rolled Structural Steel Bars, Plates, Shapes and Sheet Piling.

CHANGE SIGNIFICANCE: Section 1810.3.2.3 has been updated to improve clarity as it applies to steel foundation elements. Structural steel is defined in Section 202, and steel pipe piles and fully welded steel piles do not necessarily fit the code's definition. The new language assigns appropriate ASTM references to the applicable foundation elements. ASTM A252 applies only to steel pipe piles. ASTM A913 and ASTM A992 both apply to structural shapes and not plates; thus they were not appropriate for fully welded steel piles fabricated from plates. The latest 2011 edition of ASTM A 6 has been added as the appropriate reference for the material requirements for H-piles and sheet piling. Because ASTM A 6 includes references to the applicable ASTM standards such as A36, A572, A690, A913 and A992, duplicate references of these standards are not necessary for H-piles and sheet piling.

Table 1810.3.2.6 has been coordinated with the title change to Section 1810.3.2.3. Structural steel is defined in Section 202, and although steel pipe and fully welded steel piles do not fully meet that classification, the intent is to apply the allowable stress limits to those sections as well. Consequently, the term "structural" has been deleted.

Sections 1810.3.5.3.1 and 1810.3.5.3.2 now clarify the code's intent by separating the requirements for structural steel H-piles from those addressing fully welded steel piles fabricated from plates, while also adding the new provisions on structural steel sheet piling. A reference to ASTM A 6 for HP shapes in Section 1810.3.5.3.1 automatically satisfies the three specified dimensional limitations. Additionally, allowance is made for other structural steel H-pile shapes if they meet the three-dimensional limitations. Clarifying language is added, with new Section 1810.3.5.3.2 permitting the three-dimensional limitations to be applied to fully welded steel piles fabricated from plates. New Section 1810.3.5.3.3 has been added for structural steel sheet piling, requiring that the profiles conform to the manufacturer's specifications and the general requirements in ASTM A 6. As an additional note, the new 2011 edition of the referenced standard ASTM A 6 has been added to Chapter 35.

1901.3

Anchoring to Concrete

CHANGE TYPE: Modification

CHANGE SUMMARY: Sections 1908 and 1909 of the 2012 IBC, which contain the requirements for anchorage to concrete, have been deleted because they are obsolete and not consistent with current referenced standards. In their place, new provisions on anchoring to concrete have been added to the general provisions found in Section 1901.

2015 CODE: <u>**1901.3 Anchoring to Concrete.** Anchoring to concrete shall be in accordance with ACI 318 as amended in Section 1905, and applies to cast-in (headed bolts, headed studs and hooked J- or L-bolts) anchors and post-installed expansion (torque-controlled and displacement-controlled), undercut and adhesive anchors.</u>

~~SECTION 1908~~
~~ANCHORAGE TO CONCRETE—ALLOWABLE STRESS DESIGN~~

~~**1908.1 Scope.** The provisions of this section shall govern the *allowable stress design* of headed bolts and headed stud anchors cast in normal-weight concrete for purposes of transmitting structural loads from one connected element to the other. These provisions do not apply to anchors installed in hardened concrete or where load combinations include earthquake loads or effects. The bearing area of headed anchors shall be not less than one and one-half times the shank area. Where strength design is used, or where load combinations include earthquake loads or effects, the design strength of anchors shall be determined in accordance with Section 1909. Bolts shall conform to ASTM A 307 or an *approved* equivalent.~~

1901.3 continues

Photo Courtesy of Powers Fasteners

Adhesive anchor installation

1901.3 continued

~~**1908.2 Allowable Service Load.** The allowable service load for headed anchors in shear or tension shall be as indicated in Table 1908.2. Where anchors are subject to combined shear and tension, the following relationship shall be satisfied:~~

$$(Ps / Pt)5/3 + (Vs / Vt)5/3 \leq 1 \qquad \textbf{(Equation 19-1)}$$

~~where:~~

~~*Ps* = Applied tension service load, pounds (N).~~
~~*Pt* = Allowable tension service load from Table 1908.2, pounds (N).~~
~~*Vs* = Applied shear service load, pounds (N).~~
~~*Vt* = Allowable shear service load from Table 1908.2, pounds (N).~~

~~**TABLE 1908.2**~~
~~**ALLOWABLE SERVICE LOAD ON EMBEDDED BOLTS (pounds)**~~

~~**1908.3 Required Edge Distance and Spacing.** The allowable service loads in tension and shear specified in Table 1908.2 are for the edge distance and spacing specified. The edge distance and spacing are permitted to be reduced to 50 percent of the values specified with an equal reduction in allowable service load. Where edge distance and spacing are reduced less than 50 percent, the allowable service load shall be determined by linear interpolation.~~

~~**1908.4 Increase in Allowable Load.** Increase of the values in Table 1908.2 by one-third is permitted where the provisions of Section 1605.3.2 permit an increase in allowable stress for wind loading.~~

~~**1908.5 Increase for Special Inspection.** Where special inspection is provided for the installation of anchors, a 100- percent increase in the allowable tension values of Table 1908.2 is permitted. No increase in shear value is permitted.~~

~~**SECTION 1909**~~
~~**ANCHORAGE TO CONCRETE—STRENGTH DESIGN**~~

~~**1909.1 Scope.** The provisions of this section shall govern the strength design of anchors installed in concrete for purposes of transmitting structural loads from one connected element to the other. Headed bolts, headed studs and hooked (J- or L-) bolts cast in concrete and expansion anchors and undercut anchors installed in hardened concrete shall be designed in accordance with Appendix D of ACI 318 as modified by Sections 1905.1.9 and 1905.1.10, provided they are within the scope of Appendix D.~~

~~The strength design of anchors that are not within the scope of Appendix D of ACI 318, and as amended in Sections 1905.1.9 and 1905.1.10, shall be in accordance with an *approved* procedure.~~

CHANGE SIGNIFICANCE: Provisions for anchorage to concrete by both allowable stress design and strength design have been deleted and replaced with new provisions in Section 1901 addressing anchoring to concrete. Past provisions dealing with anchorage to concrete by allowable stress design are no longer consistent with ACI 318, AISC 360 or ASCE 7. Anchorage to concrete by strength design is covered in ACI 318 as amended in Section 1905, and includes cast-in (headed bolts, headed studs and hooked J- or L-bolts) anchors as well as post-installed expansion (torque-controlled and displacement-controlled), undercut and adhesive anchors. Anchorage to concrete is now covered in Chapter 17 of ACI 318-14, which is the referenced standard for concrete design in the 2015 IBC.

1901.4

Composite Structural Steel and Concrete Structures

Concrete-filled pipe columns

Photo Courtesy of Magnusson Klemencic Associates (MKA)

CHANGE TYPE: Modification

CHANGE SUMMARY: Section 1912 of the 2012 IBC containing specific requirements for concrete-filled pipe columns has been deleted because it is no longer necessary, as new provisions on composite structural steel and concrete structures have been added to the general provisions in Section 1901.

2015 CODE: **1901.4 Composite Structural Steel and Concrete Structures.** Systems of structural steel acting compositely with reinforced concrete shall be designed in accordance with Section 2206 of this code.

~~SECTION 1912~~
~~CONCRETE-FILLED PIPE COLUMNS~~

~~**1912.1 General.** Concrete-filled pipe columns shall be manufactured from standard, extra-strong or double-extra-strong steel pipe or tubing that is filled with concrete so placed and manipulated as to secure maximum density and to ensure complete filling of the pipe without voids.~~

~~**1912.2 Design.** The safe supporting capacity of concrete filled pipe columns shall be computed in accordance with the approved rules or as determined by a test.~~

~~**1912.3 Connections.** Caps, base plates and connections shall be of *approved* types and shall be positively attached to the shell and anchored to the concrete core. Welding of brackets without mechanical anchorage shall be prohibited. Where the pipe is slotted to accommodate webs of brackets or other connections, the integrity of the shell shall be restored by welding to ensure hooping action of the composite section.~~

~~**1912.4 Reinforcement.** To increase the safe load-supporting capacity of concrete-filled pipe columns, the steel reinforcement shall be in the form of rods, structural shapes or pipe embedded in the concrete core with sufficient clearance to ensure the composite action of the section, but not nearer than 1 inch (25 mm) to the exterior steel shell. Structural shapes used as reinforcement shall be milled to ensure bearing on cap and base plates.~~

~~**1912.5 Fire-Resistance-Rating Protection.** Pipe columns shall be of such size or so protected as to develop the required fire-resistance ratings specified in Table 601. Where an outer steel shell is used to enclose the fire protective covering, the shell shall not be included in the calculations for strength of the column section. The minimum diameter of pipe columns shall be 4 inches (102 mm) except that in structures of Type V construction not exceeding three stories above grade plane or 40 feet (12 192 mm) in building height, pipe columns used in basements and as secondary steel members shall have a minimum diameter of 3 inches (76 mm).~~

~~**1912.6 Approvals.** Details of column connections and splices shall be shop fabricated by approved methods and shall be approved only after tests in accordance with the approved rules. Shop-fabricated concrete-filled pipe columns shall be inspected by the building official or by an approved representative of the manufacturer at the plant.~~

CHANGE SIGNIFICANCE: The design and construction of concrete-filled pipe columns is a composite structural system involving structural steel and concrete, addressed by the standards referenced in Section 2206. The requirements in former Section 1912 were incomplete, not current, and have not been maintained. Section 1901.4 has been added that appropriately refers the code user to Section 2206, which covers the design of systems using structural steel elements acting compositely with reinforced concrete and references AISC 360, ACI 318, ASCE 7 and AISC 341. Concrete-filled pipe columns are a composite structural system covered by the applicable standards.

1904

Durability Requirements

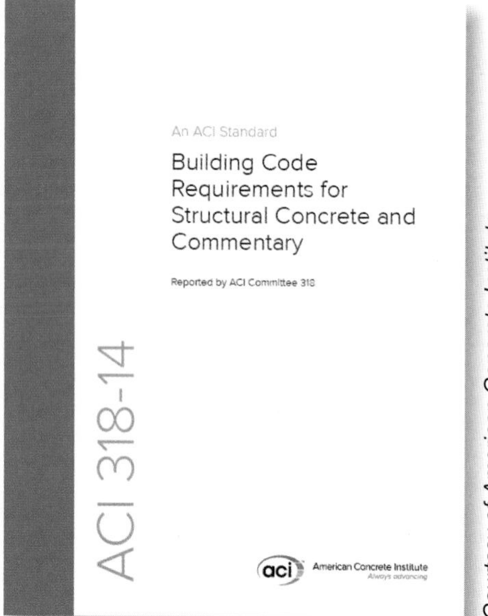

Courtesy of American Concrete Institute

2014 Edition of ACI 318

CHANGE TYPE: Modification

CHANGE SUMMARY: The durability requirements for structural concrete have been deleted from the IBC and replaced by a reference to the durability provisions in ACI 318.

2015 CODE:

SECTION 202 DEFINITIONS

NONSTRUCTURAL CONCRETE. Any element made of plain or reinforced concrete that is not part of a structural system required to transfer either gravity or lateral loads to the ground.

SECTION 1904 DURABILITY REQUIREMENTS

~~**1904.1 Exposure Categories and Classes.** Concrete shall be assigned to exposure classes in accordance with the durability requirements of ACI 318 based on:~~

~~1. Exposure to freezing and thawing in a moist condition or deicer chemicals;~~
~~2. Exposure to sulfates in water or soil;~~
~~3. Exposure to water where the concrete is intended to have low permeability; and~~
~~4. Exposure to chlorides from deicing chemicals, salt, saltwater, brackish water, seawater or spray from these sources, where the concrete has steel reinforcement.~~

1904.1 Structural Concrete. Structural concrete shall conform to the durability requirements of ACI 318.

> **Exception:** For Group R-2 and R-3 occupancies not more than three stories above grade plane, the specified compressive strength, f'_c, for concrete in basement walls, foundation walls, exterior walls and other vertical surfaces exposed to the weather shall be not less than 3000 psi (20.7 MPa).

~~**1904.2 Concrete Properties.** Concrete mixtures shall conform to the most restrictive maximum water-cementitious materials ratios, maximum cementitious admixtures, minimum air-entrainment and minimum specified concrete compressive strength requirements of ACI 318 based on the exposure classes assigned in Section 1904.1.~~

> ~~**Exception:** For occupancies and appurtenances thereto in Group R occupancies that are in buildings less than four stories above grade plane, normal-weight aggregate concrete is permitted to comply with the requirements of Table 1904.2 based on the weathering classification (freezing and thawing) determined from Figure 1904.2 in lieu of the durability requirements of ACI 318.~~

1904.2 Nonstructural Concrete. The registered design professional shall assign nonstructural concrete a freeze-thaw exposure class, as defined in ACI 318, based on the anticipated exposure of nonstructural concrete. Nonstructural concrete shall have a minimum specified compressive strength, f'_c, of 2500 psi (17.2 MPa) for Class F0; 3000 psi (20.7 MPa) for Class F1; and 3500 psi (24.1 MPa) for Classes F2 and F3. Nonstructural concrete shall be air entrained in accordance with ACI 318.

~~TABLE 1904.2~~
~~MINIMUM SPECIFIED COMPRESSIVE STRENGTH ($f'c$)~~

~~FIGURE 1904.2~~
~~WEATHERING PROBABILITY MAP FOR CONCRETE a,b,c~~

~~a. Lines defining areas are approximate only. Local areas can be more or less severe than indicated by the region classification.~~
~~b. A "severe" classification is where weather conditions encourage or require the use of deicing chemicals or where there is potential for a continuous presence of moisture during frequent cycles of freezing and thawing. A "moderate" classification is where weather conditions occasionally expose concrete in the presence of moisture to freezing and thawing, but where deicing chemicals are not generally used. A "negligible" classification is where weather conditions rarely expose concrete in the presence of moisture to freezing and thawing.~~
~~c. Alaska and Hawaii are classified as severe and negligible, respectively.~~

CHANGE SIGNIFICANCE: The durability provisions, weathering probability map, and minimum concrete strength table have been deleted and replaced with a reference to the durability requirements for structural concrete contained in Part 6 of ACI 318-14. The weathering probability map has been deleted because the concrete classes must be determined by the designer, regardless of geographic location. For clarification and usability, a new definition of non-structural concrete has been added to the IBC. An exception exempts Group R-2 and R-3 occupancies not more than three stories in height from the durability requirements by requiring not less than 3000-psi concrete in basement walls, foundation walls, exterior walls and other vertical surfaces exposed to the weather. ACI 332, *Building Code Requirements for Residential Concrete*, which is referenced in the IRC, specifies 3000-psi concrete for this condition. Because neither the IRC nor ACI 332 use the new exposure classes (F0, F1, F2, F3), the traditional "exposed to the weather" language is used in the exception. ACI 318 requires the designer to classify concrete into one of the freezing-and-thawing classes as follows:

- F0—Concrete not exposed to freezing-and-thawing cycles;
- F1—Concrete exposed to freezing-and-thawing cycles and occasional exposure to moisture;
- F2—Concrete exposed to freezing-and-thawing cycles and in continuous contact with moisture;
- F3—Concrete exposed to freezing-and-thawing cycles and in continuous contact with moisture and exposed to deicing chemicals.

The commentary to Part 6 of ACI 318-14 provides further discussion and examples to help the designer determine the appropriate class. The 2014 edition of ACI 318 is the applicable referenced standard for concrete design in the 2015 IBC.

1905.1.3

Modifications to ACI 318, Section 18.5

CHANGE TYPE: Modification

CHANGE SUMMARY: The requirements for the design of wall piers have been deleted from Section 1905 because they are now addressed in ACI 318.

2015 CODE:

SECTION 1905
MODIFICATIONS TO ACI 318

1905.1 General. The text of ACI 318 shall be modified as indicated in Sections 1905.1.1 through 1905.1.8 ~~1905.10~~.

1905.1.1 ACI 318, Section ~~2.2~~ 2.3. Modify existing definitions and add the following definitions to ACI 318, Section ~~2.2~~ 2.3.

> ~~***WALL PIER.*** *A wall segment with a horizontal length-to-thickness ratio of at least 2.5, but not exceeding 6, whose clear height is at least two times its horizontal length.*~~

1905.1.3 ACI 318, Section ~~21.4~~ 18.5. Modify ACI 318, Section ~~21.4~~ 18.5, by adding new Section 18.5.2.2 and renumbering existing Sections ~~21.4.3~~ 18.5.2.2 and 18.5.2.3 to become ~~21.4.4~~ 18.5.2.3 and 18.5.2.4, respectively. ~~and adding new Sections 21.4.3, 21.4.5, 21.4.6 and 21.4.7 to read as follows:~~

> ~~*21.4.3*~~ *18.5.2.2 - Connections that are designed to yield shall be capable of maintaining 80 percent of their design strength at the deformation induced by the design displacement or shall use Type 2 mechanical splices.*
>
> ~~*21.4.4*~~ *18.5.2.3 - Elements of the connection that are not designed to yield shall develop at least 1.5* S_y.
>
> *18.5.2.4 - In structures assigned to SDC D, E or F, wall piers shall be designed in accordance with 18.10.8 or 18.14 in ACI 318.*
>
> ~~*21.4.5 - Wall piers in Seismic Design Category D, E or F shall comply with Section 1905.1.4 of the International Building Code.*~~

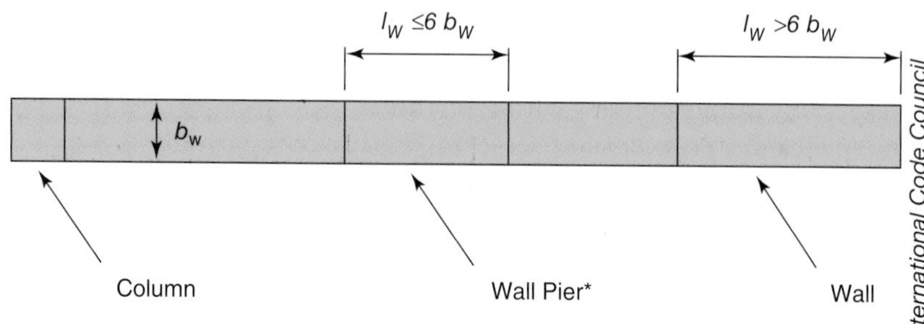

Wall pier definition in ACI 318

~~*21.4.6 - Wall piers not designed as part of a moment frame in buildings assigned to Seismic Design Category C shall have transverse reinforcement designed to resist the shear forces determined from 21.3.3. Spacing of transverse reinforcement shall not exceed 8 inches (203 mm). Transverse reinforcement shall be extended beyond the pier clear height for at least 12 inches (305 mm).*~~

~~***Exceptions:***~~

1. ~~*Wall piers that satisfy 21.13.*~~
2. ~~*Wall piers along a wall line within a story where other shear wall segments provide lateral support to the wall piers and such segments have a total stiffness of at least six times the sum of the stiffnesses of all the wall piers.*~~

~~*21.4.7 - Wall segments with a horizontal length-to thickness ratio less than 2.5 shall be designed as columns.*~~

~~**1905.1.4 ACI 318, Section 21.9.** Modify ACI 318, Section 21.9, by deleting Section 21.9.8 and replacing with the following:~~

~~*21.9.8 - Wall piers and wall segments.*~~

~~*21.9.8.1 - Wall piers not designed as a part of a special moment frame shall have transverse reinforcement designed to satisfy the requirements in 21.9.8.2.*~~

~~***Exceptions:***~~

1. ~~*Wall piers that satisfy 21.13.*~~
2. ~~*Wall piers along a wall line within a story where other shear wall segments provide lateral support to the wall piers and such segments have a total stiffness of at least six times the sum of the stiffnesses of all the wall piers.*~~

~~*21.9.8.2 - Transverse reinforcement with seismic hooks at both ends shall be designed to resist the shear forces determined from 21.6.5.1. Spacing of transverse reinforcement shall not exceed 6 inches (152 mm). Transverse reinforcement shall be extended beyond the pier clear height for at least 12 inches (305 mm).*~~

~~*21.9.8.3 - Wall segments with a horizontal length-to thickness ratio less than 2.5 shall be designed as columns.*~~

(*Because this code change deleted/revised substantial portions of Chapter 19, the entire code change text is too extensive to be included here. Refer to the* 2015 IBC Code Changes Resource Collection *for the complete text and history of code changes to Chapter 19.*)

CHANGE SIGNIFICANCE: The requirements for the design of wall piers were not included in the 2008 edition of ACI 318; therefore they were added to the 2012 IBC as an amendment to ACI 318 in Section 1905.1.3. A definition of "wall pier" has now been added, and design requirements are included in the special structural wall provisions of Section 18.10 of ACI 318-14. Thus the amendments in the IBC pertaining to the design of wall piers are no longer necessary and have been deleted from the code. It should be noted that the 2014 edition of ACI 318 is the referenced standard for concrete design in the 2015 IBC.

1905.1.8

Modifications to ACI 318, Section 17.2.3

CHANGE TYPE: Modification

CHANGE SUMMARY: Extensive modifications have been made to the concrete anchorage provisions of Section 1905.1.8 to maintain the intent regarding light-frame shear wall anchorage, while achieving consistency with Chapter 17 of the 2014 edition of ACI 318.

2015 CODE:

SECTION 1905
MODIFICATIONS TO ACI 318

~~1905.1.9~~ 1905.1.8 ACI 318, Section 17.2.3. ~~Delete ACI 318 Sections D.3.3.4 through D.3.3.7 and replace with the following:~~ Modify ACI 318 Sections 17.2.3.4.2, 17.2.3.4.3(d) and 17.2.3.5.2 to read as follows:

~~*D.3.3.4 - The anchor design strength associated with concrete failure modes shall be taken as 0.75øNn and 0.75øVn, where ø is given in D4.3 or D4.4 and Nn and Vn are determined in accordance with D5.2, D5.3, D5.4, D6.2 and D6.3, assuming the concrete is cracked unless it can be demonstrated that the concrete remains uncracked.*~~

~~*D.3.3.5 - Anchors shall be designed to be governed by the steel strength of a ductile steel element as determined in accordance with D.5.1 and D.6.1, unless either D.3.3.6 or D.3.3.7 is satisfied.*~~

17.2.3.4.2 - Where the tensile component of the strength-level earthquake force applied to anchors exceeds 20 percent of the total factored anchor tensile force associate with the same load combination, anchors and their attachments shall be designed in accordance with 17.2.3.4.3. The anchor design tensile strength shall be determined in accordance with 17.2.3.4.4.

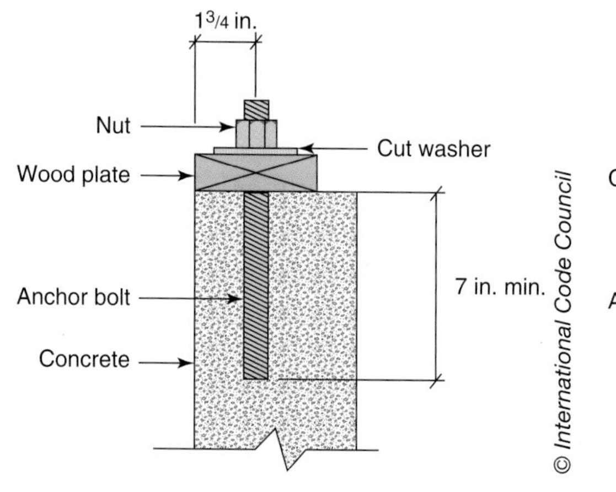

Light-frame wood

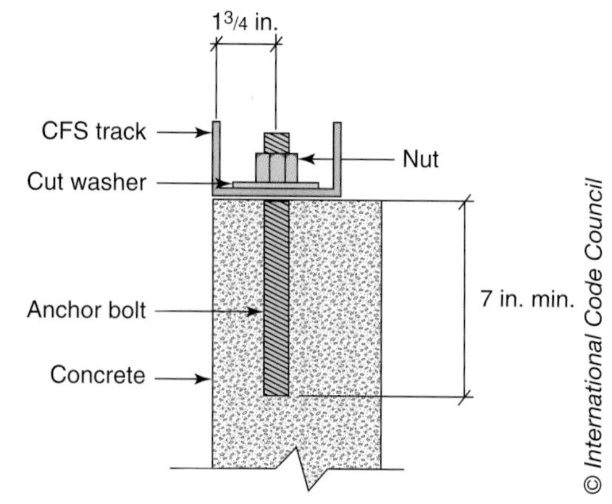

Cold-formed steel (CFS) shear wall anchorage

Exception:

~~*1.*~~ *Anchors designed to resist wall out-of-plane forces with design strengths equal to or greater than the force determined in accordance with ASCE 7 Equation 12.11-1 or 12.14-10 ~~need not~~ shall be deemed to satisfy Section ~~D.3.3.5~~ 17.2.3.4.3(d).*

~~*2. Anchors in concrete designed to support nonstructural components in accordance with ASCE 7 Section 13.4.2 need not satisfy Section D.3.3.4.3.*~~

*17.2.3.4.3(d) – The anchor or group of anchors shall be designed for the maximum tension obtained from design load combinations that include **E**, with **E** increased by Ω_0. The anchor design tensile strength shall be calculated from 17.2.3.4.4.*

17.2.3.5.2 – Where the shear component of the strength-level earthquake force applied to anchors exceeds 20 percent of the total factored anchor shear force associated with the same load combination, anchors and their attachments shall be designed in accordance with 17.2.3.5.3. The anchor design shear strength for resisting earthquake forces shall be determined in accordance with 17.5.

Exceptions:

1. ~~*D.3.3.5.3 need not apply and the design shear strength in accordance with D.6.2.1(c) need not be computed*~~ *For the calculation of the in-plane shear strength of anchor bolts attaching wood sill plates of bearing or non- bearing walls of light-frame wood structures to foundations or foundation stem walls, the in-plane shear strength in accordance with ~~D.6.2 and D.6.3~~ 17.5.2 and 17.5.3 need not be computed and ~~D.3.3.5.3~~ 17.2.3.5.3 shall be deemed to be satisfied provided all of the following are ~~satisfied~~ met:*

   ~~*2.1*~~ *1.1. The allowable in-plane shear strength of the anchor is determined in accordance with AWC NDS Table 11E for lateral design values parallel to grain.*

   ~~*2.2*~~ *1.2. The maximum anchor nominal diameter is $^5/_8$ inches (16 mm).*

   ~~*2.3*~~ *1.3. Anchor bolts are embedded into concrete a minimum of 7 inches (178 mm).*

   ~~*2.4*~~ *1.4. Anchor bolts are located a minimum of 1 $^3/_4$ inches (45 mm) from the edge of the concrete parallel to the length of the wood sill plate.*

   ~~*2.5*~~ *1.5. Anchor bolts are located a minimum of 15 anchor diameters from the edge of the concrete perpendicular to the length of the wood sill plate.*

   ~~*2.6*~~ *1.6. The sill plate is 2-inch or 3-inch nominal thickness.*

2. ~~*Section D.3.3.5.3 need not apply and the design shear strength in accordance with Section D.6.2.1(c) need not be computed*~~ *For the calculation of the in-plane shear strength of anchor bolts attaching cold-formed steel track of bearing or non-bearing walls of light-frame construction to foundations*

1905.1.8 continues

1905.1.8 continued

or foundation stem walls, the in-plane shear strength in accordance with 17.5.2 and 17.5.3 need not be computed and 17.2.3.5.3 shall be deemed to be satisfied provided all of the following are ~~satisfied~~ met:

2.1. The maximum anchor nominal diameter is $^5/_8$ inches (16 mm).

2.2. Anchors are embedded into concrete a minimum of 7 inches (178 mm).

2.3. Anchors are located a minimum of $1^3/_4$ inches (45 mm) from the edge of the concrete parallel to the length of the track.

2.4. Anchors are located a minimum of 15 anchor diameters from the edge of the concrete perpendicular to the length of the track.

2.5. The track is 33 to 68 mil designation thickness.

Allowable in-plane shear strength of exempt anchors, parallel to the edge of concrete shall be permitted to be determined in accordance with AISI S100 Section E3.3.1.

~~3. Anchors in concrete designed to support nonstructural components in accordance with ASCE 7 Section 13.4.2 need not satisfy Section D.3.3.4.3.~~

~~4~~3. In light-frame construction, bearing or nonbearing walls, shear strength of concrete anchors less than or equal to 1 inch (25 mm) in diameter ~~connecting~~ of sill plate or track to foundation or foundation stem wall need not satisfy ~~D.3.3.5.3~~ 17.2.3.5.3(a) through (c) when the design strength of the anchors is determined in accordance with ~~D.6.2.1(c)~~ 17.5.2.1(c).

~~**1905.1.10 ACI 318, Section D.4.2.2.** Delete ACI 318, Section D.4.2.2, and replace with the following:~~

~~***D.4.2.2*** *For anchors with diameters not exceeding 4 in., the concrete breakout strength requirements shall be considered satisfied by the design procedure of D.5.2 and D.6.2. For anchors in shear with diameters exceeding 4 inches, shear anchor reinforcement shall be provided in accordance with the procedures of D.6.2.9.*~~

CHANGE SIGNIFICANCE: Although the 2012 IBC references ACI 318-11 for concrete design and construction, the text of Section 1905 reflects modifications to ACI 318-08. Extensive revisions to Section 1905.1.9 allowed it to maintain the intent of previous changes to the anchorage provisions for light-frame wood and cold-formed steel (CFS) shear walls while achieving consistency with the anchorage provisions in Appendix D of the 2011 edition of ACI 318, which have undergone significant changes from ACI 318-08. The new language in Section 1905.8 of the 2015 IBC represents industry consensus among various code change proposals dealing with concrete anchorage and is intended to have a positive impact on the design community by removing confusion resulting from the inconsistency between 2012 IBC Chapter 19 and ACI 318. Anchorage to concrete is now covered in Chapter 17 of ACI 318-14. It should be noted that the 2014 edition of ACI 318 is the approved referenced standard for concrete design in the 2015 IBC.

2101.2

Masonry Design Methods

CHANGE TYPE: Modification

CHANGE SUMMARY: The references in Chapter 21 to specific sections in the Masonry Standards Joint Committee (MSJC) code have been deleted because the 2013 edition of TMS 402/ACI 530/ASCE 5 has been substantially reorganized to be more user-friendly. The charging language of Section 2101.2 has been modified to simply reference TMS 402/ACI 530/ASCE 5 or TMS 403 for the design and construction of masonry structures.

2015 CODE:

SECTION 202
DEFINITIONS

~~**ANCHOR.** Metal rod, wire or strap that secures masonry to its structural support.~~

2101.2 Design Methods. Masonry shall comply with the provisions of ~~one of the following design methods in this chapter~~ TMS 402/ACI 530/ASCE 5 or TMS 403 as well as the ~~requirements of Sections 2101 through 2104. Masonry designed by the allowable stress design provisions of Section 2101.2.1, the strength design provisions of Section 2101.2.2, the prestressed masonry provisions of Section 2101.2.3, or the direct design requirements of Section 2101.2.7 shall comply with Section 2105~~ applicable requirements of this chapter.

~~**2101.2.1 Allowable Stress Design.** Masonry designed by the allowable stress design method shall comply with the provisions of Sections 2106 and 2107.~~

~~**2101.2.2 Strength Design.** Masonry designed by the strength design method shall comply with the provisions of Sections 2106 and 2108, except that autoclaved aerated concrete (AAC) masonry shall comply with the provisions of Section 2106 and Chapter 1 and Appendix A of TMS 402/ACI 530/ASCE 5.~~

~~**2101.2.3 Prestressed Masonry.** Prestressed masonry shall be designed in accordance with Chapters 1 and 4 of TMS 402/ACI 530/ASCE 5 and Section 2106. Special inspection during construction shall be provided as set forth in Section 1705.4.~~

~~**2101.2.4 Empirical Design.** Masonry designed by the empirical design method shall comply with the provisions of Sections 2106 and 2109 or Chapter 5 of TMS 402/ACI 530/ASCE 5.~~

~~**2101.2.5 Glass Unit Masonry.** Glass unit masonry shall comply with the provisions of Section 2110 or Chapter 7 of TMS 402/ACI 530/ASCE 5.~~

~~**2101.2.6**~~ **2101.2.1 Masonry Veneer.** Masonry veneer shall comply with the provisions of Chapter 14 ~~or Chapter 6 of TMS 402/ACI 530/ASCE 5~~.

2101.2 continues

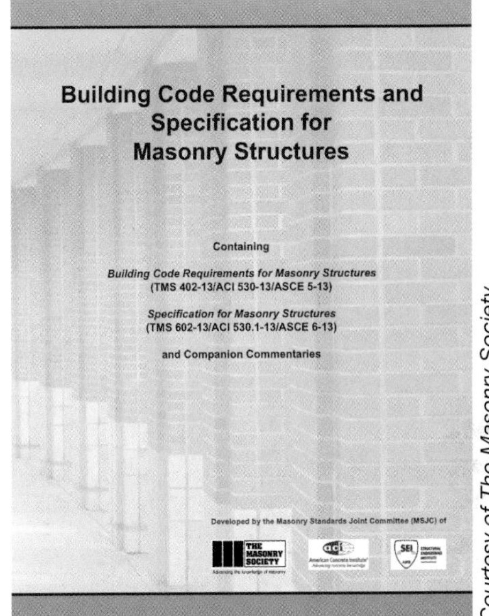

Building Code Requirements and Specification for Masonry Structures
Containing
Building Code Requirements for Masonry Structures (TMS 402-13/ACI 530-13/ASCE 5-13)
Specification for Masonry Structures (TMS 602-13/ACI 530.1-13/ASCE 6-13)
and Companion Commentaries
Developed by the Masonry Standards Joint Committee (MSJC) of
THE MASONRY SOCIETY

Courtesy of The Masonry Society

The 2013 Masonry Standards Joint Committee (MSJC) code, specification, and commentry

2101.2 continued

~~**2101.2.7 Direct Design.** Masonry designed by the direct design method shall comply with the provisions of TMS 403.~~

~~**2101.3 Construction Documents.** The construction documents shall show all of the items required by this code including the following:~~

1. ~~Specified size, grade, type and location of reinforcement, anchors and wall ties.~~
2. ~~Reinforcing bars to be welded and welding procedure.~~
3. ~~Size and location of structural elements.~~
4. ~~Provisions for dimensional changes resulting from elastic deformation, creep, shrinkage, temperature and moisture.~~
5. ~~Loads used in the design of masonry.~~
6. ~~Specified compressive strength of masonry at stated ages or stages of construction for which masonry is designed, except where specifically exempted by this code.~~
7. ~~Details of anchorage of masonry to structural members, frames and other construction, including the type, size and location of connectors.~~
8. ~~Size and permitted location of conduits, pipes and sleeves.~~
9. ~~The minimum level of testing and inspection as defined in Chapter 17, or an itemized testing and inspection program that meets or exceeds the requirements of Chapter 17~~.

2101.3 Special Inspection. The special inspection of masonry shall be as defined in Chapter 17, or an itemized testing and inspection program shall be provided that meets or exceeds the requirements of Chapter 17.

2101.3.1 2111.2 Fireplace Drawings. The construction documents shall describe in sufficient detail the location, size and construction of masonry fireplaces. The thickness and characteristics of materials and the clearances from walls, partitions and ceilings shall be indicated.

CHANGE SIGNIFICANCE: The series of references in Section 2101.2 to specific sections of the IBC and TMS 402/ACI 530/ASCE 5 (known as the Masonry Standards Joint Committee [MSJC] code) have become unnecessary over time. Because the 2013 edition of the MSJC code has been substantially reorganized, it was found to be simpler to delete the references to various sections rather than update and maintain them. The charging language of Section 2101.2 is now simplified so that it references TMS 402/ACI 530/ASCE 5 or TMS 403 for the design and construction of masonry structures. The reference to Chapter 14 for masonry veneers is maintained because Chapter 14 addresses some types of masonry veneers not covered by the MSJC code and specification, such as anchored stone veneer. Chapter 14 continues to contain a reference to Chapter 6 of the reference standard. The construction document requirements set forth in former Section 2101.3 are virtually identical to the requirements of Section 1.2.2 of the MSJC code and have therefore been deleted. A reference to Chapter 17 for special inspection is maintained in new

Section 2101.3 to facilitate compliance with the special inspection requirements for masonry structures. Former Section 2101.3.1 for fireplace drawings has been relocated to Section 2111.2, which covers requirements specific to fireplaces. The term "anchor" is deleted from Section 2102.1 because that term is used generically throughout the IBC for all types of building materials, and the definition of the term "anchor" in the MSJC code is identical to the IBC definition.

2103

Masonry Construction Materials

CHANGE TYPE: Modification

CHANGE SUMMARY: Masonry material provisions that have historically been found in Section 2103 have been deleted because they are contained in the MSJC Specification TMS 602/ACI 530.1/ASCE 6.

2015 CODE:

SECTION 202
DEFINITIONS

~~**THIN-BED MORTAR.** Mortar for use in construction of AAC unit masonry with joints 0.06 inch (1.5 mm) or less.~~

2102.1 General. The following terms are defined in Chapter 2:

~~**THIN-BED MORTAR**~~.

2103.1 Masonry Units. Concrete masonry units, clay or shale masonry units, stone masonry units, glass unit masonry and AAC masonry units shall comply with Article 2.3 of TMS 602/ACI 530.1/ASCE 6. Architectural cast stone shall conform to ASTM C1364.

~~**2103.1 Concrete Masonry Units.** Concrete masonry units shall conform to the following standards: ASTM C 55 for concrete brick; ASTM C 73 for calcium silicate face brick; ASTM C 90 for load-bearing concrete masonry units or ASTM C 744 for prefaced concrete and calcium silicate masonry units.~~

~~**2103.2 Clay or Shale Masonry Units.** Clay or shale masonry units shall conform to the following standards: ASTM C 34 for structural clay load-bearing wall tile; ASTM C 56 for structural clay nonload-bearing wall tile; ASTM C 62 for building brick (solid masonry units made from clay or shale); ASTM C 1088 for solid units of thin veneer brick; ASTM C 126 for~~

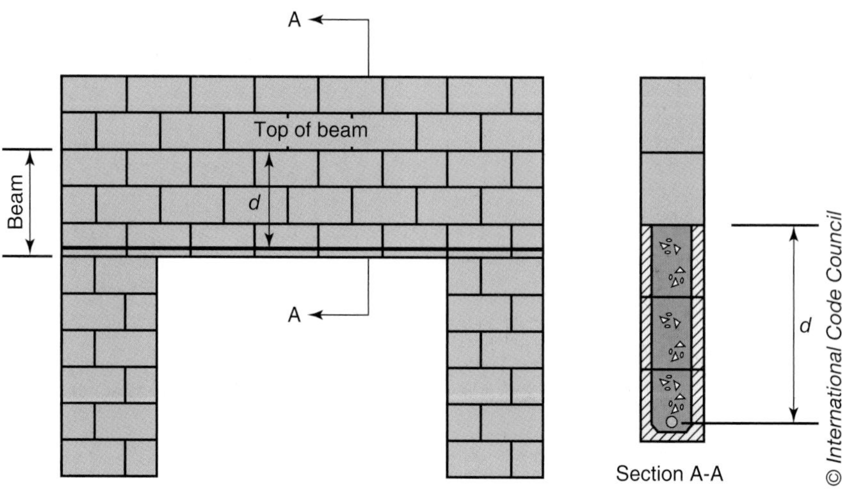

Typical "d" distance for a masonry beam per MSJC code

~~ceramic-glazed structural clay facing tile, facing brick and solid masonry units; ASTM C 212 for structural clay facing tile; ASTM C 216 for facing brick (solid masonry units made from clay or shale); ASTM C 652 for hollow brick (hollow masonry units made from clay or shale) or ASTM C 1405 for glazed brick (single-fired solid brick units).~~

> **Exception:** Structural clay tile for nonstructural use in fireproofing of structural members and in wall furring shall not be required to meet the compressive strength specifications. The fire-resistance rating shall be determined in accordance with ASTM E 119 or UL 263 and shall comply with the requirements of Table 602.

~~**2103.3 AAC Masonry.** AAC masonry units shall conform to ASTM C 1386 for the strength class specified.~~

~~**2103.4 Stone Masonry Units.** Stone masonry units shall conform to the following standards: ASTM C 503 for marble building stone (exterior); ASTM C 568 for limestone building stone; ASTM C 615 for granite building stone; ASTM C 616 for sandstone building stone; or ASTM C 629 for slate building stone.~~

~~**2103.5 Architectural Cast Stone.** Architectural cast stone shall conform to ASTM C 1364.~~

~~**2103.6 Ceramic Tile.** Ceramic tile shall be as defined in, and shall conform to the requirements of, ANSI A137.1.~~

~~**2103.7 Glass Unit Masonry.** Hollow glass units shall be partially evacuated and have a minimum average glass face thickness of 3/16 inch (4.8 mm). Solid glass-block units shall be provided when required. The surfaces of units intended to be in contact with mortar shall be treated with a polyvinyl butyral coating or latex-based paint. Reclaimed units shall not be used.~~

~~2103.8~~ 2103.1.1 Second-Hand Units. Second-hand masonry units shall not be reused unless they conform to the requirements of new units. The units shall be of whole, sound materials and free from cracks and other defects that will interfere with proper laying or use. Old mortar shall be cleaned from the unit before reuse.

~~**2103.9 Mortar.** Mortar for use in masonry construction shall conform to ASTM C 270 and Articles 2.1 and 2.6 A of TMS 602/ACI 530.1/ASCE 6, except for mortars listed in Sections 2103.10, 2103.11 and 2103.12. Type S or N mortar conforming to ASTM C 270 shall be used for glass unit masonry.~~

2103.2 Mortar. Mortar for masonry construction shall comply with Section 2103.2.1, 2103.2.2, 2103.2.3 or 2103.4.

2103.2.1 Masonry Mortar. Mortar for use in masonry construction shall conform to Articles 2.1 and 2.6 A of TMS 602/ACI 530.1/ASCE 6.

2103 continues

2103 continued

~~2103.10~~ 2103.2.2 Surface-Bonding Mortar. Surface-bonding mortar shall comply with ASTM C 887. Surface bonding of concrete masonry units shall comply with ASTM C 946.

~~2103.11~~ 2103.2.3 Mortars for Ceramic Wall and Floor Tile. Portland cement mortars for installing ceramic wall and floor tile shall comply with ANSI A108.1A and ANSI A108.1B and be of the compositions indicated in Table 2103.11.

2103.2.4 Mortar for Adhered Masonry Veneer. Mortar for use with adhered masonry veneer shall conform to ASTM C 270 for Type N or S, or shall comply with ANSI A118.4 for latex-modified Portland cement mortar.

~~**2103.12 Mortar for AAC Masonry.** Thin-bed mortar for AAC masonry shall comply with Article 2.1 C.1 of TMS 602/ ACI 530.1/ASCE 6. Mortar used for the leveling courses of AAC masonry shall comply with Article 2.1 C.2 of TMS 602/ ACI 530.1/ASCE 6.~~

~~2103.13~~ 2103.3 Grout. Grout shall comply with Article 2.2 of TMS 602/ACI 530.1/ASCE 6.

~~2103.14~~ 2103.4 Metal Reinforcement and Accessories. Metal reinforcement and accessories shall conform to Article 2.4 of TMS 602/ACI 530.1/ASCE 6. Where unidentified reinforcement is approved for use, not less than three tension and three bending tests shall be made on representative specimens of the reinforcement from each shipment and grade of reinforcing steel proposed for use in the work.

TABLE ~~2103.11~~ 2103.2.3
CERAMIC TILE MORTAR COMPOSITIONS

CHANGE SIGNIFICANCE: Many masonry material specifications that have historically been provided in IBC Section 2103 are also contained in the referenced standard, TMS 602/ACI 530.1/ASCE 6. Therefore, the code provisions have been deleted and replaced with a reference to the appropriate article in the MSJC specification. Those provisions that are not included in the MSJC code or specification continue to be maintained in Section 2103. These include architectural cast stone meeting ASTM C1364, compressive strength exemptions for structural clay tile used as fireproofing, second-hand units, surface-bonding mortar, mortars for tile, and testing of unidentified reinforcement and accessories.

2104
Masonry Construction

CHANGE TYPE: Modification

CHANGE SUMMARY: Many masonry construction provisions previously found in Section 2104 that are contained in the MSJC Specification TMS 602/ACI 530.1/ASCE 6 have been deleted and replaced with references to the specification.

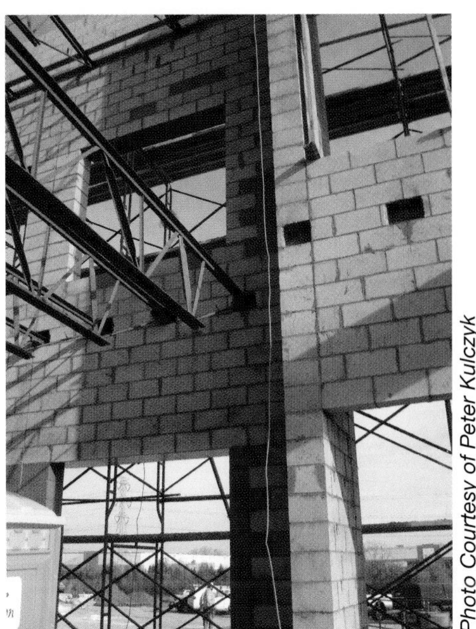

Photo Courtesy of Peter Kulczyk

Masonry construction

2015 CODE: 2104.1 Masonry Construction. Masonry construction shall comply with the requirements of Sections ~~2104.1.1 through 2104.4~~ 2104.1.1, 2104.1.2 and with TMS 602/ACI 530.1/ASCE 6.

~~**2104.1.1 Tolerances.** Masonry, except masonry veneer, shall be constructed within the tolerances specified in TMS 602/ACI 530.1/ASCE 6.~~

~~**2104.1.2 Placing Mortar and Units.** Placement of mortar, grout, and clay, concrete, glass, and AAC masonry units shall comply with TMS 602/ACI 530.1/ASCE 6.~~

~~**2104.1.3 Installation of Wall Ties.** Wall ties shall be installed in accordance with TMS 602/ACI 530.1/ASCE 6.~~

~~**2104.1.4 Chases and Recesses.** Chases and recesses shall be constructed as masonry units are laid. Masonry directly above chases or recesses wider than 12 inches (305 mm) shall be supported on lintels.~~

~~**2104.1.5 Lintels.** The design for lintels shall be in accordance with the masonry design provisions of either Section 2107 or 2108.~~

~~**2104.1.6**~~ **2104.1.1 Support on Wood.** Masonry shall not be supported on wood girders or other forms of wood construction except as permitted in Section 2304.12.

~~**2104.2 Corbeled Masonry.** Corbeled masonry shall comply with the requirements of Section 1.12 of TMS 402/ACI 530/ASCE 5.~~

~~**2104.2.1**~~ **2104.1.2 Molded Cornices.** Unless structural support and anchorage are provided to resist the overturning moment, the center of gravity of projecting masonry or molded cornices shall lie within the middle one-third of the supporting wall. Terra cotta and metal cornices shall be provided with a structural frame of approved noncombustible material anchored in an approved manner.

~~**2104.3 Cold Weather Construction.** The cold weather construction provisions of TMS 602/ACI 530.1/ASCE 6, Article 1.8 C, shall be implemented when the ambient temperature falls below 40°F (4°C).~~

~~**2104.4 Hot Weather Construction.** The hot weather construction provisions of TMS 602/ACI 530.1/ASCE 6, Article 1.8 D, shall be implemented when the ambient air temperature exceeds 100°F (37.8°C), or 90°F (32.2°C) with a wind velocity greater than 8 mph (12.9 km/hr).~~

2104 continues

2104 continued

CHANGE SIGNIFICANCE: Many masonry construction requirements previously located in Section 2103 that are contained in TMS 602/ ACI 530.1/ASCE 6 have been deleted and replaced with a reference to the appropriate article in the MSJC specification. The provisions that are not included in the MSJC code or specification are maintained in Section 2104, such as support of masonry on wood construction, and support and anchorage of molded cornices and terra cotta.

2105

Quality Assurance

CHANGE TYPE: Modification

CHANGE SUMMARY: Provisions for the quality assurance of masonry structures and related definitions have been deleted from Section 2105 and replaced with a reference to the MSJC Specification TMS 602/ACI 530.1/ASCE 6 and the special inspection and testing requirements contained in Chapter 17.

2015 CODE:

SECTION 202
DEFINITIONS

~~**COMPRESSIVE STRENGTH OF MASONRY.** Maximum compressive force resisted per unit of net cross-sectional area of masonry, determined by the testing of masonry prisms.~~

~~**PRISM.** An assemblage of masonry units and mortar with or without grout used as a test specimen for determining properties of the masonry.~~

2102.1 General. For the purposes of this chapter and as used elsewhere in this code, the following terms are defined in Chapter 2:

~~**COMPRESSIVE STRENGTH OF MASONRY.**~~
~~**PRISM.**~~

2105 continues

Photo Courtesy of Masonry Institute of America (MIA)

Inspection of cells and reinforcement location prior to grouting

2105 continued

SECTION 2105
QUALITY ASSURANCE

~~**2105.1 General.**~~
~~**2105.2 Acceptance Relative to Strength Requirements.**~~
~~**2105.2.1 Compliance with f ′m and f ′AAC.**~~
~~**2105.2.2 Determination of Compressive Strength.**~~
~~**2105.2.2.1 Unit Strength Method.**~~
~~**2105.2.2.1.1 Clay Masonry.**~~

~~**TABLE 2105.2.2.1.1**~~
~~**COMPRESSIVE STRENGTH OF CLAY MASONRY**~~

~~**2105.2.2.1.2 Concrete Masonry.**~~

~~**TABLE 2105.2.2.1.2**~~
~~**COMPRESSIVE STRENGTH OF CONCRETE MASONRY**~~

~~**2105.2.2.1.3 AAC Masonry.**~~
~~**2105.2.2.2 Prism Test Method.**~~
~~**2105.2.2.2.1 General.**~~
~~**2105.2.2.2.2 Number of Prisms Per Test.**~~
~~**2105.3 Testing Prisms from Constructed Masonry.**~~
~~**2105.3.1 Prism Sampling and Removal.**~~
~~**2105.3.2 Compressive Strength Calculations.**~~
~~**2105.3.3 Compliance.**~~

2105.1 General. A quality assurance program shall be used to ensure that the constructed masonry is in compliance with the construction documents.

The quality assurance program shall comply with the inspection and testing requirements of Chapter 17 and TMS 602/ACI 530.1/ASCE 6.

CHANGE SIGNIFICANCE: Because the quality assurance provisions found in Section 2105 of the 2012 IBC are essentially the same as those contained in the referenced MSJC code and specification, they have been deleted from the code. The revised section now simply references TMS 602/ACI 530.1/ASCE 6 for quality assurance and Chapter 17 for special inspection and testing requirements. Section 1705.4 references the quality assurance program requirements of TMS 402/ACI 530/ASCE 5 and TMS 602/ACI 530.1/ASCE 6 for special inspections and tests of masonry construction, with some exceptions.

CHANGE TYPE: Clarification

CHANGE SUMMARY: The definitions of "masonry fireplace" and "masonry chimney" have been deleted from Chapter 21 and appropriately relocated to Chapter 2. Requirements for the reinforcement and anchorage of masonry fireplaces and chimneys in Sections 2111 and 2113 have been updated and reorganized to clarify the intent.

2111, 2113

Masonry Fireplaces and Chimneys

Photo Courtesy of GeoNet

Brick chimney collapse

2015 CODE:

SECTION 2111 MASONRY FIREPLACES

~~**2111.1 Definition.** A masonry fireplace is a fireplace constructed of concrete or masonry. Masonry fireplaces shall be constructed in accordance with this section.~~

2111.1 General. The construction of masonry fireplaces, consisting of concrete or masonry, shall be in accordance with this section.

2111.4 Seismic Reinforcement. In structures assigned to Seismic Design Category A or B, seismic reinforcement ~~and seismic anchorage are~~ is not required. ~~Masonry or concrete fireplaces shall be constructed, anchored, supported and reinforced as required in this chapter.~~ In structures assigned to Seismic Design Category C or D, masonry ~~and concrete~~ fireplaces shall be reinforced and anchored as detailed in Sections 2111.3.1, 2111.3.2, and 2111.4 ~~and 2111.4.1 for chimneys serving fireplaces~~. In structures assigned to Seismic Design Category E or F, masonry ~~and concrete chimneys~~ fireplaces shall be reinforced in accordance with the requirements of Sections 2101 through 2108.

2111.5 Seismic Anchorage. Masonry ~~and concrete chimneys~~ fireplaces and foundations ~~in structures assigned to Seismic Design Category C or D shall be anchored at each floor, ceiling or roof line more than 6 feet (1829 mm) above grade, except where constructed completely within the exterior walls. Anchorage shall conform to the following requirements.~~ shall be anchored at each floor, ceiling or roof line more than 6 feet (1829 mm) above grade with two $^{3}/_{16}$-inch by 1-inch (4.8 mm by 25 mm) straps embedded a minimum of 12 inches (305 mm) into the chimney. Straps shall be hooked around the outer bars and extend 6 inches (152 mm) beyond the bend. Each strap shall be fastened to a minimum of four floor joists with two 1/2-inch (12.7 mm) bolts.

> **Exception:** Seismic anchorage is not required for the following:
>
> 1. In structures assigned to Seismic Design Category A or B.
> 2. Where the masonry fireplace is constructed completely within the exterior walls.

2111, 2113 continues

2111, 2113 continued

~~**2111.4.1 Anchorage.** Two $^{3}/_{16}$-inch by 1-inch (4.8 mm by 25.4 mm) straps shall be embedded a minimum of 12 inches (305 mm) into the chimney. Straps shall be hooked around the outer bars and extend 6 inches (152 mm) beyond the bend. Each strap shall be fastened to a minimum of four floor joists with two 1/2-inch (12.7 mm) bolts.~~

SECTION 2113 MASONRY CHIMNEYS

~~**2113.1 Definition.** A masonry chimney is a chimney constructed of solid masonry units, hollow masonry units grouted solid, stone or concrete, hereinafter referred to as "masonry." Masonry chimneys shall be constructed, anchored, supported and reinforced as required in this chapter.~~

2113.1 General. The construction of masonry chimneys consisting of solid masonry units, hollow masonry units grouted solid, stone or concrete shall be in accordance with this section.

2113.3 Seismic Reinforcement. ~~Masonry or concrete chimneys shall be constructed, anchored, supported and reinforced as required in this chapter.~~ In structures assigned to Seismic Design Category A or B, seismic reinforcement is not required. In structures assigned to Seismic Design Category C or D, masonry ~~and concrete~~ chimneys shall be reinforced and anchored as detailed in Sections 2113.3.1, 2113.3.2 and 2113.4. ~~In structures assigned to Seismic Design Category A or B, reinforcement and Seismic anchorage is not required~~. In structures assigned to Seismic Design Category E or F, masonry ~~and concrete~~ chimneys shall be reinforced in accordance with the requirements of Section~~s~~ 2113.4 ~~2101 through 2108~~.

2113.4 Seismic Anchorage. Masonry ~~and concrete~~ chimneys and foundations ~~in structures assigned to Seismic Design Category C or D shall be anchored at each floor, ceiling or roof line more than 6 feet (1829 mm) above grade, except where constructed completely within the exterior walls. Anchorage shall conform to the following requirements.~~ shall be anchored at each floor, ceiling or roof line more than 6 feet (1829 mm) above grade with two $^{3}/_{16}$-inch by 1-inch (4.8 mm by 25 mm) straps embedded a minimum of 12 inches (305 mm) into the chimney. Straps shall be hooked around the outer bars and extend 6 inches (152 mm) beyond the bend. Each strap shall be fastened to a minimum of four floor joists with two ½-inch (12.7 mm) bolts.

> **Exception:** Seismic anchorage is not required for the following;
>
> 1. In structures assigned to Seismic Design Category A or B.
> 2. Where the masonry fireplace is constructed completely within the exterior walls.

~~**2113.4.1 Anchorage.** Two $^{3}/_{16}$-inch by 1-inch (4.8 mm by 25 mm) straps shall be embedded a minimum of 12 inches (305 mm) into the chimney. Straps shall be hooked around the outer bars and extend 6 inches (152 mm) beyond the bend. Each strap shall be fastened to a minimum of four floor joists with two ½-inch (12.7 mm) bolts.~~

CHANGE SIGNIFICANCE: Chapter 21 deals with masonry and not concrete; thus the references to concrete fireplaces and chimneys have been deleted in both Sections 2111 and 2113, respectively. Although Section 2111 of the 2012 IBC addresses masonry fireplaces, Section 2111.3 refers to chimneys. Due to this inconsistency, the text implies that fireplaces in Seismic Design Categories (SDC) C and D are required to be anchored even though they are not required to be anchored in SDCs E and F. Because of these issues, the provisions in both Sections 2111 and 2113 have been revised to remove inconsistencies and reorganized to better clarify the intent. In an additional change, the definitions of "masonry fireplace" and "masonry chimney" have been appropriately relocated to Chapter 2.

2210

Cold-Formed Steel

Steel deck

Photo Courtesy of the Steel Deck Institute www.sdi.org

CHANGE TYPE: Modification

CHANGE SUMMARY: A new Steel Deck Institute (SDI) standard addressing the design and construction of composite concrete slabs and steel decks has been added to IBC Chapter 35.

2015 CODE:

SECTION 2210
COLD-FORMED STEEL

2210.1 General. The design of cold-formed carbon and low-alloy steel structural members shall be in accordance with AISI S100. The design of cold-formed stainless-steel structural members shall be in accordance with ASCE 8. Cold-formed steel light-frame construction shall also comply with Section 2211. Where required, the seismic design of cold-formed steel structures shall be in accordance with the additional provisions of Section 2210.2.

2210.1.1 Steel Decks. The design and construction of cold-formed steel decks shall be in accordance with this section.

2210.1.1.3 Composite Slabs on Steel Decks. Composite slabs of concrete and steel deck shall be permitted to be designed and constructed in accordance with SDI-C.

CHAPTER 35

SDI. SDI-C-2011 Standard for Composite Steel Floor Deck Slabs.

CHANGE SIGNIFICANCE: Previous editions of the IBC contained no specific provisions for the design of composite slabs on steel decks. The previous ASCE 3 standard referenced in the 2009 IBC was deleted from the 2012 IBC, which resulted in designers and building officials having to use the alternate materials and methods of construction provisions of Section 104.11 for these common structural systems. The 2011 Standard for Composite Steel Floor Deck Slabs (SDI-C) is an update of the previous 2006 version developed and approved through a consensus process under ANSI guidelines. The standard is available for download from the SDI website at www.sdi.org.

2211

Cold-Formed Steel Light-Frame Construction

CHANGE TYPE: Modification

CHANGE SUMMARY: A new American Iron and Steel Institute standard, AISI S220, is now referenced for the construction of cold-formed steel light-frame non-structural products.

2015 CODE:

SECTION 2211
COLD-FORMED STEEL LIGHT-FRAME CONSTRUCTION

2211.1 General. The design and installation of structural members and nonstructural members utilized in cold-formed steel light-frame construction where the specified minimum base steel thickness is ~~between 0.0179 inches (0.455 mm) and~~ not greater than 0.1180 inches (2.997 mm) shall be in accordance with AISI S200 and Sections 2211.2 through 2211.7, or AISI S220, as applicable.

2211.3.3 Trusses Spanning 60 Feet or Greater. The owner or the owner's authorized agent shall contract with a registered design professional for the design of the temporary installation restraint/bracing and the permanent individual truss member restraint/bracing for trusses with clear spans 60 feet (18 288 mm) or greater. Special inspection of trusses over 60 feet (18 288 mm) in length shall conform to Section 1705.

2211.4 Structural Wall Stud Design. Structural wall studs shall be designed in accordance with either AISI S211 or AISI S100.

2211 continues

© William Britten/E+/Getty Images

Cold-formed steel construction

2211 continued

CHAPTER 35

AISI. AISI S220—11 North American Standard for Cold-formed Steel Framing-Nonstructural Members.

CHANGE SIGNIFICANCE: The American Iron and Steel Institute (AISI), ASTM Committees C11 and A05, and the steel framing industry worked together on a "code synchronization" effort, the goal of which was to organize and maintain building code requirements for cold-formed steel light-frame construction products. Interested parties and stakeholders such as the Steel Framing Industry Association (SFIA), the Steel Stud Manufacturers Association (SSMA), the Association of the Wall and Ceiling Industry (AWCI), and the Gypsum Association (GA) all participated in this effort. The result has been the development of a new standard that includes requirements for cold-formed steel light-frame non-structural members. The scope of Section 2201.1 now includes cold-formed steel light-frame non-structural products. Section 2211.1 has been modified because with the addition of the reference for non-structural cold-formed steel framing, the lower limit of the minimum base thickness has been deleted. Section 2211.4 has been modified so the charging language reflects the distinction between AISI S211 and AISI S220. Section 2203 references AISI S220, Section A6.5, which includes requirements for the identification and protection of non-structural cold-formed steel framing. Table 2506.2 has been modified to refer to "structural" and "non-structural" CFS studs and track, and AISI S200 and AISI S220 are incorporated into the table as primary references. Modifications to Table 2507.2 have also been made to coordinate the table entries with changes to Table 2506.2.

2303.1.4

Structural Glued Cross-Laminated Timber

CHANGE TYPE: Addition

CHANGE SUMMARY: A new definition for a wood-based product identified as cross-laminated timber (CLT) has been added to Chapter 2. The new manufacturing standard ANSI/APA PRG 320 is now referenced in Chapter 23 and has been added to Chapter 35.

2015 CODE:

SECTION 202
DEFINITIONS

CROSS-LAMINATED TIMBER. A prefabricated engineered wood product consisting of at least three layers of solid-sawn lumber or structural composite lumber where the adjacent layers are cross-oriented and bonded with structural adhesive to form a solid wood element.

2303.1.4 Structural Glued Cross-Laminated Timber. Cross-laminated timbers shall be manufactured and identified as required in ANSI/APA PRG 320.

CHAPTER 35

APA. ANSI/APA PRG 320-2011 Standard for Performance-Rated Cross-Laminated Timber.

2303.1.4 continues

Photo Courtesy of FP Innovations

Cross-laminated timber

2303.1.4 continued

CHANGE SIGNIFICANCE: First developed in Europe 20 years ago, cross-laminated timber (CLT) has been used extensively there as large-section structural timber. A new North American product manufacturing standard, ANSI/APA PRG 320-2011, *Standard for Performance-Rated Cross-Laminated Timber*, provides requirements and test methods for qualification and quality assurance for performance-rated cross-laminated timber, which is manufactured from solid-sawn lumber or structural composite lumber. This large-section, engineered wood product is now defined in the code, ANSI/APA PRG 320 is referenced in Section 2303.1.4, and the newly developed consensus manufacturing standard has been added as a referenced standard to Chapter 35.

2303.1.13

Engineered Wood Rim Board

CHANGE TYPE: Addition

CHANGE SUMMARY: A new definition for engineered wood rim board has been added to Chapter 2 and two new standards are now referenced in Chapter 23 and have been added to Chapter 35.

2015 CODE:

SECTION 202
DEFINITIONS

ENGINEERED WOOD RIM BOARD. A full-depth structural composite lumber, wood structural panel, structural glued laminated timber, or prefabricated wood I-joist member designed to transfer horizontal (shear) and vertical (compression) loads, provide attachment for diaphragm sheathing, siding and exterior deck ledgers, and provide lateral support at the ends of floor or roof joists or rafters.

2303.1.13 Engineered Wood Rim Board. Engineered wood rim boards shall conform to ANSI/APA PRR 410 or shall be evaluated in accordance with ASTM D 7672. Structural capacities shall be in accordance with ANSI/APA PRR 410 or established in accordance with ASTM D 7672. Rim boards conforming to ANSI/APA PRR 410 shall be marked in accordance with that standard.

CHAPTER 35

APA. ANSI/APA PRR 410-2011 Standard for Performance-Rated Engineered Wood Rim Boards.

ASTM. ASTM D 7672-2011e1 Standard Specifications for Evaluating Structural Capacities of Rim Board Products and Assemblies.

Photo Courtesy of APA – The Engineered Wood Association

Engineered rim board

CHANGE SIGNIFICANCE: An engineered rim board is an important structural element in many engineered wood floor applications where both structural load path through the perimeter member and dimensional change compatibility are important design considerations. Two new standards address products intended for engineered wood rim board applications. Both ANSI/APA PRR 410 and ASTM D7672 address the fundamental requirements for the testing and evaluation of engineered rim board. ASTM D7672 is applicable in the determination of product-specific rim board performance (i.e., structural capacities) for engineered wood products that may be recognized in a manufacturer's product evaluation reports. The PRR 410 standard also includes performance categories for engineered wood products used in engineered rim board applications. Under PRR 410, products are assigned a grade based on performance category (e.g., categories based on structural capacity) and must bear a mark in accordance with the grade.

2304.6

Exterior Wall Sheathing

CHANGE TYPE: Modification

CHANGE SUMMARY: Section 2304.6 has been modified to establish minimum structural performance requirements and clarify that wall sheathing on the outside of exterior walls, as well as connection of sheathing to framing, must be capable of resisting wind pressures in accordance with Section 1609, which in turn references ASCE/SEI 7-10. The new term "gable" has been added to clarify that exterior wall sheathing requirements for out-of-plane wind resistance are equally applicable to the gable area at end walls.

2015 CODE:

SECTION 202
DEFINITIONS

GABLE. The triangular portion of the wall beneath a dual-slope, pitched, or mono-slope roof.

2304.6 Exterior Wall Sheathing. ~~Except as provided for in Section 1405 for weatherboarding or where stucco construction that complies with Section 2510 is installed, enclosed buildings shall be sheathed with one of the materials of the nominal thickness specified in Table 2304.6 or any other approved material of equivalent strength or durability~~ Wall sheathing on the outside of exterior walls, including gables, and the connection of sheathing to framing shall be designed in accordance with the general provisions of this code and shall be capable of resisting wind pressures in accordance with Section 1609.

Photo Courtesy of SBE Builders

Inspection of exterior wall sheathing

TABLE ~~2304.6~~ 2308.5.11 Minimum Thickness of Wall Sheathing

Sheating Type	Minimum Thickness	Maximum Wall Stud Spacing
Diagonal wood boards	5/8 inch	24 inches on center
Structural fiberboard	½ inch	16 inches on center
Wood structural panel	In accordance with Tables 2308.6.3(2) and 2308.6.3(3)	—
M-S "Exterior Glue" and M-2 "Exterior Glue" Particleboard	In accordance with Section 2306.3 and Table 2308.6.3(4)	—
Gypsum sheathing	½ inch	16 inches on center
~~Gypsum wallboard~~	~~½ inch~~	~~24 inches on center~~
Reinforced cement mortar	1 inch	24 inches on center
Hardboard panel siding	In accordance with Table 2308.6.3(5)	—

For SI: 1 inch = 25.4 mm.

2304.6.1 Wood Structural Panel Sheathing. Where wood structural panel sheathing is used as the exposed finish on the outside of exterior walls, it shall have an exterior exposure durability classification. Where wood structural panel sheathing is used elsewhere, but not as the exposed finish, it shall be of a type manufactured with exterior glue (Exposure 1 or Exterior). ~~Wood structural panel wall sheathing or siding used as structural sheathing shall be capable of resisting wind pressures in accordance with Section 1609. Maximum wind speeds for wood~~ Wood structural panel sheathing, ~~used to resist wind pressures~~ connections and framing spacing shall be in accordance with Table 2304.6.1 for the applicable wind speed and exposure category when used with enclosed buildings with a mean roof height not greater than 30 feet (9144 mm) and a topographic factor (K_{zt}) of 1.0.

CHANGE SIGNIFICANCE: Unless the prescriptive conventional wood frame construction provisions of Section 2308 are used, wall sheathing installed on the outside of exterior walls, including gable ends, and the connection of sheathing to the framing system must be designed in accordance with the general provisions of the code and be capable of resisting wind pressures in accordance with Section 1609. Because Table 2304.6 in the 2012 IBC is essentially a prescriptive table, it has been appropriately relocated to Section 2308. Section 2304.6 has been rewritten to establish minimum structural performance requirements and clarify that wall sheathing on the outside of exterior walls, as well as the connection of sheathing to framing, must be capable of resisting wind pressures in accordance with Section 1609. The term "gable" is included to clarify that exterior wall sheathing requirements for out-of-plane wind resistance are also applicable to the gable area at end walls.

2304.6 continues

2304.6 continued

Modifications to Section 2304.6.1 coordinate with the minimum structural performance requirements of Section 2304.6. Prior language dealing with design for out-of-plane wind resistance has been deleted because it is now addressed in the modified language of Section 2304.6. Gypsum wallboard has been removed from the table to make clear that the table only applies to exterior wall sheathing. A new Section 2308.5.11 has been added that requires the outside of exterior walls, including gables, of enclosed buildings to be sheathed with materials specified in Table 2308.9.3 and fastened in accordance with Table 2304.10.1, or as an alternative, be designed in accordance with accepted engineering practice.

2304.10.6
Load Path

CHANGE TYPE: Modification

CHANGE SUMMARY: The minimum required thickness of steel straps used to splice discontinuous framing members has been modified to be consistent with the standard thickness established in the new AISI Product Data Standard, S201.

2015 CODE: 2304.10.6 Load Path. Where wall framing members are not continuous from foundation sill to roof, the members shall be secured to ensure a continuous load path. Where required, sheet metal clamps, ties or clips shall be formed of galvanized steel or other approved corrosion-resistant material not less than 0.0329 inch (0.0836 mm) base metal ~~0.040 inch (1.01 mm) nominal~~ thickness.

CHANGE SIGNIFICANCE: The minimum required 0.040-inch thickness of sheet metal steel connections as set forth in the 2012 IBC is not consistent with standard thicknesses in the new AISI Product Data Standard, S201-12. AISI S201 provides criteria, material requirements and product requirements for structural and non-structural members utilized in cold-formed steel framing applications where the specified minimum base steel thickness is between 18 mils (0.0179 inches) and 11 mils (0.1180 inches). The minimum thickness specified in the code represented galvanized 20-gage steel. The term "gage" is no longer a steel thickness designation and is now designated as 33 mils. According to the AISI Product Standard S201, the base metal thickness that corresponds to 33 mils is 0.0329 inches. The modified minimum required base metal thickness of 0.0329 inches for steel straps used to splice discontinuous framing members is now consistent with the standard thickness in the new AISI standard.

Photo Courtesy of Simpson Strong-Tie Company Inc.

Coil strap

2304.12

Protection Against Decay and Termites

CHANGE TYPE: Modification

CHANGE SUMMARY: Modifications to Section 2304.12 identify exactly where waterborne preservatives are required and where they are not required.

2015 CODE: 2304.~~11~~12 Protection Against Decay and Termites. Wood shall be protected from decay and termites in accordance with the applicable provisions of Sections 2304.12.1 through ~~2304.11.9~~ 2304.12.7.

~~**2304.11.1 General.** Where required by this section, protection from decay and termites shall be provided by the use of naturally durable or preservative-treated wood.~~

~~2304.11.2 Wood Used Above Ground~~ 2304.12.1 Locations Requiring Water-Borne Preservatives or Naturally Durable Wood. Wood used above ground in the locations specified in Sections ~~2304.11.2.1~~ 2304.12.1.1 through ~~2304.11.2.7~~ 2304.12.1.5, 2304.12.3 ~~2304.11.3~~ and 2304.12.5 shall be naturally durable wood or preservative-treated wood using water-borne preservatives, in accordance with AWPA U1 ~~(Commodity Specifications A or F)~~ for above-ground use.

~~2304.11.2.1~~ 2304.12.1.1 Joists, Girders and Subfloor. Where wood joists or the bottom of a wood structural floor without joists are closer than 18 inches (457 mm), or wood girders are closer than 12 inches (305 mm) to the exposed ground in crawl spaces or unexcavated areas located within the perimeter of the building foundation~~, the floor construction (including posts, girders, joists and subfloor)~~ shall be of naturally durable or preservative-treated wood.

~~2304.11.2.2~~ 2304.12.1.2 Wood Supported by Exterior Foundation Walls. Wood framing members, including wood sheathing, that ~~rest on~~ are in contact with exterior foundation walls and are less than 8 inches (203 mm) from exposed earth shall be of naturally durable or preservative-treated wood.

Photo Courtesy of BB&S Treated Lumber of New England

Treated lumber

~~2304.11.2.3~~ 2304.12.1.3 Exterior Walls Below Grade. Wood framing members and furring strips ~~attached directly to~~ in direct contact with the interior of exterior masonry or concrete walls below grade shall be of naturally durable or preservative-treated wood.

2304.12.2 Other Locations. Wood used in the locations specified in Sections 2304.12.2.1 through 2304.12.2.5 shall be naturally durable wood or preservative treated wood in accordance with AWPA U1. Preservative treated wood used in interior locations shall be protected with two coats of urethane, shellac, latex epoxy or varnish unless waterborne preservatives are used. Prior to application of the protective finish, the wood shall be dried in accordance with the manufacturer's recommendations.

~~2304.11.2.7~~ 2304.12.2.2 Posts or Columns. Posts or columns supporting permanent structures and supported by a concrete or masonry slab or footing that is in direct contact with the earth shall be of naturally durable or preservative-treated wood.

Exceptions:

~~1.~~ Posts or columns that are ~~either~~ not exposed to the weather ~~or located in basements or cellars,~~ are supported by concrete piers or metal pedestals projected at least 1 inch (25 mm) above the slab or deck and ~~6~~ 8 inches (152 mm) above exposed earth, and are separated ~~therefrom~~ by an impervious moisture barrier.

~~2. Posts or columns in enclosed crawl spaces or unexcavated areas located within the periphery of the building, supported by a concrete pier or metal pedestal at a height greater than 8 inches (203 mm) from exposed ground, and are separated therefrom by an impervious moisture barrier.~~

2304.12.2.5. Supporting Members for Permeable Floors and Roofs. Wood structural members that support moisture-permeable floors or roofs that are exposed to the weather, such as concrete or masonry slabs, shall be of naturally durable or preservative-treated wood unless separated from such floors or roofs by an impervious moisture barrier.

~~2304.11.4~~ 2304.12.3 Wood in Contact with the Ground or Fresh Water. Wood used in contact with the ground (exposed earth) ~~in the locations specified in Sections 2304.11.4.1 and 2304.11.4.2~~ shall be naturally durable (species for both decay and termite resistance) or preservative treated ~~using water-borne preservatives~~ in accordance with AWPA U1 ~~(Commodity Specifications A or F)~~ for soil or fresh water use.

Exception: Untreated wood is permitted where such wood is continuously and entirely below the groundwater level or submerged in fresh water.

~~2304.11.4.1~~ 2304.12.3.1 Posts or Columns. Posts and columns supporting permanent structures that are embedded in concrete that is ~~in direct contact with the earth, embedded in concrete that is~~ exposed to the weather or in direct contact with the earth shall be of preservative-treated wood.

2304.12 continues

2304.12 continued

~~**2304.11.4.2 Wood Structural Members.** Wood structural members that support moisture-permeable floors or roofs that are exposed to the weather, such as concrete or masonry slabs, shall be of naturally durable or preservative-treated wood unless separated from such floors or roofs by an impervious moisture barrier.~~

~~**2304.11.7**~~ **2304.12.5 Wood Used in Retaining Walls and Cribs.** Wood installed in retaining or crib walls shall be preservative treated in accordance with AWPA U1 ~~(Commodity Specifications A or F)~~ for soil and fresh water use.

(Sections not shown remain unchanged.)

CHANGE SIGNIFICANCE: Requirements for protection against decay and termites in Section 2304.12 have been reorganized, and technical changes clarify where waterborne preservatives are required and where they are not. References to commodity specifications in the U1 standard have been deleted because Section 6 of AWPA's U1 standard provides a listing of all specifications for treated wood commodities. It also provides information on the listed preservative systems and species/species groupings that can be treated under AWPA Standards for each use category (use exposure condition). The use category descriptions are given in Section 2 of the U1 standard.

By rewording Section 2304.12.1.1 and deleting "the floor construction (including posts, girders, joists and subfloor)," it has been clarified that only the floor elements within the specified proximity to exposed ground need to be protected. The new Section 2304.12.2 references the subsections for specific locations where other-than waterborne preservatives are permitted under certain circumstances provided treatment is in accordance with the AWPA U1 standard. The requirement that waterborne preservatives be used exclusively has been deleted from Section 2304.12.3 because the specific locations where water-borne preservatives must be used is clearly specified in Section 2304.12.1.

CHANGE TYPE: Modification

CHANGE SUMMARY: Section 2308, which contains prescriptive requirements for conventional wood frame construction, has been reformatted and reorganized in its entirety. Significant changes include the introduction of new designations for wall bracing methods similar to those in the IRC as shown in new Table 2308.6.3(1), and reformatted wall bracing requirements set forth in Table 2308.6.1.

2308

Conventional Light-Frame Construction

2015 CODE:

SECTION 2308 CONVENTIONAL LIGHT-FRAME CONSTRUCTION

2308.1 General. The requirements of this section are intended for conventional light-frame construction. Other construction methods are permitted to be used, provided a satisfactory design is submitted showing compliance with other provisions of this code. Interior non-load-bearing partitions, ceilings and curtain walls of conventional light-frame construction are not subject to the limitations of Section 2308.2. Detached one- and two-family dwellings and multiple single-family dwellings (townhouses) not more than three stories above grade plane in height with a separate means of egress and their accessory structures shall comply with the *International Residential Code.*

2308.1.1 Portions Exceeding Limitations of Conventional Light-Frame Construction. When portions of a building of otherwise conventional light-frame construction exceed the limits of Section 2308.2, those portions and the supporting load path shall be designed in accordance with accepted engineering practice and the provisions of this code. For the purposes of this section, the term "portions" shall mean parts of buildings containing volume and area such as a room or a series of rooms. The extent of such design need only demonstrate compliance of the

2308 continues

Photo Courtesy of Peter Kulczyk

Conventional wood frame construction

2308 continued

non-conventionally light-framed elements with other applicable provisions of this code and shall be compatible with the performance of the conventional light-framed system.

2308.1.2 Connections and Fasteners. Connectors and fasteners used in conventional construction shall comply with the requirements of Section 2304.10.

2308.2 Limitations. Buildings are permitted to be constructed in accordance with the provisions of conventional light-frame construction, subject to the limitations in Sections 2308.2.1 through 2308.2.6.

2308.2.1 Stories. Structures of conventional light-frame construction shall be limited in story height according to Table 2308.2.1.

2308.2.2 Allowable Floor-to-Floor Height. Maximum floor-to-floor height shall not exceed 11 feet, 7 inches (3531 mm). Exterior bearing wall and interior braced wall heights shall not exceed a stud height of 10 feet (3048 mm).

2308.2.3 Allowable Loads. Loads shall be in accordance with Chapter 16 and shall not exceed the following:

1. Average dead loads shall not exceed 15 psf (718 N/m^2) for combined roof and ceiling, exterior walls, floors and partitions.

   Exceptions:

   1. Subject to the limitations of Section 2308.6.9.2, stone or masonry veneer up to the lesser of 5 inches (127 mm) thick or 50 psf (2395 N/m^2) and installed in accordance with Chapter 14 is permitted to a height of 30 feet (9144 mm) above a noncombustible foundation, with an additional 8 feet (2438 mm) permitted for gable ends.
   2. Concrete or masonry fireplaces, heaters and chimneys shall be permitted in accordance with the provisions of this code.

2. Live loads shall not exceed 40 psf (1916 N/m^2) for floors.
3. Ground snow loads shall not exceed 50 psf (2395 N/m^2).

TABLE 2308.2.1 Allowable Story Height

Seismic Design Category	Allowable Story Above Grade Plane
A and B	Three stories
C	Two Stories
D and E[a]	One story

a. For the purposes of this section, for buildings assigned to *Seismic Design Category* D or E, cripple walls shall be considered to be a *story* unless cripple walls are solid blocked and do not exceed 14 inches in height.

TABLE 2308.6.1 Wall Bracing Requirements

Seismic Design Category	Story Condition (See Section 2308.2)	Maximum Spacing of Braced Wall Lines	Braced Panel Location, Spacing (o.c.) and Minimum Percentage (x) — Bracing Method			Maximum Distance of Braced Wall Panels from Each End of Braced Wall Line
			LIB	DWB WSP	SFB PBS PCP HPS GB,[c,d]	
A and B		35′-0″	Each end and ≤25′-0″ o.c	Each end and ≤25′-0″ o.c	Each end and ≤25′-0″ o.c	12′-6″
		35′-0″	Each end and ≤25′-0″ o.c	Each end and ≤25′-0″ o.c	Each end and ≤25′-0″ o.c	12′-6″
		35′-0″	NP	Each end and ≤25′-0″ o.c	Each end and ≤25′-0″ o.c	12′-6″
C		35′-0″	NP	Each end and ≤25′-0″ o.c	Each end and ≤25′-0″ o.c	12′-6″
		35′-0″	NP	Each end and ≤25′-0″ o.c.(min. 25% of wall length)[e]	Each end and ≤25′-0″ o.c. (min. 25% of wall length)[e]	12′-6″
D and E		25′-0″	NP	Sds < 0.50: Each end and ≤25′-0″ o.c. (min. 21% of wall length)[e] 0.5 ≤ Sds < 0.75: Each end and ≤25′-0″ o.c. (min. 32% of wall length)[e] 0.75 ≤ Sds ≤ 1.00: Each end and ≤25′-0″ o.c. (min. 37% of wall length)[e] Sds > 1.00: Each end and ≤25′-0″ o.c. (min. 48% of wall length)[e]	Sds < 0.50: Each end and ≤25′-0″ o.c. (min. 43% of wall length)[e] 0.5 ≤ Sds < 0.75: Each end and ≤25′-0″ o.c. (min. 59% of wall length)[e] 0.75 ≤ Sds ≤ 1.00: Each end and ≤25′-0″ o.c. (min. 75% of wall length)[e] Sds > 1.00: Each end and ≤25′-0″ o.c. (min. 100% of wall length)[e]	8′-0″

For SI: 1 inch = 25.4 mm, 1 foot = 304.8 mm.
NP = Not Permitted

a. This table specifies minimum requirements for *braced wall panels* along interior or exterior *braced wall lines*.
b. See Section 2308.6.2 for full description of bracing methods.
c. For method GB, gypsum wallboard applied to framing supports that are spaced at 16 inches on center.
d. The required lengths shall be doubled for gypsum board applied to only one face of a braced wall panel.
e. Percentage shown represents the minimum amount of bracing required along the building length (or wall length if the structure has an irregular shape).

2308 continues

2308 continued

2308.2.4 Ultimate Wind Speed. V_{ult} shall not exceed 130 miles per hour (57 m/s) (3-second gust).

Exceptions:

1. V_{ult} shall not exceed 140 mph (61.6 m/s) (3-second gust) for buildings in Exposure Category B that are not located in a hurricane-prone region.
2. Where V_{ult} exceeds 130 mph (3-second gust), the provisions of either AWC WFCM or ICC 600 are permitted to be used.

2308.2.5 Allowable Roof Span. Ceiling joist and rafter framing constructed in accordance with Section 2308.7 and trusses shall not span more than 40 feet (12 192 mm) between points of vertical support. A ridge board in accordance with Section 2308.7 or 2308.7.3.1 shall not be considered a vertical support.

2308.2.6 Risk Category Limitation. The use of the provisions for conventional light-frame construction in this section shall not be permitted for Risk Category IV buildings assigned to Seismic Design Category B, C, D or E.

(Because Code Change S273-12 deleted and replaced Section 2308 in its entirety, the entire code change text is too extensive to be included here. Refer to the 2015 IBC Code Changes Resource Collection *for the complete text and history of the code change.)*

CHANGE SIGNIFICANCE: The majority of changes to Section 2308 consist of reorganization and reformatting, although several substantive changes also occurred. The IRC wall bracing method designations are introduced in new Table 2308.6.3(1) instead of the traditional bracing method numbering system of Method 1 (let-in braces), Method 2 (diagonal wood boards), Method 3 (wood structural panel) and so on. The new designations for bracing methods are shown in the following table.

2012 IBC Bracing Method	2015 IBC Bracing Method	Description
1	LIB	Let-in brace
2	DWB	Diagonal wood board
3	WSP	Wood structural panel
4	SFB	Structural fiberboard
5	GB	Gypsum board
6	PBS	Particleboard sheathing
7	PCP	Portland cement plaster
8	HPS	Hardboard panel siding
Alternate braced wall panel	ABW	Alternate braced wall panel
Alternate braced wall panel adjacent to an opening	PFH	Portal frame with hold-downs

The alternate braced wall panel is now designated ABW and the figure from the IRC has been introduced. Similarly, the alternate braced wall panel adjacent to a door or window opening is renamed "portal frame with hold-downs (PFH)" as in the IRC. Because this figure was already in the IBC, the title was changed to reflect the new name PFH.

The specific wall bracing requirements based on Seismic Design Category have been reformatted and are shown in tabular form in Table 2308.6.1, which is similar to the presentation approach in the IRC.

2308.2.5

Allowable Roof Span

CHANGE TYPE: Modification

CHANGE SUMMARY: Provisions related to limitations on roof span have been clarified as a part of the reformatting and reorganization of Section 2308.

2015 CODE: **2308.2.5 Allowable Roof Span.** ~~5. Roof trusses and~~ Ceiling joist and rafter~~s~~ framing constructed in accordance with Section 2308.7 and trusses shall not span more than 40 feet (12 192 mm) between points of vertical support. A ridge board in accordance with Section 2308.7 or 2308.7.3.1 shall not be considered a vertical support.

CHANGE SIGNIFICANCE: The 2012 IBC indicates that "roof trusses and rafters shall not span more than 40 feet between points of vertical support," allowing for possible misinterpretation in regard to rafter spans. The revision ensures that a ridge board will not be misconstrued as being a point of vertical support. A conventionally framed roof with a ridge board, rafters, and ceiling joists will function similar to trusses, spanning from wall to wall, with the rafter in compression and the ceiling joist in tension. A ridge beam, on the other hand, is designed to support vertical loads (reactions) from the rafters and, as such, is considered a point of vertical support.

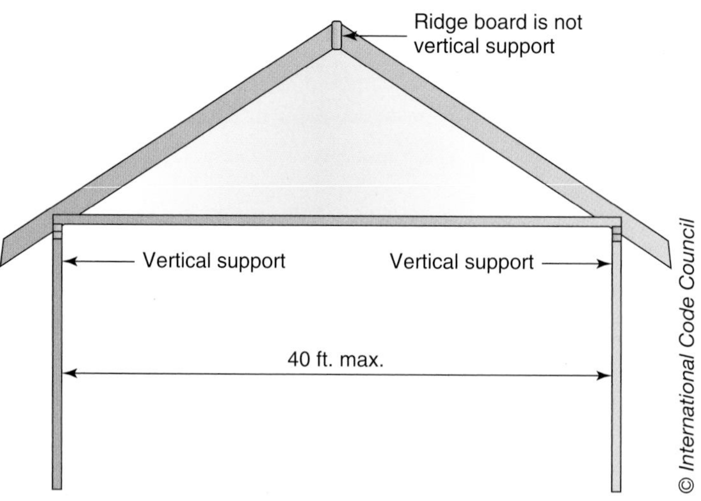

Rafter, ceiling joist, and ridge board framing

2308.7

Roof and Ceiling Framing

CHANGE TYPE: Modification

CHANGE SUMMARY: Ceiling joist and rafter span tables from the IRC have been incorporated into the conventional construction provisions of the IBC.

2015 CODE: **2308.7 Roof and Ceiling Framing.** The framing details required in this section apply to roofs having a minimum slope of three units vertical in 12 units horizontal (25-percent slope) or greater. Where the roof slope is less than three units vertical in 12 units horizontal (25-percent slope), members supporting rafters and ceiling joists such as ridge board, hips and valleys shall be designed as beams.

~~2308.10.2~~ 2308.7.1 Ceiling Joist Spans. ~~Allowable s~~Spans for ceiling joists shall be in accordance with Table 2308.7.1(1) or 2308.7.1(2). For other grades and species, and other loading conditions, refer to the ~~*AF&PA*~~ *AWC STJR*.

~~TABLE 2308.10.2(1)~~ TABLE 2308.7.1(1)
CEILING JOIST SPANS FOR COMMON LUMBER SPECIES (Uninhabitable attics without storage, live load = 10 psf, L/Δ = 240)
(Table R802.4(1) from the International Residential Code)

~~TABLE 2308.10.2(2)~~ TABLE 2308.7.1(2)
CEILING JOIST SPANS FOR COMMON LUMBER SPECIES (Uninhabitable attics with limited storage, live load = 20 psf, L/Δ = 240)
(Table R802.4(2) from the International Residential Code)

~~2308.10.3~~ 2308.7.2 Rafter Spans. ~~Allowable s~~Spans for rafters shall be in accordance with Table 2308.7.2(1), 2308.7.2(2), 2308.7.2(3), 2308.7.2(4), 2308.7.2(5) or 2308.7.2(6). For other grades and species and other loading conditions, refer to the ~~*AF&PA*~~ *AWC STJR*. The span of each rafter shall be measured along the horizontal projection of the rafter.

2308.7 continues

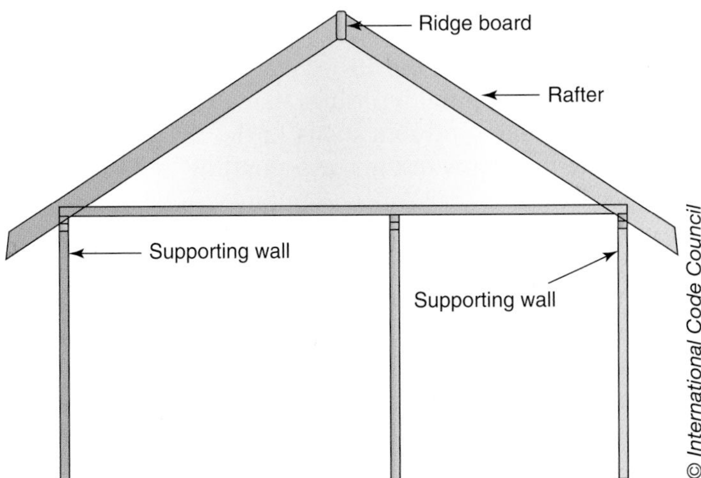

Typical rafter and ceiling joist framing

2308.7 continued

~~TABLE 2308.10.3(1)~~ TABLE 2308.7.2(1)
RAFTER SPANS FOR COMMON LUMBER SPECIES
(Roof Live Load = 20 pounds per square foot, Ceiling Not Attached to Rafters, L/Δ = 180)
(Table R802.5.1(1) from the International Residential Code)

~~TABLE 2308.10.3(2)~~ TABLE 2308.7.2(2)
RAFTER SPANS FOR COMMON LUMBER SPECIES
(Roof Live Load = 20 pounds per square foot, Ceiling Attached to Rafters, L/Δ = 240)
(Table R802.5.1(2) from the International Residential Code)

~~TABLE 2308.10.3(3)~~ TABLE 2308.7.2(3)
RAFTER SPANS FOR COMMON LUMBER SPECIES
(Ground Snow Load = 30 pounds per square foot, Ceiling Not Attached to Rafters, L/Δ = 180)
(Table R802.5.1(3) from the International Residential Code)

~~TABLE 2308.10.3(4)~~ TABLE 2308.7.2(4)
RAFTER SPANS FOR COMMON LUMBER SPECIES
(Ground Snow Load = 50 pounds per square foot, Ceiling Not Attached to Rafters, L/Δ = 180)
(Table R802.5.1(4) from the International Residential Code)

~~TABLE 2308.10.3(5)~~ TABLE 2308.7.2(5)
RAFTER SPANS FOR COMMON LUMBER SPECIES
(Ground Snow Load = 30 pounds per square foot, Ceiling Attached to Rafters, L/Δ = 240)
(Table R802.5.1(5) from the International Residential Code)

~~TABLE 2308.10.3(6)~~ TABLE 2308.7.2(6)
RAFTER SPANS FOR COMMON LUMBER SPECIES
(Ground Snow Load = 50 pounds per square foot, Ceiling Attached to Rafters, L/Δ = 240)
(Table R802.5.1(6) from the International Residential Code)

CHANGE SIGNIFICANCE: The spans for joists and rafters based on the conventional wood frame construction provisions of the IBC and IRC are identical, and buildings that meet the restrictions and limitations for conventional construction in the IBC have loading limitations that are commensurate with residential buildings that are within the scope of the IRC. Adjustments to the allowable spans in the IRC tables will now automatically update the spans in the corresponding tables in the IBC, and the species-specific spans for joists and rafters will be automatically correlated between the two codes. The provisions containing the charging text have been modified to read in the same manner as the corresponding provisions in the IRC. Footnotes to the IRC tables that are not applicable have been excluded from the IBC tables. It is important to note that the span tables in the 2015 IRC are modified for some lumber species and grades. For a complete discussion of these modifications, refer to *Significant Changes to the International Residential Code—2015 Edition.*

2309
Wood Frame Construction Manual

Courtesy of American Wood Council

The 2015 Edition of the *Wood Frame Construction Manual*

CHANGE TYPE: Addition

CHANGE SUMMARY: Section 2309 has been added to reference the American Wood Council's (AWC) *Wood Frame Construction Manual* (WFCM) for structural design of wood frame buildings assigned to Risk Category I or II.

2015 CODE: 2301.2 General Design Requirements. The design of structural elements or systems, constructed partially or wholly of wood or wood-based products, shall be in accordance with one of the following methods:

1. Allowable stress design in accordance with Sections 2304, 2305 and 2306.
2. Load and resistance factor design in accordance with Sections 2304, 2305 and 2307.
3. Conventional light-frame construction in accordance with Sections 2304 and 2308.

   Exception: ~~Buildings designed in accordance with the provisions of the AF&PA WFCM shall be deemed to meet the requirements of the provisions of Section 2308.~~
4. <u>AWC WFCM in accordance with Section 2309.</u>

~~4.~~<u>5</u>. The design and construction of log structures shall be in accordance with the provisions of ICC 400.

<u>**SECTION 2309**</u>
<u>**WOOD FRAME CONSTRUCTION MANUAL**</u>

<u>**2309.1 Wood Frame Construction Manual.** Structural design in accordance with the AWC WFCM shall be permitted for buildings assigned to Risk Category I or II subject to the limitations of Section 1.1.3 of the AWC WFCM and the load assumptions contained therein. Structural elements beyond these limitations shall be designed in accordance with accepted engineering practice.</u>

CHANGE SIGNIFICANCE: The *Wood Frame Construction Manual* (WFCM) is an ANSI standard developed by American Wood Council (AWC) technical committees that contains both engineering criteria and engineered prescriptive provisions for wood frame buildings that may be used for the design of wood frame structures within its scope. Although the WFCM provisions are intended primarily for detached one- and two-family dwellings due to the floor live load assumption associated with those occupancies, many of the provisions for specific geographic wind, seismic and snow loads are applicable to other types of buildings. For example, wind provisions are specifically addressed in the WFCM in regard to roof sheathing/wall sheathing sizing, the fastening schedule, uplift straps, shear anchorage, shear wall lengths and wall studs for out-of-plane wind loads, and are applicable to other occupancies within the load limitations of the WFCM tables. Similarly, roof rafter size and spacing for

2309 continues

2309 continued heavy snow loads, as well as shear wall lengths and anchorage for seismic considerations, are applicable within the load limitations of the WFCM tables. Any conditions that are outside the scope of the WFCM limitations or tabulated requirements, such as floor joist design for 60-psf loading and the design of supporting gravity elements for the additional floor live load, are beyond the applicability of the WFCM and must be designed in accordance with accepted engineering practice. The reference to the WFCM as an alternative in Section 2308.1 has been deleted because it may lead to confusion about the applicability. It is more appropriate to use the applicability limits within the WFCM itself rather than the limits for conventional construction in Section 2308.2.

2406.4.7

Safety Glazing Adjacent to Bottom Stair Landing

CHANGE TYPE: Modification

CHANGE SUMMARY: The height criteria for regulating glazing at the landing at the bottom of a stair has been revised and the method for measuring the horizontal distance has been clarified, now generally requiring safety glazing if located less than 60 inches above the bottom landing of a stair.

2015 CODE: 2406.4.7 Glazing Adjacent to the Bottom Stairway Landing. Glazing adjacent to the landing at the bottom of a stairway where the glazing is less than ~~36 inches (914 mm)~~ 60 inches (1524 mm) above the landing and within a 60 inch~~es~~ (1524 mm) horizontal~~ly of~~ arc that is less than 180 degrees (3.14 rad) from the bottom tread nosing shall be considered a hazardous location.

2406.4.7 continues

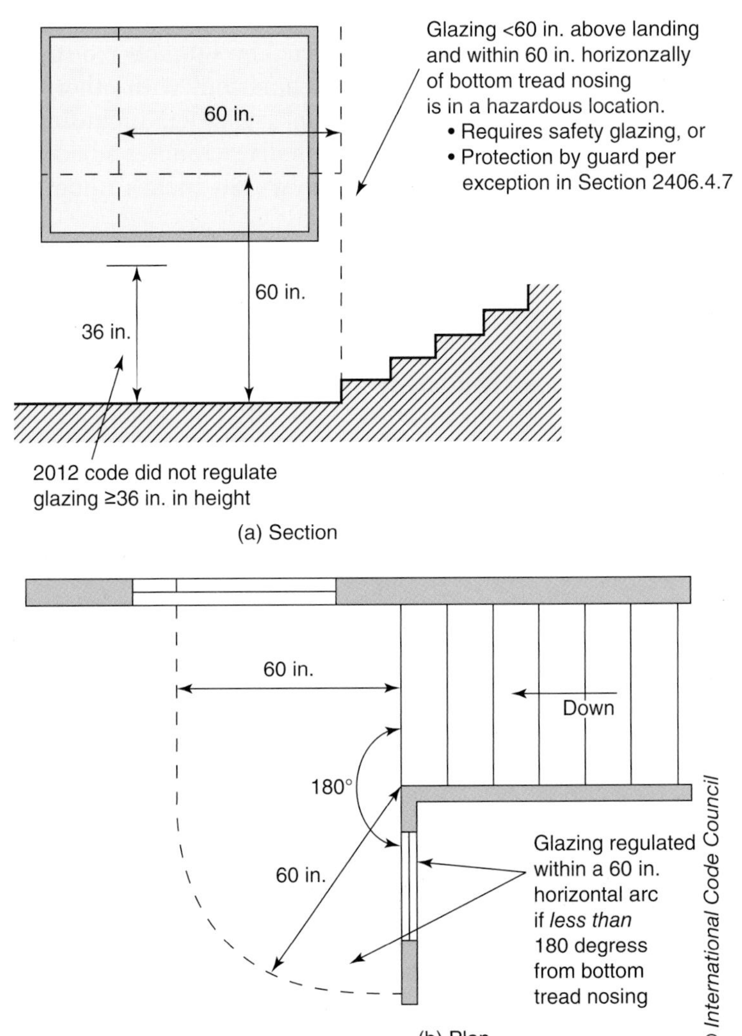
Glazing <60 in. above landing and within 60 in. horizonzally of bottom tread nosing is in a hazardous location.
• Requires safety glazing, or
• Protection by guard per exception in Section 2406.4.7
60 in.
60 in.
36 in.
2012 code did not regulate glazing ≥36 in. in height
(a) Section
60 in.
Down
180°
60 in.
Glazing regulated within a 60 in. horizontal arc if *less than* 180 degress from bottom tread nosing
(b) Plan
© International Code Council

Requirement for safety glazing adjacent to bottom stairway landing

2406.4.7 continued

Exception: Glazing that is protected by a guard complying with Sections 1015 and 1607.8 where the plane of the glass is greater than 18 inches (457 mm) from the guard.

CHANGE SIGNIFICANCE: The revised provisions for safety glazing application related to glazing adjacent to the bottom stair landing are generally consistent with the 2009 IBC. The modifications reverse a change made in the 2012 code in order to make the IBC and IRC consistent. Where glazing is located adjacent to the bottom landing of a stairway, safety glazing is now required if it is less than 60 inches above the landing. Under the 2012 IBC, the height threshold was reduced to 36 inches in order to coordinate with a similar provision in the IRC. In the 2015 IBC, the 60-inch height limitation has been restored and the bottom landing requirements are more consistent with those of Section 2406.4.6 addressing glazing adjacent to stairways and ramps.

Whether an individual is falling at the bottom landing as addressed by Section 2406.4.7 or at an intermediate landing per Section 2406.4.6, the height of the regulated glazing should be consistent because the person does not fall any differently. A person falling at the bottom of the stairway may impact the glazing located adjacent to the landing at virtually any height, be it with the entire body or simply with a hand or arm that is reaching out to grasp for support. The 60-inch requirement was selected for this location because it is consistent with other slip or fall locations where the code requires safety glazing, including Sections 2406.4.2, 2406.4.5 and 2406.4.6. Additionally, windows adjacent to walking surfaces are regulated to heights above 36 inches under Item 3 of Section 2406.4.3.

Glazing is also now regulated if it is within a 180-degree arc of the bottom tread, which is more consistent with those earlier editions that regulated the glazing "within any direction" when applying the 60-inch horizontal distance and determining the potential impact zone. This 180-degree requirement regulates glazing facing the tread, glazing running parallel to the direction of stair travel, as well as glazing that is splayed at an angle to the direction of travel. The provision recognizes that it is unlikely a person could turn around a corner and impact the glazing. The 180-degree angle should be measured using the line of the bottom tread nosing.

Chapter 25

Gypsum Panel Products

Gypsum panel products
Photo Courtesy of Gypsum Association

CHANGE TYPE: Addition

CHANGE SUMMARY: The definition of "gypsum board" has been revised and a new definition for "gypsum panel product" has been added to Chapter 2. Multiple references to gypsum panel products have also been added to Chapter 25 where applicable.

2015 CODE:

SECTION 202
DEFINITIONS

GYPSUM BOARD. The generic name for a family of sheet products consisting of a noncombustible core primarily of gypsum with paper surfacing. Gypsum wallboard, gypsum sheathing, gypsum base for gypsum veneer plaster, exterior gypsum soffit board, predecorated gypsum board and water-resistant gypsum backing board complying with the standards listed in Tables 2506.2, 2507.2 and Chapter 35 are types of gypsum board.

GYPSUM PANEL PRODUCT. The general name for a family of sheet products consisting essentially of gypsum.

CHAPTER 25
GYPSUM BOARD, GYPSUM PANEL PRODUCTS, AND PLASTER

2501.1 General. Provisions of this chapter shall govern the materials, design, construction and quality of gypsum board, gypsum panel products, lath, gypsum plaster and cement plaster.

2501.2 Performance. Lathing, plastering, gypsum board, and gypsum panel product construction shall be done in the manner and with the materials specified in this chapter, and when required for fire protection, shall also comply with the provisions of Chapter 7.

2503.1 Inspection. Lath, ~~and~~ gypsum board and gypsum panel products shall be inspected in accordance with Section 110.3.5.

2504.1 Scope. The following requirements shall be met where construction involves gypsum board, gypsum panel products, or lath and plaster in vertical and horizontal assemblies.

2504.1.1 Wood Framing. Wood supports for lath, ~~or~~ gypsum board, or gypsum panel products, as well as wood stripping or furring, shall not be less than 2 inches (51 mm) nominal thickness in the least dimension.

> **Exception:** The minimum nominal dimension of wood furring strips installed over solid backing shall not be less than 1 inch by 2 inches (25 mm by 51 mm).

Chapter 25 continues

Chapter 25 continued

2504.1.2 Studless Partitions. The minimum thickness of vertically erected studless solid plaster partitions of ⅜-inch (9.5 mm) and ¾-inch (19.1 mm) rib metal lath or ½-inch thick (12.7 mm) long-length gypsum lath, ~~and~~ gypsum board, or gypsum panel product partitions shall be 2 inches (51 mm).

2505.1 Resistance to Shear (Wood Framing). Wood-framed shear walls sheathed with gypsum board, gypsum panel products, or lath and plaster shall be designed and constructed in accordance with Section 2306.3 and are permitted to resist wind and seismic loads. Walls resisting seismic loads shall be subject to the limitations in Section 12.2.1 of ASCE 7.

2505.2 Resistance to Shear (Steel Framing). Cold-formed steel-framed shear walls sheathed with gypsum board or gypsum panel products, and constructed in accordance with the materials and provisions of Section 2211.6 are permitted to resist wind and seismic loads. Walls resisting seismic loads shall be subject to the limitations in Section 12.2.1 of ASCE 7.

SECTION 2506
GYPSUM BOARD AND GYPSUM PANEL PRODUCT MATERIALS

TABLE 2506.2 Gypsum Board Materials and Accessories

Material	Standard
~~Steel studs, load-bearing~~ Cold-formed steel studs and track, structural	AISI S200 and ASTM C955, Section 8
~~Steel studs, nonload-bearing~~ Cold-formed steel studs and track, nonstructural	AISI S220 and ASTM C645, Section 10

(Portions of table not shown remain unchanged.)

2506.2.1 Other Materials. Metal suspension systems for acoustical and lay-in panel ceilings shall conform with ASTM C 635 listed in Chapter 35 and Section 13.5.6 of ASCE 7 for installation in high seismic areas.

SECTION 2507
LATHING AND PLASTERING

TABLE 2507.2 Lath, Plastering Materials and Accessories

Material	Standard
~~Steel studs, load bearing~~ Cold-formed steel studs and track, structural	~~ASTM C 645~~ AISI S200 and: ASTM C955, Section 8
Cold-formed steel studs and track, nonstructural	AISI S220 and ASTM C 645, Section 10
Hydraulic Cement	ASTM C 1157; C 1600

(Portions of table not shown remain unchanged.)

AISI. AISI S220—11 North American Standard for Cold-formed Steel Framing-Nonstructural Members.

CHANGE SIGNIFICANCE: "Gypsum panel product" is a term that has been created by the gypsum manufacturing industry to describe gypsum sheet products that are manufactured unfaced or with a facing other than paper. Glass mat-faced and unfaced gypsum sheet materials are examples of gypsum panel products. Some gypsum application standards referenced by the code, such as GA 216, ASTM C 840, and ASTM C 1280, are used to define application requirements for both board and panel products, a dual role that was not reflected in prior code text. In addition, while the ASTM manufacturing standards for many gypsum panel products (ASTM C1278, C1178, C1658 and C1177) were incorporated into Chapter 25 during the past decade, the general text of Chapter 25 was not previously updated to reflect the incorporation of the new standards. The new code provisions address both issues. The tables in Section 2508 have been modified to indicate where the application standards may function as an application reference standard for either a board or a panel product. The term "gypsum panel product" has been added throughout Chapter 25 where appropriate. The definition for "gypsum panel product" in Chapter 2 has been extracted verbatim from ASTM C 11, *Standard Terminology Relating to Gypsum and Related Building Materials and Systems.*

The material column in Table 2506.2 has also been updated to refer to "structural" and "nonstructural" CFS studs and track, and references to AISI S200 and AISI S220 have been incorporated into the table. In addition, Table 2507.2 has been updated and coordinated with Table 2506.2.

2612

Plastic Composites

CHANGE TYPE: Addition

CHANGE SUMMARY: New definitions and applicable test standards now address the use of plastic composites for use as exterior deck boards, stair treads, handrails and guards.

2015 CODE:

SECTION 202
DEFINITIONS

PLASTIC COMPOSITE. A generic designation that refers to wood/plastic composites and plastic lumber.

PLASTIC LUMBER. A manufactured product made primarily of plastic materials (filled or unfilled) which is generally rectangular in cross-section.

WOOD/PLASTIC COMPOSITE. A composite material made primarily from wood or cellulose-based materials and plastic.

2601.1 Scope. These provisions shall govern the materials, design, application, construction and installation of foam plastic, foam plastic insulation, plastic veneer, interior plastic finish and trim, ~~and~~ light-transmitting plastics, and plastic composites, including plastic lumber. See Chapter 14 for requirements for exterior wall finish and trim.

Photo Courtesy of Peter Kulczyk

Plastic composite decking

SECTION 2612
PLASTIC COMPOSITES

2612.1 General. Plastic composites shall consist either of wood/plastic composites or of plastic lumber. Plastic composites shall comply with the provisions of this code and with the additional requirements of Section 2612.

2612.2 Labeling and Identification. Packages and containers of plastic composites used in exterior applications shall bear a *label* showing the manufacturer's name, product identification and information sufficient to determine that the end use will comply with code requirements.

2612.2.1 Performance Levels. The label for plastic composites used in exterior applications as deck boards, stair treads, handrails and guards shall indicate the required performance levels and demonstrate compliance with the provisions of ASTM D 7032.

2612.2.2 Loading. The label for plastic composites used in exterior applications as deck boards, stair treads, handrails and guards shall indicate the type and magnitude of the load determined in accordance with ASTM D 7032.

2612.3 Flame Spread Index. Plastic composites shall exhibit a flame spread index not exceeding 200 when tested in accordance with ASTM E 84 or UL 723 with the test specimen remaining in place during the test.

> **Exception:** Materials determined to be noncombustible in accordance with Section 703.5.

2612.4 Termite and Decay Resistance. Plastic composites containing wood, cellulosic or any other biodegradable materials shall be termite and decay resistant as determined in accordance with ASTM D 7032.

2612.5 Construction Requirements. Plastic composites shall be permitted to be used as exterior deck boards, stair treads, handrails and guards in buildings of Type VB construction.

2612.5.1 Span Rating. Plastic composites used as structural components of exterior deck boards shall have a span rating determined in accordance with ASTM D 7032.

2612.6 Plastic Composite Decking, Handrails and Guards. Plastic composite decking, handrails and guards shall be installed in accordance with this code and the manufacturer's instructions.

(*Sections 2612 and 2613 have been renumbered as Sections 2613 and 2614, respectively.*)

2612 continues

2612 continued

SECTION 1410
PLASTIC COMPOSITE DECKING

1410.1 Exterior deck boards, stair treads, handrails and guardrail systems constructed of plastic composites, including plastic lumber, shall comply with Section 2612.

CHANGE SIGNIFICANCE: Composite plastics are in common use, and with the addition of these new provisions the code will now provide guidance on where the materials are permitted and what standards the materials must meet. As addressed by the definitions, plastic composites can be plastic lumber or wood plastic composites. Both types of products are made of plastic materials with added fibrous materials to provide stiffness. Wood plastic composites contain wood or cellulosic materials (normally over 50 percent) as the primary fiber that provides the material its stiffness. On the other hand, plastic lumber contains primarily plastic (normally over 50 percent) and uses a variety of materials such as fiberglass to provide stiffness.

Materials of composite plastics are commonly used for the construction of decks throughout the United States but historically neither the IBC nor the IRC provided any details or references for them. ICC Evaluation Services recognizes both of the materials under Acceptance Criteria AC 174, *Acceptance Criteria for Deck Board Span Ratings and Guardrail Systems*. Because these materials are currently being used and performing adequately, placing the requirements within the IBC will eliminate the need for the building official to accept them as alternate materials or methods of construction.

As can be seen by the various subsections of Section 2612, the composite plastic materials must be provided with a label indicating information regarding the manufacturer (Section 2612.2) and showing compliance with the performance and loading criteria from Sections 2612.2.1 and 2612.2.2. These new requirements will limit the flame spread of the plastic composite materials (Section 2612.3) and limit the use of the materials to exterior deck boards, stair treads, handrails and guards to buildings of Type VB construction (Section 2612.5). Because these products can vary based upon the individual manufacturer and product, Section 2612.6 requires the material to be installed in accordance with both the code and the manufacturer's installation instructions, and the material's span rating shall be evaluated per Section 2612.5.1.

PART 7

Building Services, Special Devices, and Special Conditions

Chapters 27 through 34

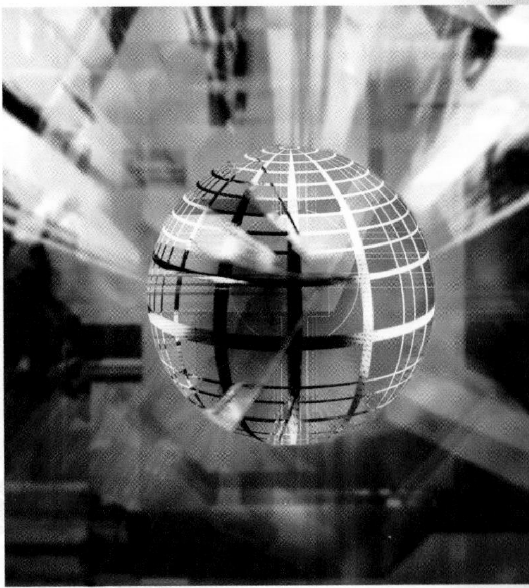

- **Chapter 27** Electrical No changes addressed
- **Chapter 28** Mechanical Systems No changes addressed
- **Chapter 29** Plumbing Systems
- **Chapter 30** Elevators and Conveying Systems
- **Chapter 31** Special Construction No changes addressed
- **Chapter 32** Encroachments into the Public Right-of-Way No changes addressed
- **Chapter 33** Safeguards during Construction No changes addressed
- **Chapter 34** Existing Structures

Although building services such as electrical systems (Chapter 27), mechanical systems (Chapter 28) and plumbing systems (Chapter 29) are regulated primarily through separate and distinct codes, limited provisions are set forth in the *International Building Code*. Chapter 30 regulates elevators and similar conveying systems to a limited degree, as most requirements are found in American Society of Mechanical Engineers (ASME) standards. The special construction provisions of Chapter 31 include those types of elements or structures that are not conveniently addressed in other portions of the code. By "special construction," the code is referring to membrane structures, pedestrian walkways, tunnels, awnings, canopies, marquees, and similar building features that are unregulated elsewhere. Chapter 32 governs the encroachment of structures into the public right-of-way, and Chapter 33 addresses safety during construction and the protection of adjacent public and private properties. ■

2902.3

Public Toilet Facilities

3004

Elevator Hoistway Venting

3006

Elevator Lobbies

CHAPTER 34

Existing Structures

2902.3

Public Toilet Facilities

CHANGE TYPE: Modification

CHANGE SUMMARY: Limited-size quick-service tenant spaces are no longer required to provide toilet facilities for the public customers.

2015 CODE: 2902.3 Required Public Toilet Facilities. Customers, patrons and visitors shall be provided with public toilet facilities in structures and tenant spaces intended for public utilization. The number of plumbing fixtures located within the required toilet facilities shall be provided in accordance with Section 2902.1 for all users. Employees shall be provided with toilet facilities in all occupancies. Employee toilet facilities shall be either separate or combined employee and public toilet facilities.

> **Exceptions:** Public toilet facilities shall not be required in:
>
> 1. Open or enclosed parking garages~~. Toilet facilities shall not be required in parking garages~~ where there are no parking attendants.
> 2. Structures and tenant spaces intended for quick transactions, including take out, pick up and drop off, having a public access area less than or equal to 300 square feet (28 m^2).

CHANGE SIGNIFICANCE: The purpose of Section 2902.3 has always been to provide public toilet facilities for the comfort and convenience of any customer or visitor who may spend a period of time within a facility that is open to the public. Where individuals are not in the facility for an

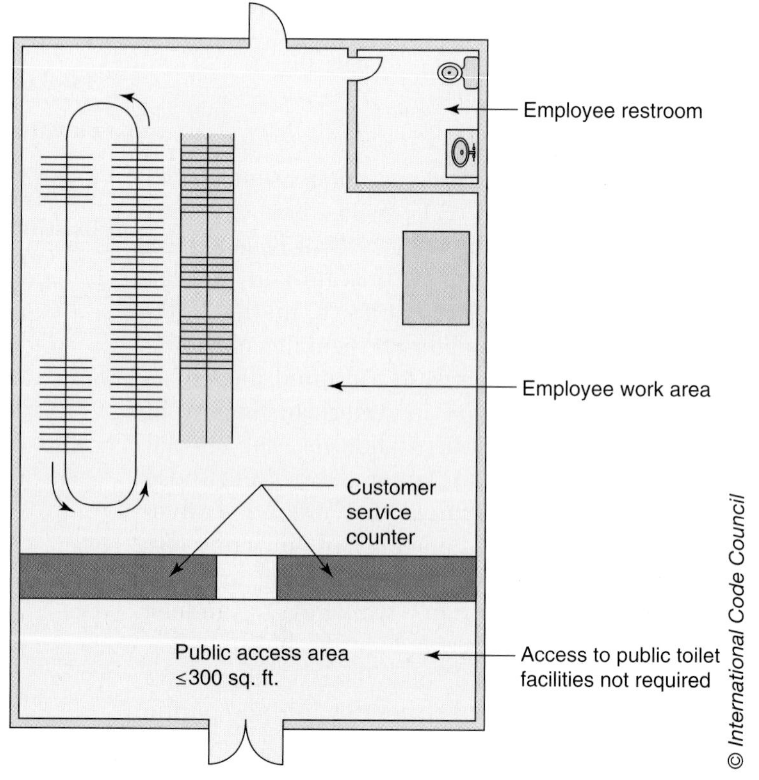

Toilet facilities for the public are not required

© International Code Council

Quick transaction tenant space

extended length of time, the need for available public toilets is minimized. With the added restriction that the exception is limited to a maximum area of 300 square feet for the public area, the limited occupant load combined with the limited length of time the customers are within the facility makes it unlikely that a patron would need access to toilet facilities.

Quick-transaction tenant spaces are unique in that patrons spend a short period of time completing a transaction and then depart. Therefore, they tend to not have a need for toilet facilities and typically do not have an expectation that public toilets are available within such small spaces. Examples of these types of spaces include take-out food locations such as pizza carry-outs, dry cleaners, ATM facilities, shoe repair shops, newspaper stands, and so forth. These are just a few examples; there are numerous possible tenant spaces that might be eligible to use the exception. There are also likely examples of where the uses may not always be appropriate. The key issue in the exception is that it is "intended for quick transactions" and the public access area must be limited to 300 square feet in floor area. Therefore, the examples given within the code should not be viewed as limiting the exception to those uses or as being an exhaustive list.

It is the public-access area and not the entire business that is regulated by the area limitation. Generally, the public areas will be separated by counters, walls, or other means that clearly designate the public area from work spaces, stock areas or other business functions. In the case of a small shoe repair shop or a carry-out pizza tenant, it would generally be a safety hazard or a health hazard to allow the public to have access to and travel through the working space in order to reach designated public toilet facilities. Therefore, if restrooms were required in these small facilities they would probably be located in the front area where the public has access. It would then require the size of the public area to be increased or possibly create a visual obstruction at the storefront and affect the street or sidewalk visibility of the business.

3004

Elevator Hoistway Venting

CHANGE TYPE: Deletion

CHANGE SUMMARY: The elevator hoistway venting provisions of Section 3004 have been deleted; such hoistways are no longer required to be vented to the exterior.

2015 CODE:

~~SECTION 3004~~
~~HOISTWAY VENTING~~

~~**3004.1 Vents Required.** Hoistways of elevators and dumbwaiters penetrating more than three stories shall be provided with a means for venting smoke and hot gases to the outer air in case of fire.~~

> ~~**Exception:** Venting is not required for the following elevators and hoistways:~~
>
> ~~1. In occupancies of other than Groups R-1, R-2, I-1, I-2 and similar occupancies with overnight sleeping units, where the building is equipped throughout with an approved automatic sprinkler system installed in accordance with Section 903.3.1.1 or 903.3.1.2.~~
> ~~2. Sidewalk elevator hoistways.~~
> ~~3. Elevators contained within and serving open parking garages only.~~
> ~~4. Elevators within individual residential dwelling units.~~

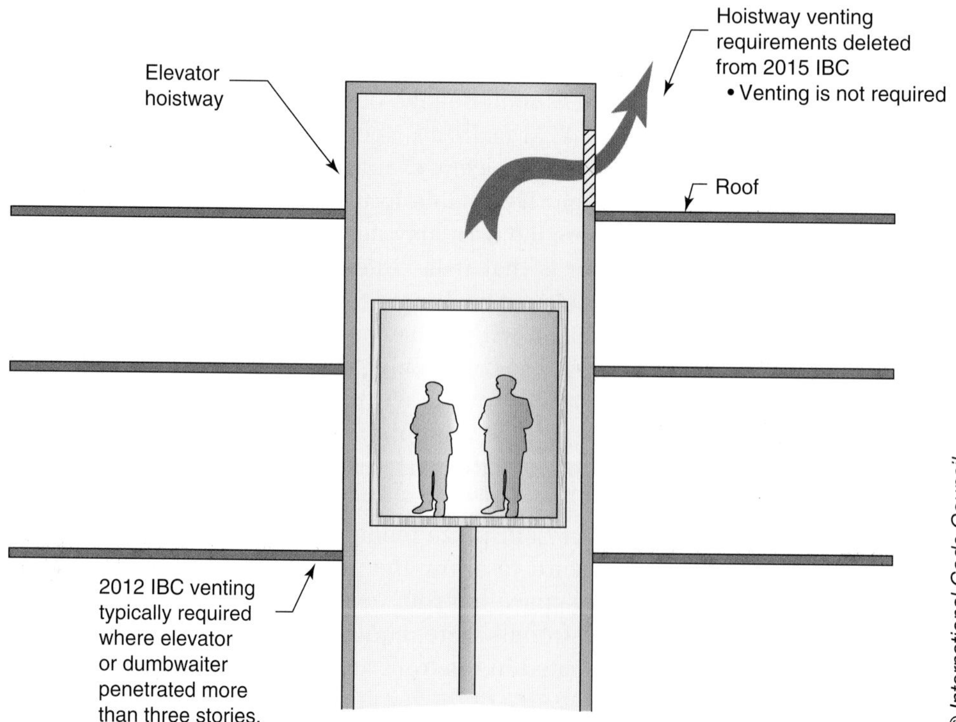

Hoistway venting requirements deleted from 2015 IBC
• Venting is not required
Elevator hoistway
Roof
2012 IBC venting typically required where elevator or dumbwaiter penetrated more than three stories.
© International Code Council

Hoistway venting is not required

~~**3004.2 Location of Vents.** Vents shall be located at the top of the hoistway and shall open either directly to the outer air or through noncombustible ducts to the outer air. Noncombustible ducts shall be permitted to pass through the elevator machine room, provided that portions of the ducts located outside the hoistway or machine room are enclosed by construction having not less than the fire-resistance rating required for the hoistway. Holes in the machine room floors for the passage of ropes, cables or other moving elevator equipment shall be limited as not to provide greater than 2 inches (51 mm) of clearance on all sides.~~

~~**3004.3 Area of Vents.** Except as provided for in Section 3004.3.1, the area of the vents shall be not less than 3½ percent of the area of the hoistway nor less than 3 square feet (0.28 m²) for each elevator car, and not less than 3½ percent nor less than 0.5 square feet (0.047 m²) for each dumbwaiter car in the hoistway, whichever is greater. Of the total required vent area, not less than one-third shall be permanently open. Closed portions of the required vent area shall consist of openings glazed with annealed glass not greater than ⅛ inch (3.2 mm) in thickness.~~

> ~~**Exception:** The total required vent area shall not be required to be permanently open where all the vent openings automatically open upon detection of smoke in the elevator lobbies or hoistway, upon power failure and upon activation of a manual override control. The manual override control shall be capable of opening and closing the vents and shall be located in an approved location.~~

~~**3004.3.1 Reduced Vent Area.** Where mechanical ventilation conforming to the *International Mechanical Code* is provided, a reduction in the required vent area is allowed provided that all of the following conditions are met:~~

1. ~~The occupancy is not in Group R-1, R-2, I-1 or I-2 or of a similar occupancy with overnight sleeping units.~~
2. ~~The vents required by Section 3004.2 do not have outside exposure.~~
3. ~~The hoistway does not extend to the top of the building.~~
4. ~~The hoistway and machine room exhaust fan is automatically reactivated by thermostatic means.~~
5. ~~Equivalent venting of the hoistway is accomplished.~~

~~3004.4~~ 3002.9 Plumbing and Mechanical Systems. Plumbing and mechanical systems shall not be located in an elevator hoistway enclosure.

> **Exception:** Floor drains, sumps and sump pumps shall be permitted at the base of the hoistway enclosure provided they are indirectly connected to the plumbing system.

CHANGE SIGNIFICANCE: Elevator hoistways have been required to vent to the exterior of the building for decades by the IBC as well as its legacy codes. Over the years, numerous changes have occurred in areas such as elevator lobbies, energy conservation, automatically operated

3004 continues

3004 continued

dampers, better smoke control and more sprinklered buildings. However, the venting requirements have remained and undergone minor revisions, although the exact purpose or need for the vents is no longer clear. Based on the lack of a specific detailed need for the venting and recognizing that the requirement has been removed from the 2010 edition of the ASME A17.1 *Safety Code for Elevators and Escalators*, the venting requirement has been deleted.

Although it appears that the original intent was focused more upon fire-fighting and post-fire overhaul, the operation of the vents or their being opened for relieving shaft pressures can also lead to smoke movement up through the shaft and on to other floors. In addition, the amount of conditioned air lost through the vent or unconditioned air allowed to enter into the building greatly affects energy conservation.

The only provision from the 2012 IBC that has been retained is the prohibition of installing plumbing and mechanical systems within the hoistway enclosure. This provision is still appropriate and therefore has been relocated to Section 3002.9 addressing other hoistway enclosure requirements.

3006

Elevator Lobbies

CHANGE TYPE: Modification

CHANGE SUMMARY: The elevator lobby requirements have been relocated from Section 713.14.1, where they were previously included with the general shaft enclosure requirements, to Chapter 30, which addresses elevators.

2015 CODE:

SECTION 3006
ELEVATOR LOBBIES AND
HOISTWAY OPENING PROTECTION

3006.1 General. Elevator hoistway openings and enclosed elevator lobbies shall be provided in accordance with the following:

1. Where hoistway opening protection is required by Section 3006.2 such protection shall be in accordance with Section 3006.3.
2. Where enclosed elevator lobbies are required for underground buildings such lobbies shall comply with Section 405.4.3.
3. Where an area of refuge is required and an enclosed elevator lobby is provided to serve as an area of refuge, the enclosed elevator lobby shall comply with Section 1009.6.

3006 continues

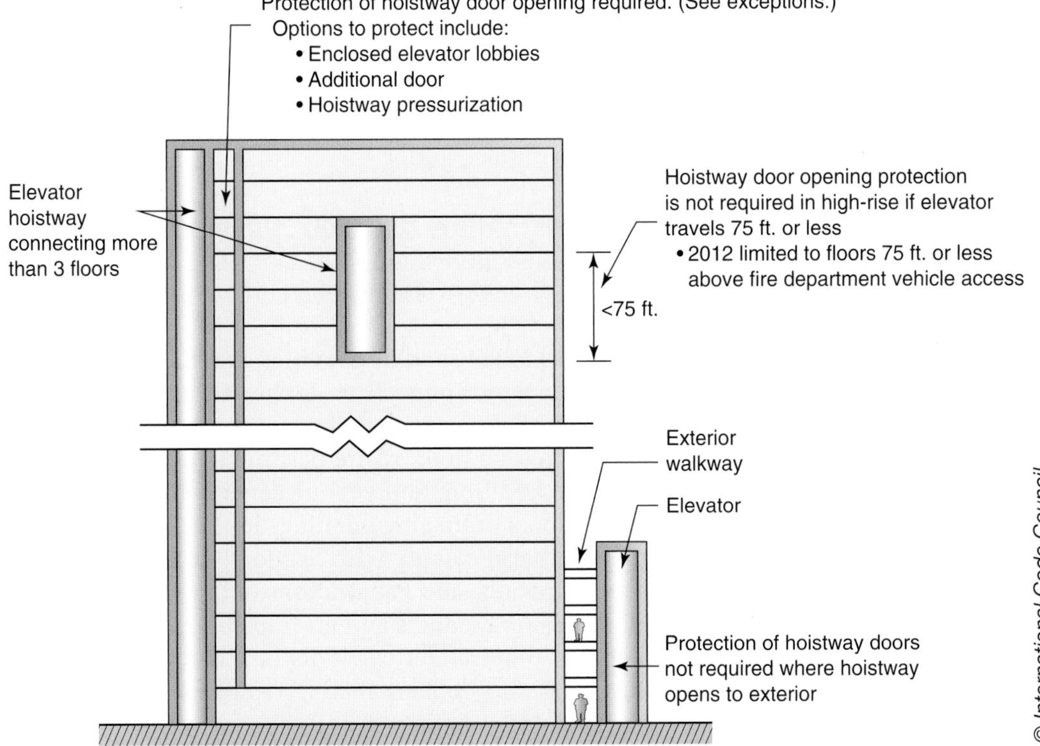

The elevator lobby provisions have been relocated to Chapter 30

3006 continued

4. Where fire service access elevators are provided, enclosed elevator lobbies shall comply with Section 3007.6.
5. Where occupant evacuation elevators are provided, enclosed elevator lobbies shall comply with Section 3008.6.

3006.2 Hoistway Opening Protection Required. Elevator hoistway door openings shall be protected in accordance with Section 3006.3 where an elevator hoistway connects more than three stories, is required to be enclosed within a shaft enclosure in accordance with Section 712.1.1 and any of the following conditions apply:

1. The building is not protected throughout with an *automatic sprinkler system* in accordance with Section 903.3.1.1 or 903.3.1.2;
2. The building contains a Group I-1 Condition 2 occupancy;
3. The building contains a Group I-2 occupancy;
4. The building contains a Group I-3 occupancy;
5. The building is a high rise and the elevator hoistway is more than 75 feet (22 860 mm) in height. The height of the hoistway shall be measured from the lowest floor to the highest floor of the floors served by the hoistway.

Exceptions:

1. Protection of elevator hoistway door openings is not required where the elevator serves only open parking garages in accordance with Section 406.5.
2. Protection of elevator hoistway door openings is not required at the level(s) of exit discharge, provided the level(s) of exit discharge is equipped with an *automatic sprinkler system* in accordance with Section 903.3.1.1.
3. Enclosed elevator lobbies and protection of elevator hoistway door openings are not required on levels where the elevator hoistway opens to the exterior.

~~713.14.1~~ 3006.3 ~~Elevator Lobby.~~ Hoistway Opening Protection. Where Section 3006.2 requires protection of the elevator hoistway door opening, the protection shall be provided by one of the following:

1. An enclosed elevator lobby shall be provided at each floor ~~where an elevator shaft enclosure connects more than three stories. The lobby enclosure shall~~ to separate the elevator hoistway shaft enclosure doors from each floor by fire partitions in accordance with Section 708. In addition, ~~to the requirements in Section 708 for fire partitions,~~ doors protecting openings in the elevator lobby enclosure walls shall ~~also~~ comply with Section 716.5.3 as required for corridor walls. ~~and~~ Penetrations of the enclosed elevator lobby ~~enclosure~~ by ducts and air transfer openings shall be protected as required for corridors in accordance with Section 717.5.4.1. ~~Elevator lobbies shall have at least one means of egress complying with Chapter 10 and other provisions within this code.~~

~~**Exceptions:**~~

~~1. Enclosed elevator lobbies are not required at the level(s) of exit discharge, provided the level(s) of exit discharge is equipped with an automatic sprinkler system in accordance with Section 903.3.1.1.~~

~~2. Elevators not required to be located in a shaft in accordance with Section 712.1 are not required to have enclosed elevator lobbies.~~

2. An enclosed elevator lobby shall be provided at each floor to separate the elevator hoistway shaft enclosure doors from each floor by smoke partitions in accordance with Section 710 where the building is equipped throughout with an *automatic sprinkler system* installed in accordance with Section 903.3.1.1 or 903.3.1.2. In addition, doors protecting openings in the smoke partitions shall comply with Sections 710.5.2.2, 710.5.2.3 and 716.5.9. Penetrations of the enclosed elevator lobby by ducts and air transfer openings shall be protected as required for corridors in accordance with Section 717.5.4.1.

3. ~~Enclosed elevator lobbies are not required where an~~ Additional doors ~~are~~ shall be provided at each elevator ~~the~~ hoistway door opening in accordance with Section 3002.6. Such door shall comply with the smoke and draft control door assembly requirements in Section 716.5.3.1 when tested in accordance with UL 1784 without an artificial bottom seal.

~~4. Enclosed elevator lobbies are not required where the building is protected by an automatic sprinkler system installed in accordance with Section 903.3.1.1 or 903.3.1.2. This exception shall not apply to the following:~~

~~4.1. Group I-2 occupancies;~~

~~4.2. Group I-3 occupancies; and~~

~~4.3. Elevators serving floor levels over 75 feet above the lowest level of fire department vehicle access in high-rise buildings.~~

~~5. Smoke partitions shall be permitted in lieu of fire partitions to separate the elevator lobby at each floor where the building is equipped throughout with an automatic sprinkler system installed in accordance with Section 903.3.1.1 or 903.3.1.2. In addition to the requirements in Section 710 for smoke partitions, doors protecting openings in the smoke partitions shall also comply with Sections 710.5.2.2, 710.5.2.3, and 716.5.9 and duct penetrations of the smoke partitions shall be protected as required for corridors in accordance with Section 717.5.4.1.~~

4. ~~6. Enclosed elevator lobbies are not required where~~ The elevator hoistway ~~is~~ shall be pressurized in accordance with Section 909.21.

~~7. Enclosed elevator lobbies are not required where the elevator serves only open parking garages in accordance with Section 406.5.~~

3006.4 Means of Egress. Elevator lobbies shall be provided with at least one means of egress complying with Chapter 10 and other provisions in this code. Egress through an elevator lobby shall be permitted in accordance with Item 1 of Section 1016.2.

3006 continues

3006 continued

(Because this code change affected substantial portions of Sections 713.14.1 and the new Section 3006, the entire code change text is too extensive to be included here. Refer to Code Changes FS-61, FS-67, FS-69, FS-70, FS-71 and E-110 in the 2015 IBC Code Changes Resource Collection *for the complete text and history of the code change.)*

CHANGE SIGNIFICANCE: The major aspect of this revision is the fact that the elevator lobby provisions have been consolidated into a new section in Chapter 30 with all of the other elevator requirements, as opposed to being located in Section 713 with the general shaft enclosure requirements. A new charging section has been created in Section 3006.1 that directs code users to specific sections throughout the code that contain elevator lobby requirements or affect the construction details for them. By referencing users to the appropriate provisions, it will be clear that there are several distinct types of lobbies that differ in their purpose. With these section references and brief descriptions, users will be able to quickly determine if more than one set of requirements may be applicable to a specific building.

When dealing with high-rise buildings, the code no longer requires the hoistway door opening protection for every installation, but only for those hoistways that connect a large number of stories. Previously, a hoistway that only connected two stories required elevator lobby protection if it happened to be located above the 75-foot elevation. Under the 2015 IBC, the door opening protection is only required where the elevation within the hoistway exceeds 75 feet when measured from the lowest floor to the highest floor served by that hoistway.

Per new Exception 3 of Section 3006.2, on levels where an elevator hoistway opens to the exterior there is no need for either an elevator lobby or to protect the door openings because smoke accumulation will not occur. Conceptually, this exemption is similar to Exception 1, which provides a similar exemption within an open parking garage.

Although it is not directly a part of Section 3006 or of the related code changes, code users should be aware of revisions made to the accessible means-of-egress elevator access provisions in new Section 1009.6.2 (previously 1007.6). In an effort to coordinate the accessible means of egress provisions with the revised elevator lobby provisions, the requirements for elevator lobbies that are used as an area of refuge were also modified. Previously, the code required elevator lobbies used as an area of refuge to be protected in accordance with the smokeproof enclosure provisions. It was deemed that the reference was inappropriate because the smokeproof enclosure requirements are more focused on stairways and are not practical for an elevator lobby. Therefore, the provisions for areas of refuge simply rely on the separation requirements of Section 1009.6.4 instead of the smoke-proof enclosure provisions.

The elevator hoistway pressurization provisions of Section 909.21.1 have also been modified to provide additional guidance on how the system is to be designed and operated. The revisions should help to make compliance with the pressurization provisions easier and ultimately improve performance and code compliance.

Chapter 34
Existing Structures

CHANGE TYPE: Deletion

CHANGE SUMMARY: Chapter 34 has been deleted from the IBC in its entirety, and existing buildings will now be solely regulated by the *International Existing Building Code* (IEBC).

2015 CODE:

~~CHAPTER 34~~
~~EXISTING STRUCTURES~~
(Chapter has been deleted in its entirety.)

101.4.7 Existing Buildings. The provisions of the *International Existing Building Code* shall apply to all matters governing the repairs, alterations, change of occupancy, addition to and relocation of existing buildings.

CHANGE SIGNIFICANCE: The deletion of Chapter 34 from the IBC and the reliance on the IEBC will help code users by providing a more consistent and coordinated document to deal with existing buildings. With the removal of Chapter 34, there will now be a single specific document dealing with existing buildings as opposed to the partial requirements that were previously found in Chapter 34.

Under the 2012 IBC, Section 3401.6 accepted the IEBC as an alternative compliance option and it was "deemed to comply" with the provisions of Chapter 34. Under the 2015 IBC, the IEBC will become a referenced code in Section 101.4, similar to how the IECC is used for energy efficiency in Chapter 13 and Section 101.4.6. The IEBC takes a more comprehensive approach to existing buildings than what was previously included within the IBC. Instead of adding additional language into the IBC to deal with this complex issue, the decision was made to rely on the IEBC as the only document for the regulation of existing buildings. The requirements found within Chapter 34 were previously replicated in Chapters 4 and 14 of the IEBC. Eliminating Chapter 34 from the IBC will therefore remove the duplication and yet still provide code users with the same tools and requirements to regulate existing buildings.

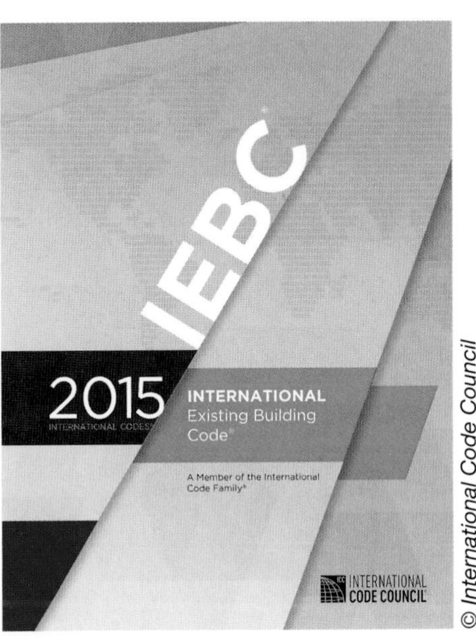

© International Code Council

The 2015 *International Existing Building Code* (IEBC)

© International Code Council

Existing buildings are regulated by the IEBC

Index

G

U

V

W

INTERNATIONAL CODE COUNCIL
People Helping People Build a Safer World
Growing your career is what ICC Membership is all about
As your building career grows, so does the need to expand your code knowledge and job skills. Whether you're seeking a higher level of certification or professional quality training, Membership in ICC offers the best in I-Code resources and training for growing your building career today and for the future.
Learn new job skills to prepare for a higher level of responsibility within your organization
Improve your code knowledge to keep pace with the latest International Codes (I-Codes)
Achieve additional ICC Certifications to open the door to better job opportunities
Plus, an affordable ICC Membership provides exclusive Member-only benefits including:
Free code opinions from I-Code experts
Free I-Code book(s) to new Members*
Access to employment opportunities in the ICC Career Center
Discounts on professional training & Certification renewals
Savings of up to 25% off on code books & training materials
Free benefits - Governmental Members: Your staff can receive free ICC benefits too*
And much more!
Join the International Code Council (ICC) and start growing your building career now!
Visit our Member page at www.iccsafe.org/membership for an application.
*Some restrictions apply. Speak with an ICC Member Services Representative for details.
14-09142